McNAUGHTON

VOLUME 3 1944-1966

McNAUGHTON

VOLUME 3 1944–1966

John Swettenham

Second edition:

Editor and Publisher

IEEE Canada, 2022

ISBN: 978-0-9692316-5-3

First edition copyright page:

PRINTED AND BOUND IN CANADA BY THE RYERSON PRESS
TORONTO WINNIPEG VANCOUVER

SBN 7700 0299 4
Library of Congress Catalog Number: 70-375263

Published as a Centennial of Canadian Confederation project. The preparation of the manuscript was subsidized by the Centennial Commission.

Publié en célébration du Centenaire de la Confédération Canadienne, et rédigé grâce à une subvention de la Commission du Centenaire.

All picture credits are to the Public Archives of Canada unless otherwise noted.

Contents

PART FIVE: RECESSION AND RESURGENCE (1944-1966)

Illustrations

Maps

McNAUGHTON

VOLUME 3 1944-1966

PART FIVE

RECESSION AND RESURGENCE (1944-1966)

Chapter 1

From Soldier to Civilian

General and Mrs. McNaughton reached New York at noon on 2 February, 1944. A party of waiting Canadians, including Dr. C. J. Mackenzie (representing the Prime Minister), Colonel G. S. Currie (Ralston's deputy) and Major-General Maurice Pope (to whom Mackenzie King had confided, "We must do nothing to prejudice our relations with General McNaughton")[1] came aboard the troopship as soon as the gang plank was down. The couple were rushed to the home of the Canadian representative in New York (Consul General H. D. Scully) where they remained, secluded from reporters, until the evening. Then, accompanied by a captain of the United States security police who "remained in close contact with the General until he boarded the train," the McNaughtons left Grand Central Station for Canada.[2]

McNaughton had decided to spend a few days at a place he loved—the Seigniory Club, which, backed by hills, overlooks a fine stretch of the Ottawa River about forty miles east of the capital. There, surrounded by the peaceful countryside of rural Quebec, he hoped that he would be able to see the feverish events of the past two months in their true perspective. In the early afternoon of 3 February the special train pulled into the small station at Montebello. The McNaughtons moved out of the heated car, down onto the snowy platform, and the General looked about him. He was back on Canadian ground. It was cold but brilliantly sunny and all around stretched the rolling fields and hills, dazzling white, under hard blue cloudless skies. He filled his lungs with the keen air and smiled with pleasure.[3]

Then newspaper reporters, who had been waiting at the station, came up. McNaughton was reluctant to make any statement but they pressed him. How was his health after the voyage? McNaughton, having been cooped up aboard ship for days, again in New York and on the stuffy train, took another deep breath of the exhilarating air: "Never felt better in my life!" he said briskly. Well, they admitted, he certainly looked fitter than he had on his last visit to Canada two years before; but surely the reason given for his replacement had been ill-health? The General refused to qualify his statement: he was absolutely fit. According to the press summaries, he added: "It was up to the persons who had given those reasons to make the explanation"; he would not.[4]

Nevertheless at a further interview that evening in his suite at the Seigniory Club, McNaughton explained his remark. There had been rumours, he said, that he was chronically ill. That was not the case. There was "absolutely nothing organically" the matter with him, but he had "suffered a severe attack of influenza and was forced to have a short rest after which he expected to be as fit as ever." He wanted a holiday in fresh Canadian air with his children and grandchildren but if there should be "a call to serve" at some later date he would answer it.[5] As might be expected, it was his earlier statement that made the headlines—to the consternation of the government. Ralston, who had greeted McNaughton's arrival with a telegram ("I fervently hope that you will find here recuperation and complete recovery"),[6] was particularly incensed.

After reading the evening papers Mackenzie King called for one of the reporters (John Connolly of the British United Press) who had interviewed McNaughton the previous evening. Connolly described what McNaughton had said at the later interview and King had been very relieved after the shock which, he said, the newspaper reports had given him; from these he had been afraid that "General McNaughton was referring to him when he said that those who had spread the rumours of his illness would have to explain them."[7] King telephoned McNaughton personally to invite him and his wife to luncheon on 7 February at Laurier House. There, for two hours before luncheon and for two hours afterwards in King's study, the Prime Minister and McNaughton discussed the affairs of the Canadian Army and McNaughton's removal; McNaughton did not blame King for that and there was no rancour. Nevertheless, he made it quite clear, as he was to reiterate the following September, that he did not feel that it was in the public interest that he should be asked to serve again under Ralston in whom he had "no confidence."[8] The question, however, did not immediately arise as McNaughton was on leave until June. Photographs that appeared in the newspapers of a smiling King waving goodbye to his guests at the door of Laurier House did much to silence rumours of a rift between McNaughton and the government; the Ottawa *Journal*, on 8 February, went so far as to predict "McNaughton will enter King's Cabinet as Minister of Reconstruction."[9]

King had, however, expressed concern over the earlier press reports that there was nothing wrong with McNaughton's health, and after McNaughton had shown the Prime Minister the draft of an explanation he proposed to make, it was agreed that Norman Robertson (Under-Secretary of State for External Affairs) should visit the General at the Seigniory Club next day to explore how best these could be corrected. Robertson arrived before luncheon on the 8th, and after a meal in McNaughton's suite the two considered various formulae. There seemed to be three possibilities: one, to continue to base the reasons for McNaughton's recall on grounds of health; second, to admit the controversies between McNaughton, the Defence Department and the British and to ask, now that a decision had been reached, that these be put aside; and third, to hold a complete investigation. After McNaughton's admission that there was very little wrong with him the first seemed impractical. The third was not feasible because "too many leaders of other nations" would become involved; strategy would be discussed and such disclosures could be useful to the enemy. The second course seemed the right one to follow and a draft press release, based on the one McNaughton had shown to King, was drawn up to the satisfaction of both Robertson and McNaughton. The release would be subject to Mackenzie King's approval. Robertson drove back to Ottawa and returned to Montebello late that night after seeing King. He suggested that McNaughton should categorically deny having made the statement that had caused the controversy, which McNaughton refused to do. Shortly before midnight Robertson, in Ottawa again, telephoned McNaughton to say that it was now felt to be advisable to make no statement whatsoever and to let the opposition make what capital they could; with that McNaughton agreed.[10]

This was undoubtedly, from the government's point of view, a wise decision. In the opinion of one correspondent, had McNaughton made a further statement on the lines tentatively agreed by himself and Robertson,* that would have made the General "more than ever the nation's hero." He would be pressed to present his own version of his removal, "which would, according to the view of observers on Parliament Hill, possibly result in an election." And, it was considered, the General would be quite justified in taking this stand after all he has been through." Nevertheless McNaughton was loyal to King; silence was in the government's interest.[11]

On 13 February the McNaughtons moved to Ottawa where they stayed at the Chateau Laurier until they could find a suitable apartment. Their home on Chapel Street had been rented to the Polish Embassy. The Chateau proved to be far too public, and they were relieved when, at the end of the month, they found a furnished apartment on Rideau Terrace. In the House, meanwhile, demands for the real reasons behind McNaughton's removal had been raised, which Mackenzie King staved off until Ralston should have finished presenting his departmental estimates during the debate on the War Appropriation Bill. The mood of the country also demanded an explanation. McNaughton, "the great scientist soldier," had been publicized "on the high, wide and handsome scale" in Canada, Britain and the United States and then, "as if he had never amounted to a row of beans, out goes McNaughton and all that Canada hears is that the man is in ill health. The story simply does not hold water." There was speculation that Canada was "taking orders from Whitehall and playing a second-class role in the councils of the Commonwealth;" that McNaughton "was shoved out and that the shoving was done in Whitehall."

> Around the club cars, the hotel lobbies and the pubs they say even uglier things. They say McNaughton was fired because he still insisted that the Canadian Army ought to fight as such and under Canadian leadership.[12]

---

* He suggested some dramatic headlines: "McNaughton Urges No Further Debate"; "General Asks Contentious Issue Be Forgotten"; "Says Government's Decisions should be Loyally Obeyed at Crucial Time"; "Willing to Forget Personal Views in Common Interest"; "A Nation at War Must Act as One, Says General."

Mackenzie King handled the matter with consummate skill. He knew that any reference by Ralston to having been told by the British that McNaughton was not fit to command in the field would bring "a terrible explosion from McNaughton." McNaughton was sure to demand a full-scale inquiry and there is little doubt that an analysis of SPARTAN would have vindicated him. And what then? Certainly it would have done nothing for Anglo-Canadian relations at the height of an all-out war. King persuaded a reluctant Ralston to "in no way, show any sign of bitterness" and to get everything in a form "devoid of provocation." As for himself, the Prime Minister admitted that he had a very difficult time not to "say anything that could injure McNaughton or injure Ralston's feelings." Then, on 14 February, King announced that it would not be in the public interest to make disclosures in the House, as he would be bound to do if he were to answer questions; he therefore invited the leaders of the other parties to meet him, when the correspondence would be put before them. This they agreed to do. The correspondence was read next morning and we may be sure that Stuart's "sequence-of-events" summary played a prominent part, for King leaned on it heavily. He had agreed with Stuart's version, thus showing Ralston "that I would be solidly behind him on the facts of the case." That done, the Prime Minister recorded: "I think all who came were impressed with the unwisdom of having a public disclosure of what the despatches revealed, and the effect this would have upon the morale of the troops, and on the United Nations effort, comfort to the enemy etc." And so *l'affaire McNaughton* petered out, the opposition switching, "and rightly so," King said, to the criticism "that it did not appear that the Canadian Government itself had sufficient voice in planning and control over their own troops."[13]

With the spring the McNaughtons moved to a summer home at Eagle Lake, Parham, in Ontario. The General fished, but there was much else to keep him occupied. He maintained close touch with the NRC; he made speeches; visitors were constantly at the door and his correspondence was heavy. One letter from Admiral Lord Louis Mountbatten (then Supreme Allied Commander in Southeast Asia, whom McNaughton had known as Chief of Combined Operations), particularly pleased him. "During my time as C.C.O.," Mountbatten wrote in mid-February when the news that McNaughton had left the

Army had reached him, "no other senior officer took such a lively interest in Combined Operations or contributed so much to the technical development of our equipment."[14] Morgan, despite the feverish pace with which planning for the invasion now proceeded, found time to write, as, of course, did Dill from Washington. Dill hoped to see him in Canada if he could get away: "I want to see how you both are and to hear all your news. . . . My love to you both and may 1944 bring you real happiness. I know what a sad year 1943 has been for you."[15] Field-Marshal Lord Birdwood, of ANZAC fame, wrote "as an old soldier–the senior in the British Army." He felt he must tell McNaughton how greatly he regretted the fact that the General was no longer in command:

I of course know no details of any sort so cannot speak but having at the back of my mind a feeling that you may not have had an altogether square deal I feel I must write to tell you how much I admired all I heard of your very strenuous and successful work in organizing and commanding your excellent force. May another turn of the wheel bring you good fortune.[16]

Letters streamed in also from persons of every walk of life in Canada, written with full hearts, and even from men in the Canadian ranks in England and in Italy. McNaughton was the prisoner of his correspondence which he conscientiously acknowledged. And his letters of acknowledgment bred others: "I merely want to thank you for your trouble in acknowledging my letter. It is most gratifying to realize that those who rank so high above us can still find time for those little amenities of life which mean so much."[17] Patiently he wrote again. The tone of all these letters was extremely friendly, as was McNaughton's reception by the public. In May he toured the West after receiving an honorary degree at the University of Saskatchewan as "Saskatchewan's distinguished son," and wherever he went flags were out, bands, honour guards and crowds of people. At his birthplace, Moosomin, the town took a holiday; the townsfolk surged around "to hear him speak and to shake his hand if possible."[18]

On 6 June the landings in Normandy began. McNaughton listened to the radio news reports with close attention but gave no indication of his feelings; his thoughts undoubtedly were of his men, now that the time for waiting was at last over. At least one man, tensely awaiting orders to cross to France, found time to think of him. "I thought particularly of General McNaughton, who had given everything he had to this Canadian force overseas and who was not here for the fulfilment of his plans and dreams. This was going to be his kind of war. A war of guns and machines operated by brave men. A war of science, too."[19] And Crerar, who now commanded the Army, said: "This Canadian army, which I have the high honour to command, is his creation. In the hard-fought battles and victory which lie ahead we will none of us forget the debt which we Canadians owe him."[20] A little later an anonymous "Letter to 'Andy,' " published in the *Maple Leaf,* reported on successes in Italy and Normandy in these words:

The battle's far from over
But we're keen to play our part;
We're still the blade you termed us,
Edging close to Berlin's heart.
The tempered steel of Canada's men
Has smitten the Nazi horde;
We'd like to take time out to thank
The man who forged the sword.[21]

His task had been important—that was recognized—but he would have dearly loved to see it through. "My row with Ralston and Brooke carried me out of that picture," he said, and there was no turning back.[22] He was looking towards the future.

The Progressive Conservatives had already approached him although it was July before any serious discussions took place. They hoped he would join a select group to work out a dynamic postwar reconstruction programme for the party to adopt; McNaughton, they said, would provide a forceful lead. The alternatives for him to consider were: it was proposed that a national trust should be set up to control certain of the country's natural resources, on a vast scale, and that he should be chairman of the board with a salary that would make him, and his family after his death, independent; or he should become the leader of the party. He was interested but, as he pointed out, he was merely on leave from the Army and bound by regulations not to take any public part in politics as long as the connection remained. He was also asked by the Kingston Allied Labor Committee* if he would consider running as a Labor candidate for Kingston in the next federal election; but, again, he was still in the army.[23]

In June Ralston sent for McNaughton, who saw the Minister at his office on Saturday, 24 June, and the outcome was that McNaughton's leave was extended for three more months from 1 July. The meeting was cold but courteous. Ralston conducted McNaughton to the map room where intelligence officers brought the General up to date on operations in Italy and in Normandy. He was invited to visit the room whenever he was in Ottawa. The summer wore on and McNaughton heard little from Mackenzie King; however, the Prime Minister wrote on 1 September to pass along the good news that the Canadians "have entered Dieppe today." Though he had not seen McNaughton for some time, King went on, "I have had you much in my thoughts and never more so than during these past few weeks—weeks in which the Canadian Army has manifested to all the world the fruits of the splendid training which they had under your leadership and direction."[24] In his reply McNaughton mentioned that he had spent the summer at Eagle Lake, that he was fit, and "looking forward to taking up active work shortly."

Towards the end of September, Ralston (who had become alarmed about reinforcements for the Canadian Army and had decided to see for himself the situation in Italy) told McNaughton that he was going away soon and expected to be abroad for some time; McNaughton's second period of leave was nearly up and he thought they should have a chat before he left. The message reached McNaughton at Echo Beach, where he was engrossed in serious discussions with the Conservatives, and he drove in to Ottawa and met Ralston on the 21st. Ralston mentioned McNaughton's leave: what did he wish to do about it? "With the march of events in Europe," McNaughton replied, there was no longer any need "to continue to hold myself available for return to the Army."

I wished to be free to take up work which might be useful. . . . I had made no commitments but many people of many walks in life and many political persuasions had spoken to me and urged that I should

---

* The committee was made up of Kingston trade unionists whose unions were affiliated with The Congress of Industrial Organizations and The Canadian Congress of Labor.

give expression to my views and give a lead as to what I thought should be done. I mentioned that I had been offered a seat by Labour and that I had sympathy of long standing with Organized Labour. I mentioned also that some of my contacts had been with those who might be considered to represent vested interests. There might be some opportunity of reconciling these extremes.[25]

Ralston, thinking perhaps that successes in Italy and in France might have modified McNaughton's views, turned to the Army. Did not McNaughton get a thrill out of its performance? It would have performed better had it been employed as a whole, McNaughton returned: "Canada had made a grave mistake in forcing dispersion." Ralston admitted there were difficulties and cited Montgomery's rudeness to Crerar in Italy earlier in the year. This provoked McNaughton into another forthright statement. "Montgomery"—even in the United Kingdom—"had deliberately set out to capture the Canadians for his own aggrandizement." He recalled "the impossible position" in which he had been placed when Montgomery had denied him access to Canadian troops in Sicily and added, "I had no support from the Government of Canada." Ralston brought the meeting to an end; he would report the conversation to the Prime Minister and McNaughton might expect to hear from King later in the day.[26]

The Prime Minister wasted no time; his reaction was to arrange for an intermediary, Norman Robertson, to sound McNaughton out that evening. Robertson first discussed the National Research Council. McNaughton gave his view that Mackenzie had earned the right to remain as president. Then, said Robertson, there was a need to maintain Canada's position in Europe. Would McNaughton consider becoming Canada's High Commissioner in the occupied countries? McNaughton thought not, indicating that he would have difficulty in working satisfactorily with British officers in view of his strong stand for the autonomy of the Canadian Army; he had not been popular with many in England who sought to treat Canada as a "Crown Colony." Robertson then proposed the highest appointment in the gift of Canada, that of governor-general. According to Mackenzie King's papers, he had had some thought of choosing a Canadian to fill this high office before submitting to King George VI the name of the Earl of Athlone in March, 1940. He would have chosen the man who had acted as Administrator after Tweedsmuir's death—Chief Justice Sir

Lyman Duff–had he been married; and, after the Athlones, "if Lapointe wanted the job he could have had it."[27] Lapointe (Minister of Justice) had died in November, 1941, and after that date Mackenzie King had fastened on McNaughton to succeed Athlone. Robertson, McNaughton wrote,

> revived a suggestion which the PM had made during my visit in February 1942 that I should become [Governor-General].* The PM had hoped that I would accept another couple of months leave to ensure that my health was completely restored and to give time for the consideration of this. I said that I thought that this office might give useful scope for my endeavours but that I did not wish to remain longer in the Army. . . .[28]

Mackenzie King, after hearing Robertson's report, made an appointment to see McNaughton on Saturday, 23 September. He was most friendly. At the meeting he spoke of the building and the training of the Army, which he considered of "greater significance than the employment of the Army in the later operations." He referred to a newspaper report that Crerar was to be made a full general and said that McNaughton's work should be recognized first. He intended to announce McNaughton's promotion with the news of his retirement. McNaughton, though he thanked him for this "generous tribute," said he was "not much concerned with recognition." He was sure that what he had done was right and that the Army "used as a whole" would have been more effective. The Prime Minister thought a report prepared by Stuart–"Which," McNaughton interjected, "I have never seen"–came "as close to being correct as was possible." McNaughton said the matter would be analysed some day; he now wished to give his whole attention to the future. He had "no anxiety as to what would be said when all the facts were known. They could not be revealed now because of the danger to inter-imperial relations." King turned to the future. As Robertson had done, he talked of NRC. McNaughton "expressed the highest confidence in Mackenzie under whom the nucleus of pre-war organization had blossomed into flower in useful

---

* In his papers McNaughton invariably left a blank for this appointment but there is no doubt about what was meant. Though King discussed the governor-generalship with McNaughton in 1942, as is confirmed by the McNaughton Papers, Norman Robertson told the writer that the Prime Minister first broached the matter with the General in London during August, 1941.

service to the State." Mackenzie was best qualified "to direct this organization in the transition from war to peace." McNaughton had spent much time with him in talking over these problems and had attended the Review Committee meetings where postwar organization had been discussed.

Like Robertson, King turned to the possibility of McNaughton serving Canada abroad and again McNaughton offered his objections. In the occupied countries Canada's stake would be small and it would be necessary "to defer to the greater interests of the United Kingdom." Having "stood for Canada" he might not be welcome to the British Foreign Office as High Commissioner. The Prime Minister's personal inclination, he said, was to ask McNaughton to join his government but cabinet unity was important. McNaughton, with Ralston in mind, replied that he "was well aware there were certain members with whom I could not work." Well, said King, what about the Senate? That would provide a forum for McNaughton to develop his thoughts for Canada. If he became a senator, McNaughton predicted, he would almost certainly be submerged.[29] Then, said McNaughton,

> he referred to the suggestion he had made to me, some two years ago, that I might become Governor General and I indicated that I thought well of this and that it might . . . give me the opportunity I sought to contribute to the consolidation of Canada's nationhood. It was left that we would meet again early next week to explore this possibility. The difficulty was that the term of the present incumbent does not expire until next spring and I was anxious that my participation in public affairs should not be unduly delayed.[30]

He had received many approaches, he told King, and the time was fast nearing when he must make a choice. While he felt that King's opinions and his own were close concerning what should be done in Canada, they might have to work "with different groups and by different methods"; perhaps he should "throw his hat into the ring in an attempt to crystallize the best elements in the situation for forward thought and progress."[31] King made no comment other than to arrange for the talk to be continued in a few days.

McNaughton's interest in the governor-generalship had relieved King's mind: "No two men shared more in common than he and I did in the constitutional position and relationship," the Prime Minister recorded, and he would "feel every confidence in his appointment to that position."

. . . the Governor Generalship fits in with my feeling I have come to have about the desirability of standing for a new nation, getting away from this subsidiary position and to have a Canadian Governor General. To my mind I have been thinking over Massey and his wife, Vanier and his wife, and last night quite a bit of McNaughton and his wife. Of the three I really believe McNaughton would be the best and it would be a fine recognition of his services as a General at the beginning of the war. Churchill has said to me he would be prepared, if he were in England, to offer him a Peerage.[32]

Mackenzie King then sent for Ralston, who was obviously uneasy about what might have been discussed. The Minister, King wrote, "not knowing what was coming . . . was ashen pale and motionless, like that of a man . . . having his own fate decided. As I went on with the discussion, his calmness came back a little, and when I mentioned about the Governor Generalship he was at once immensely relieved."[33]

The same afternoon McNaughton, at the request of Ralston, discussed with the Minister what form McNaughton's written request to leave the Army should take. Only with Ralston, McNaughton's bitterness showed and it did again on this occasion. He warned the Minister about the future:

I believed in the British Empire but not on the basis of providing raw materials and detachments. I had great anxiety about future relations, U.S. and U.K., with Russia. I thought our commitment to maintain troops in occupied Europe should be cut down as soon as possible. They were hostages for our compliance with British schemes. . . . We had lost the potential position we might have had for our Government . . . if we had had a real army their view would have carried weight. In insisting on sending 1st Corps [to Italy] . . . the Government of Canada had run counter to agreed plans (COSSAC) and played into the hands of Brooke *et al* who wished to break up the Canadian Army for political reasons. We were too independent for them.[34]

Ralston said that McNaughton might be in a position to do something about that himself. King had mentioned the possibility of McNaughton's becoming Governor-General. Ralston hoped it would come about. This would give McNaughton an opportunity to carry on the work of consolidating Canada's constitutional position within the British Empire.

They parted in a fairly friendly way. McNaughton had never doubted Ralston's sincerity of purpose. They had differed in methods but these matters were past and he was prepared to let history decide. Ralston recalled a phrase that McNaughton had once used about him, that "he had a heart of gold."*[35] They shook hands and hoped to look forward into the future and not back.[36]

Later that day Mackenzie King wished Ralston—about to leave on his fateful mission to Italy and Northwest Europe—Godspeed and a safe return. " . . . and he left with the lightest heart that I think he has had for many a day." The Prime Minister himself, with not a cloud to mar the blue of the sky, turned his thoughts to the next general election:

> If between now and that time it should be possible to have the Canadian flag made the flag of Canada; or [a Canadian] like McNaughton announced as the next Governor General, and then go to the country on a straight social reform programme, and the importance of my being the one to serve in the making of peace, I believe it would be possible to carry the country.[37]

Three days later Hugh Keenleyside of the Department of External Affairs telephoned McNaughton. Had he any thought of going to a university as president? McNaughton told him that he had been offered McGill in the past and a recent approach had been made by Toronto. Would he, Keenleyside asked, consider the University of British Columbia? McNaughton did not think the job would appeal to him but nevertheless Mackenzie King, despite his pleasure that McNaughton was at least considering the governor-generalship, raised the matter again at their next meeting.[38]

On 30 September McNaughton's retirement was announced with promotion to full general, the third member of the Canadian forces—after Currie and Otter—to achieve that rank. The Montreal *Star* announced that in the prime of intellectual life, he was too good a man to lose. "It would be more than a pity if this man's brain, his executive

---

* The limitations of documents are demonstrated here. When tackled with this McNaughton said: "He chose to forget the other half of the words I used to him. What I actually said was, "Ralston, you have a heart of gold but a misplaced mind."

ability, his capacity for leadership and his gift for inspiring confidence were lost to Canada which is already far in his debt."[39] When McNaughton met King at his office on 3 October the General mentioned that he was under very heavy pressure to enter public life—from the Conservatives, some Liberals, Labor and the CCF. He did not like some elements of the Conservative Party, notably those represented by Premier George Drew of Ontario who, in June, had made a speech in Winnipeg urging the adoption of conscription for overseas service at once. Though conscription had been the law of the land since 1942, there were dangers about imposing it. Quebec's opposition to conscription could not be ignored. If the need for adequate reinforcement of the Army required it, then conscription should be imposed but the case presented so far was anything but conclusive. To introduce conscription at that time would set the stage for racial schism which might result in the failure of Confederation and the disappearance of the Canada they had known. "It could well be, if the ill-feeling against Quebec was developed, blood might flow and then we would have a problem for a thousand years."[40]

It had been suggested, McNaughton said, that he might be able to eradicate these policies and that he had a duty to try. At any rate he believed in tolerance and co-operation. He was talking thus freely to King, he continued, because of their old friendship, and because on major issues their views were close. Would he be right to accept the high office that had been proposed or would he serve the public better by "throwing his hat in the ring," as it had been put to him? If he should accept the governor-generalship, when would there be a public announcement? It was essential to know because of the mounting pressure.[41]

Athlone, Mackenzie King told him, wished to stay the full term which expired the following May, but if McNaughton accepted he would approach the Governor-General at once and then His Majesty The King. It would be best to make no change until after the election, which would probably coincide with the end of the war in Europe. The life of Parliament extended until April and McNaughton's appointment would be announced two months ahead of that. Then, said McNaughton, he would give it serious thought; and though King wanted no immediate decision, he would like to know his answer in a couple of weeks. They both commented on their frank conversation. In the "bitter partisanship" which marked Canadian politics "the conversation was remarkable." There was friendship behind the scenes between politicians of opposite parties in England but that had not yet developed in Canada.[42]

McNaughton then gave the most careful consideration to what he should do. He balanced the proposal which Mackenzie King had made against the proposition that he should join the Tories. Against the "difficulties of the position proposed by Mr. King" he weighed his own inclination "to get out into the open and take part in the struggle." If he were to be Governor-General there would be "things I would have to do in that office that I would not like." There would be "financial difficulties," for governors-general had been men of independent means, and "at the end of office I would have nothing left except my present pension." On the other hand, from the Tories, there was the offer of independent means without open political commitment. But in the final conclusion "after much anxious thought" and "going over matters point by point with my wife I decided that whatever the upshot and whatever the difficulties I could not refuse to do the task proposed by Mr. King and I was comforted in this by the thought that in the long-term view I might have made a better contribution to Canada. I thought too that my anxieties over Drew and his stand against Quebec, and so the unity of Canada, could be liquidated. . . ."

He was finally convinced on 10 October by Canon Brosseau, a family friend and the only man in whom McNaughton confided fully. The Canon advised him not to enter political life. He could do better work for Canada by "remaining neutral as between the parties and speaking as often as I could to give direction to the thoughts of the people." Brosseau spoke of the affection that McNaughton enjoyed, and the great influence that his words commanded.[43] That evening McNaughton's mind was clear, and he at once informed the Progressive Conservatives "that he could not refuse a task proposed by the Government" and that he hoped they would have "faith that my decision was based on my duty as I saw it; that there were no political considerations; that nothing except a national emergency could alter my decision."[44] On the 11th he wrote individual letters to every one of the Conservative leaders who had approached him; he was not free to go on "with you and your associates in the work you have undertaken "but felt sure" that some months hence when the facts are public I will have your full sympathy with the decision I have now taken." One of them wrote that Mackenzie King might be "manoeuvring" to keep McNaughton out of public life, for he had overheard a conversation in which the Prime Minister had said that McNaughton was the only person he was afraid of as a rival; but apart from that there were expressions of disappointment, certainly, but complete under standing.[45]

On 13 October McNaughton wrote to Mackenzie King: he now held himself available "to undertake the office you have suggested."

I have told those who have been urging me to enter the political arena of my decision though I have given no indication of the offer contemplated; only that it is strictly non-political in character. . . .

May I say that it is only the observations which you yourself have made to me which have given me confidence even to think of undertaking this work and to feel that through it I might do something which would be of value to Canada.[46]

Next day the Prime Minister telephoned him from Kingsmere. He was "delighted" with the decision. Dr. Mackenzie had been appointed to the NRC and, King said, it seemed to be always McNaughton's job "to put things into shape and then move on." This new office, the very highest, would give him one more opportunity for that.[47]

That day Mackenzie King wrote in his diary:

This means that Canada will have as her Governor General or rather as the representative of the King, one of her own citizens and he, one of the most distinguished in Canada's history. The appointment will create something of a sensation in Britain and in Canada but I am sure will be overwhelmingly approved in Canada. It will help to remove a certain badge of colonialism but will have to be accompanied by a change in the title . . . from Governor General to King's representative. I am certain that the time has long since gone by when we should continue to have as the King's representative in Canada members of the nobility from England as if from among our own citizens, men of equal distinction and ability could not be found and trusted. To McNaughton is owing the highest honour which Canada can confer. He is a great man and has already served Canada and the Empire and the world in a great way through his contribution to the work of research and in the formation and training of the Canadian army overseas. His resignation for the reasons given, as President of the National Research Council, was a mark of his noble character. He is a true Canadian. In visiting other parts, his name, record and appearance will be an inspiration along the right lines to the younger generation who will begin to look up to their own citizens, instead of being taught to worship as a superior breed those who come to Canada from the privileged positions in the Old Land. His appointment will be in accord with the spirit of the times and of the Canadian people, which aims at a true democracy.[48]

The Prime Minister then notified Athlone of his choice of McNaughton. "I was interested to see how His Excellency would receive this word," he admitted. "While it seemed to occasion a moment's surprise in his mind, his reaction seemed to me to be distinctly favourable rather than otherwise. He raised no question but supplemented what I had said about McNaughton possessing a fine character and being well suited for the position. He said nothing about it being any break in the tradition. He added that his wife was a fine woman."[49]

So the matter rested until, less than three weeks later, a national emergency, which became known as the conscription crisis of 1944, forced McNaughton to think again.

Chapter 2

The Conscription Crisis

The conscription, or reinforcement, crisis of 1944 came about through a temporary shortage of general duty infantrymen. Thus infantry units fighting in Italy and Northwest Europe found themselves unable to make good their losses because reinforcements were not immediately available, whereas other units were up to strength and easily reinforced. How had this situation come about? There is no one simple answer, and only the main factors can be dealt with here.*

The root of the problem is to be found in Canadian adherence to British "wastage" rates. These were designed to meet global needs, and not only those in the two theatres of heavy fighting–Italy and Northwest Europe–where the Canadians found themselves. Thus the British forecasts of the number of infantry reinforcements that would periodically be required to replace casualties proved far too low for Canadian needs; at the same time their estimates in the other arms and services were too high. Another factor was the increase in the administrative services (the "tail") required to maintain Canadian fighting troops (the "teeth") in *two* theatres, whereas the original conception had been that the Canadian Army would fight as a whole in a single theatre.

With foresight the crisis could have been avoided. Late in 1943 a warning came from the field in Italy that the British wastage figures did

* The writer summarized for General McNaughton, in thirty foolscap pages, the various factors leading up to the crisis. He found the account "interesting and convincing" and suggested that it be filed with his papers, which has now been done. There is no space, however, to include it here.

not match the Canadian experience. Had it been acted upon *at that time* recruiting streams could have been diverted into the correct channel, and enough surplus troops in the other branches of the Army could have been remustered and trained as infantry. But none of this was carried out drastically enough and the crisis struck. Then, to avoid conscription, it became necessary to induce NRMA soldiers, called up under the National Resources Mobilization Act for service in Canada only, to volunteer for overseas service, to fill the ranks of the depleted infantry units. Had they done so the crisis would have passed, but they did not.

The maximum expansion of the Army in Britain took place in 1942, when the First Canadian Army was formed. On 3 December, 1941, an assurance was given to the Cabinet War Committee by the recently appointed CGS (Stuart) that the programme could be completed by voluntary enlistments and that, in foreseeable circumstances, there would be an adequate number of reinforcements. During March of 1942, as we have seen, McNaughton informed the Cabinet War Committee in Ottawa that, in his opinion, the Army programme represented the limit of the force that Canada could deploy and maintain in a war of long duration; no major increases would or should be made. At McNaughton's direction, on his return to England, a study of the manpower required for a well-balanced army was undertaken during the late spring and early summer of 1942. He decided that, in the absence of direct experience, the British estimate of wastage resulting from battle casualties during intensive* operations should remain as the basis for providing reinforcements for the Canadian Army Overseas.

McNaughton's manpower study brought to light a disquieting situation. At 30 June, 1942, there was a deficiency of 704 officers and 10,257 other ranks against the officers and men who were required to keep the units up to strength and as reinforcements. Stuart came to England and with McNaughton visited the war Office where they learned that a cross-Channel invasion would be possible in 1943 only if there should be a break in German morale. In such circumstances battle reserves would have little significance; the landings would meet light resistance and the full scale of reinforcements would not be

* Intense rates were based on the conditions of battle (offensive or defensive) against a first-class enemy of strength not greatly less than the Canadian, with all forces, land and air, engaged.

required. Therefore, with the urgency removed, it was agreed to begin by completing the units required for the First Canadian Army; the provision of reinforcements could wait.

Shipping difficulties then arose. Space allotted up to the end of April, 1943, made no provision for the full complement of men for the Army, lines of communications and base units, and though McNaughton would have accepted philosophically a reduction in the Army programme to fit the available manpower, it was decided that the programme should stand, with the British providing up to nine thousand troops per Canadian division. Thus, on 6 January, 1943, the Cabinet War Committee approved a reorganization of the First Canadian Army, which was as follows:

An Army of three Infantry Divisions, two Armoured Divisions and three Army Tank Brigades together with Corps troops and a Canadian quota of Army, Line of Communications units, augmented by British troops on a scale of 9,000 per Division; the completion of the pool of reinforcements based upon three months [casualties] overseas at the intense rate; and, on completion of the foregoing, the despatch abroad of 5,000 men per month commencing 1 September 1943 for the maintenance of this force, allowing for battle casualties on the scale of 6 months intense, 3 months normal, and 3 months at the "no activity" rate.

This programme, it must be emphasized, imposed a ceiling* on the First Canadian Army which could in no way be exceeded; and that ceiling was approved at a time when the employment of the Army as a *whole* in *one theatre* was envisaged. Though the Army was to be split and employed in two widely separated theatres at the instigation of the Canadian government, the government did not correspondingly raise the manpower ceiling which had been established in very different circumstances.

In Britain, in the fall of 1942, Canadian infantry reinforcement units were short of men. This was aggravated by losses at Dieppe and the adoption of revised British rates of wastage which called for the holding of a larger number of reinforcements than before. McNaughton ordered the Army Service Corps and Forestry Corps to comb their units for men suitable for service with the infantry; that

* Of 226,000, revised by the Cabinet War Committee on 11 March, 1943 to a figure of 232,100.

began in April 1943. He also ordered the reclassification of surplus tradesmen throughout the Army. In May he asked Stuart to give the highest priority to the provision of infantry reinforcements; he sought, and obtained, a number of formed infantry battalions from Canada. Three of these, the Lincoln and Welland Regiment, the Argyll and Sutherland Highlanders of Canada and the New Brunswick Rangers, found places in the reorganized order of battle of the 4th Canadian Armoured Division, but another battalion, Les Voltigeurs de Québec, was disbanded and its personnel used to reduce by half the shortage of 2,282 French-speaking infantrymen. with the target date for the cross-Channel invasion then set at 1 May, 1944–giving the Canadian Army time to prepare itself for that–McNaughton thought he had several months during which the wastage rate would be far less than the five thousand reinforcements promised monthly. By May the Army would be in good shape. But in 1943 Italy absorbed a corps and McNaughton's concept of an all-Canadian army fell apart. Far from having the time he had expected to prepare the Army for the assault on France, he saw an additional administrative "tail" provided for the Mediterranean.* And thousands of men were of course "lost" in the long transit "pipelines" from Canada to Britain and Britain to Italy at any given time. No allowance was made for this. The manpower ceiling, as has been said, represented a maximum figure which could not be exceeded.

The fighting in Sicily provided the first opportunity to test the operational forecasts against the British wastage rates that had been adopted for the Canadian Army. Doubts of their validity were raised.† It was stated from Italy on 1 December, 1943, that Canadian infantry casualties in Sicily had been higher than predicted. It is not surprising. The British figures were based on the experience of the North African campaign and this was not typical of the fighting in Sicily. In North Africa the broad face of the desert gave few opportunities for linear defence; wide-sweeping enemy armoured thrusts were just as likely to overrun support and service units in the rear as the forward infantry. British wastage rates, therefore, were relatively high, for example, in Service Corps units; infantry wastage, correspondingly, was lower. But

---

* On 18 December, 1943, General Montague pointed out to Crerar that *since* base installations had been set up new demands for an additional 228 Canadian officers and 1,342 other ranks had been received from Italy.

† By Brigadier A. W. Beament.

In the summer of 1944 he could relax

The nomination at the Grey North Liberal convention, 16 December, 1944

The retreat from Grey North

in Sicily, mountains confined movement to a few heavily defended roads. The fighting became linear in character and the infantry took the worst of that. It was to be the same in Italy.

That the wastage rates were erroneous for Sicily and Italy, however, would be far more obvious to the Canadians than to the British. A larger force, containing formations that would seldom be equally engaged at one time, could carry on with a smaller proportion of reinforcements; a smaller force, such as the Canadian, might suffer severely in a single engagement. The warning from Sicily alerted Canadian Military Headquarters but caused no alarm. Nobody bothered McNaughton with it at that time, when rumours of his replacement were becoming more pronounced. Gradually, as time went on, forecasts of a purely Canadian trend could be based on the experience of further fighting. That was a fortnight before the grim battle for Ortona.

Stuart succeeded McNaughton in an acting capacity on 26 December, 1943, and, in addition, became Chief of Staff, CMHQ. At that time the CMHQ estimate of casualties for 1944 was 75,000 all ranks. Some, not seriously wounded, would be "recovered," leaving 45,000 to come from Canada. These men were to be shipped at the rate of 5,000 a month (as had been agreed) for the first six months and 3,000 during each succeeding month–a total of 48,000, as had been accepted by the Adjutant General (Major-General H. F. G. Letson) in Ottawa. Stuart changed that. Ottawa proposed that the rate of supply should be 4,000 reinforcements each month (subject to reduction in the late months of 1944, should the actual casualties prove lower than the estimates). CMHQ considered that this proposal would result in reinforcement holdings becoming dangerously low in June, July and August; far better to insure during the early months and, if necessary, to cut down later. But on 19 January, 1944, Stuart, in his new and powerful appointments, came out in support of the Ottawa view and brushed aside the recommendations made in London. He was satisfied that the amended rate of flow would cover requirements and Ralston, thus reassured by Stuart, made a statement in the House which lulled the nation into a false sense of security concerning the reinforcement situation,[1] so that it came as a great shock when, towards the end of 1944, the crisis materialized.

There were plenty of other warnings of the impending crisis but these were either ignored or acted upon too late. In February, for example, General Montague went to Italy to see the situation for himself; he reported to Stuart by cable that the casualties of general duty infantry were disproportionately high. Larger holdings of reinforcements moreover were required in Italy (in addition to those occasioned by casualty figures); convoys now sailed at thirty-six-day intervals instead of monthly, and thus men were lost in the pipeline for longer periods. But because of the shortage of reinforcements in Britain, holdings in Italy could not be stepped up; and as from March operational priority was accorded the Northwest Europe theatre. The fears of CMHQ that the reinforcement pool would fall dangerously low in June, 1944, were on the way to being realized. Nevertheless, at the end of February Stuart still found the picture favourable, with the possible exception of the infantry. New graphs would be drawn to show Canadian trends as soon as the February casualty figures were known.

The numbers of infantry reinforcements held in Britain were some nine thousand below acceptable figures. Stuart, while willing to remuster men (who were surplus in other corps) to infantry, would not recommend that action yet, and he suggested a review on 1 May. By the middle of March, however, Montgomery*—who was studying the overall manpower situation within 21 Army Group, in readiness for the Normandy invasion—called for Canadian reinforcement figures and forced Stuart's hand. Montgomery suggested that remustering to infantry and retraining should be started as soon as possible. Even after the arrival of the next reinforcement draft from Canada, there would still be a shortage of five thousand infantrymen. Stuart noted that future drafts included 1,750 reinforcements for the Armoured Corps, 820 for the Field Artillery and 800 engineers. He asked that these be replaced by general duty infantrymen but Ottawa informed him that it would be impossible at that late date. Letson, very disturbed, suggested that they should be remustered to infantry on arrival and, in addition, he authorized the remustering of one thousand gunners, five hundred engineers and five hundred men from the Armoured Corps who were already in England. As well as this, the Adjutant General agreed to send the 13th Brigade (composed largely of NRMA men assembled for home defence) from Pacific Command in Canada if the NRMA men could be induced to volunteer. To

---

* Prompted by Brigadier A. W. Beament.

encourage them to do this he arranged for all general servicemen in the 6th Division in Canada to be posted to this brigade, rather than overseas in smaller drafts, so that the NRMA men might be fired by their example.

The campaign of inducement began on 30 March and its progress was closely watched. If NRMA men in the 13th Brigade would offer themselves freely for overseas service it could reasonably be expected that NRMA men in other home defence units would do the same. Voluntary methods would then survive until the end of the war and the unity of the country need not be tested. The attitude of the 13th Brigade is, therefore, important to our later story and may be briefly described here.

Brigadier W. H. S. Macklin, the brigade commander, toured the battalions and addressed the men. Regimental officers then interviewed the members of the companies and platoons. The chaplains worked diligently. But resistance was amazingly strong and the response disappointing. Only a handful of men volunteered each day in the various units. The general servicemen, it seemed, mentally ranged themselves on one side, and the NRMA, with the attitude that if the government wanted them it would take them, on the other; they would not ease the government's task by going of their own free will. The attitude of the general servicemen, who, like everyone else, referred to the NRMA as "zombies" (the walking dead), helped to widen the gulf between them. Then drafts of general servicemen began to arrive from elsewhere in the 6th Division and the climate changed. Convinced now that the brigade would move overseas whether they volunteered or not, many NRMA men cheerfully took the plunge. When the drive ended 800 men had enlisted and these, with the general servicemen, brought the "active" strength of the brigade to 2,432. Even the most optimistic, however, could not have described the result, in this the most hopeful of all the brigades, as more than a limited success.

In Italy, meanwhile, infantry reinforcements were dwindling fast. The level of the reinforcement pool, sufficient to cover two months' wastage at intense rates as set by McNaughton, had been cut in half by Stuart at the end of March. At the beginning of May Acting Lieutenant-General E. L. M. Burns, who had succeeded Crerar as corps commander, noted that if the wastage rates in June and July continued to be intense, reinforcements would be exhausted before further drafts reached Italy. He decided to remuster 800 reinforcements of other corps to infantry but this, necessitating three to six weeks of training, would be no quick remedy. The Liri Valley fighting that brought the fall of Rome cost Canada, in May and the first week of June, 3,700 casualties. During that period the reinforcement pool in Italy was bled of 3,300 officers and men to build up units in the field. Urgent appeals to England brought a trickle of infantrymen—127—and remustering was put in hand on an increasing scale.

By the beginning of June the reinforcement crisis was steadily building up. The situation in Italy was critical. In England, despite remustering, it was no better; there the scales of reinforcements were insufficient to meet the requirements of the First Canadian Army in its invasion role, to say nothing of the 3rd Canadian Infantry Division in the actual assault. And the dispatch from Canada of the under-strength 13th Brigade—in which so much hope had been placed—would certainly not solve the problem.

Then, while the battle for Rome was still continuing, Burns advocated the *addition of one infantry brigade* and a squadron of engineers to the 5th Canadian Armoured Division to conform to British practice. These troops would have to be found in England, and Stuart, on the eve of the cross-Channel invasion, turned the proposal down. Burns then offered, by converting the units of other corps to infantry, to raise the brigade himself. Stuart determined to deal with the matter on the spot in Italy but before doing so he asked Ottawa if an additional brigade could be reinforced. He was told that the infantry reinforcement pool would have to be increased by 90 officers and 1,578 men and that the proposal was unlikely to be accepted unless militarily necessary.

In mid-July Stuart reached Italy where Generals Alexander and Leese quickly convinced him that the step was indeed operationally necessary, and Stuart instructed Burns to go ahead with the brigade on the understanding that Ottawa would have the final say. The decision in Ottawa was without doubt influenced by Stuart's report from Italy on 15 July:

Reinforcements situation here is good. Units all up to strength and two months supply at intense rates are available in reinforcement units.

This situation will improve as result of whole corps being held in general reserve for period of about two months.[2]

At the beginning of August Stuart returned to Canada to report on the war situation. He appeared before the War Committee and confirmed the impression that all was well in Europe. Mackenzie King recorded: "He made clear that we had plenty of reserves. Presented a request for some additional men for use in Italy but emphasized that this would not lead to any pressure for further men."[3] On 3 August the Cabinet War Committee approved the reorganization of the 5th Armoured Division in Italy. Three weeks later the new 12th Canadian Infantry Brigade, after little better than a month's training as such, plunged into the stern fighting for the Gothic Line.

In Northwest Europe relatively light Canadian losses on D-Day and for the rest of June (301 officers and 3,142 other ranks) helped stave off the relentlessly approaching crisis for a little while longer. Battle casualties for July were higher (605 officers and 7,213 other ranks) but the delay that had occurred in committing all the formations of First Canadian Army had put back the date when wastage would reach a peak. Stuart appeared to be well satisfied with the position as it appeared to him at the end of July: "On 30 Jul 44, after over twelve months fighting in the Middle East and two months in France we have a reinforcement pool overseas with a strength in excess of three months reinforcement requirements at intense rates. This is a most satisfactory situation. . . ."[4]

But all was not well. What Stuart would still not admit was that it was reinforcements for corps *other than infantry* that were piling up; infantry reinforcements were running out. Montague pointed out to Stuart that reinforcements, other than infantry, were, in fact, surplus holdings and must be remustered to infantry at once; this was the last moment for decision. General Crerar, writing on 4 August*[5] from France, pointed to a shortage of nearly two thousand infantry and by

* On 24 October, 1944 (after the conscription crisis had broken), Ralston revealed at a meeting of the War Committee that Stuart had received a telegram dated 4 August in which Crerar had expressed concern about the state of reinforcements; the Minister had seen it for the first time that day. "It was a strange thing," Mackenzie King commented acidly, "that the Cabinet had not had that telegram of Crerar's read to us. That early in August, we were told by Stuart, the Chief of Staff, there was no need to be concerned about reinforcements. Indeed it was only now after the middle of October that we were told for the first time there was such a need. . . . It was strange too the Minister should not have known of it until today."

15 August the deficiency had grown to 2,644. Stuart, on receiving Crerar's message of the 4th, began to grow alarmed. On the 4th he ordered 2,045 men to be remustered in England and to begin infantry training; orders went out to remuster a further 1,077 eleven days later.

The reason for Stuart's refusal to face up to the situation can only be guessed at. Did he keep the matter from Ralston because he feared the Minister would then regret having got rid of McNaughton? Or was he gambling on the war ending in 1944? The latter is far more likely.

In a telegram to the CGS on 26 August Stuart, though an air of reassurance still crept in, at last gave warning of the true position. He asked to use, at his discretion and in an emergency only, infantry *tradesmen* as general duty infantrymen (in other words, as fighting soldiers), for about three weeks.

> The present situation is not a manpower problem in a true sense. We have the men. It is not a problem of general supply. It is a problem of detailed distribution. We have taken British [War Office] rates as a guide. Experience of particular conditions of war in France has shown that [War Office] rates for infantry are too low and for practically all other arms are too high. In addition we did not anticipate that practically all inf casualties would be in general duty personnel. We are now going through a period of adjustment. In three weeks to a month we will be alright [*sic*]. I need assistance to bridge this intervening period. My recommendations . . . represent the only way you can help me to bridge this gap. . . .[6]

The CGS put Stuart's telegram in front of Ralston but, apart from initialling it, the Minister took no action. This is scarcely surprising. Though Stuart had by now penetrated to the heart of the problem, he had wrapped it around with such a cloud of reassuring words that its implications and the urgency of a vigorous solution quite escaped Ralston.* On 29 November, long after Stuart's roseate vapours had been dispersed and the government felt the bleak wind of crisis, Ralston admitted in the House that he had been remiss:

> . . . If I remember correctly it was a Sunday morning, and I simply initialled the telegram and handed it back to the Chief of the General

---

* McNaughton believed that Stuart was still gambling on a speedy end to the war. A German collapse, after the hammering they had received at Falaise, appeared not unlikely but it was highly imprudent to count on this.

Staff. The telegram concluded by stating that everything would be all right in three weeks. As a matter of fact, if I had known as much then as I know now I would have realized that this meant that it would be unlikely that the pools would be built up, although if it had come to my mind at all I would have thought that the men who were being trained in England would be coming in three weeks, and this was simply a situation to be tided over at that time.[7]

Ralston's willingness to consider this a minor tremor in a zone whose stability had always been uncertain meant that the time for action had run out; and yet, had the significance of Stuart's cable struck him or had he asked for clarification, the situation might have been remedied even at that eleventh hour.

The Adjutant General and Stuart indulged in correspondence about how best to protect the trades pay of the men who, as Stuart had suggested, should augment the rifle-strength of the battalions. It was not before 23 September that CMHQ authorized surplus tradesmen to be remustered–and that was six weeks ahead of Ottawa's final approval. Meanwhile, at the end of August, in a long overdue step, the Canadians at last revised wastage rates to fit their battle experience. The remustering of reinforcements surplus to the new figures then began; but this step, again, had been left until far too late.

In September the Canadians in Italy began to fight their way through the Gothic Line. Infantry casualties were heavy. Burns reported that, despite remustering, the future employment of the 1st Canadian Corps would be jeopardized unless a greater number of general duty infantrymen could be sent out with the October draft. Stuart, under heavy pressure from Crerar, could not comply. In an appeal to Ottawa, Stuart in vain requested the direct dispatch of a thousand infantrymen from Canada to Italy. Italy, he was told, could only be reinforced at the expense of the pools in England.

This plea from Stuart at last aroused Ralston. Something must be wrong. Two days later he was in Italy to find out for himself, and on his return he visited England and Northwest Europe. What he saw appalled him:

In Italy I visited the reinforcement units and most of the Infantry battalions. I talked with the men themselves, and time and again I got the same story–about having to go back to the line repeatedly, or not having leave, or being short-handed, when trained N.R.M.A. men could be made available to give them relief.

At the end of September, just before I left Italy, the pool of trained infantry reinforcements was down to between ten and fifteen per cent of what had been planned, and infantry reinforcements were so scarce in the United Kingdom that the only addition they were getting was the monthly convoy to arrive in about two weeks, bringing not even enough trained infantrymen to meet the casualties which would have been incurred in those two weeks.[8]

Thus Ralston faced the truth. Since the beginning of the year Stuart had allowed infantry reinforcement holdings to decline until units, at grips with the enemy in two theatres, could no longer maintain their strength. The Minister desperately sought an answer. In England he ordered a manpower study; the preliminary finding was disappointing. He cabled Canada for figures of available reinforcements. On 8 October he flew to Brussels and conferred with the acting army commander (Simonds).* He visited Canadian formations, units, reinforcement pools and hospitals. He saw Montgomery and, in Paris, Eisenhower. But you cannot shout at arithmetic and expect it to go away; a more accurate manpower study awaited him in England. In both theatres, it seemed, after allowing for casualties at generous rates, there would be a surplus of fourteen thousand reinforcements in the other arms during the balance of the year; in the infantry there would be a shortage of two thousand. And so, on 13 October, Ralston notified the Prime Minister of his apprehensions. Two days later CMHQ reached the conclusion that, at the end of the year, there would be no infantry reinforcements left. To restore the pool to the level of approximately two months' wastage at intense rates would require the bulk provision of fifteen thousand general duty infantrymen. On 16 October Ralston left by air for Canada, taking Stuart with him. On his return he lit the fire that, fanned by the opposition and the press, flared unchecked into the conscription crisis.

The Minister, on his arrival two days later, called on Mackenzie King. He outlined the seriousness of the position regarding reinforcements and said there was great unrest in the Army Overseas

* Crerar was sick.

about "a standing army in Canada doing nothing" while they were under strength. King tackled him about the request for an additional brigade in Italy, which had been sanctioned in August; had the Prime Minister not been assured then that this could be done without difficulty and without resorting to conscription? "Ralston gave some explanation about the nature of the brigade," King wrote, "but I could not follow him." The Minister "attempted at one point a sort of defence at having separated the army; also as to sending part of it to Italy, saying we had done honour to our country in giving it a place in world affairs by doing this or some such expression. I imagine he begins to see that McNaughton and Churchill were not unwise in wishing to keep the Canadian army intact for final operations in Germany."[9]

There were more than enough reinforcements other than in the infantry, Mackenzie King reflected later. He spoke of "the strange position" in which the government found itself, for thousands of airmen—"some ten thousand of Power's men"—were to be demobilized at a time when there was talk of conscripting fifteen thousand infantry soldiers. This indicated (as indeed it did) poor planning between the different services and for that, the Prime Minister concluded, Ralston as Minister of Defence must be held responsible.[10] He agreed with Dr. J. H. King (the leader of the government in the Senate) "that Ralston, who as Minister of Defence, should have supervised the Navy and the Air Force, had simply allowed Angus Macdonald and Power* to run away with everything."[11]

Ralston reported to the Cabinet war Committee on 19 October. He placed before the members a report made by Stuart in which, as McNaughton commented to Mackenzie King when he saw it on 1 November, the Chief of Staff in his conclusion exceeded his position and passed "from a factual statement of Army requirements to the advocacy of a political method of raising these forces."[12] Stuart recommended that "steps be taken to impose conscription for overseas service of persons called up under the National Resources Mobilization Act" if reinforcements could not be found elsewhere.[13] Ralston had already let Mackenzie King know that "he would have to

---

* Minister for Naval Services, and C. G. Power, Minister for Air.

support Stuart, if necessary to the length of resigning,"[14] and he proposed that fifteen thousand NRMA men be shipped overseas early in December. The Minister for Air has described the shock that this pronouncement brought: "it was realized by all that the strong position the government had enjoyed had become precarious" and that the government itself was likely to find itself "split down the middle" by a decision on conscription which, up to now, it had been possible to evade.[15] And this was the fall of 1944, when the war was almost won.

Stuart next appeared before the committee. He frankly admitted past mistakes but King, seeking his support, did not chastise him. Although he agreed that Stuart had given the wrong information, he said he hoped Stuart would do all he could to help the government out of the present situation. Winter was approaching. Surely, the Prime Minister thought, the men would be entitled to a rest? Stuart "came back with a statement they would want to . . . play a big part, and said something about ending up in glory"; that the strategy was "to press on" throughout the winter.

I asked if we had ever been consulted about strategy. He said something about Crerar and others keeping in touch. I asked had the Government of Canada approved strategy of this kind? I said this was the first time the Cabinet had heard strategy involved action of this kind. That where lives of Canadians were concerned, we were entitled to know what was involved in the strategy. . . . This point had never been apparently thought of either by Stuart or any members of the Cabinet or Ralston, though Ralston in previous years has been very strong on what disposition was of all our forces. . . .[16]

The fact was, as King reflected earlier, "that any proposal made, short of conscription of NRMA men, meets with instant rejection [by the Defence Department]. There is not an attitude of trying to work out the situation"; and this despite Stuart being worried, "as he may well be, for if the Government falls it will be due to the Army and the way they have . . . misled the Government."[17]

The government verged on disintegration:

Politically-minded men like Gardiner and Mackenzie could see in this only the possibility of disaster to the Liberal Party. Howe would have liked to see the conclusion of the war without popular outcries for or against conscription detracting from his magnificent achievement.

Others, such as Crerar and Ilsley,* felt that their position would be utterly untenable unless they supported the emphatic view, existing in their sections of the country, that the zombies should be immediately sent overseas. Popular agitation for this quickly assumed large proportions. . . . Angus Macdonald, who had been strongly in favour of conscription since 1942, of course persisted in his view.[18]

Power himself saw the larger picture, "that a policy of conscription imposed now would have the effect of almost dividing the Canadian nation in two," and had made up his mind to resign if the measure were imposed. St. Laurent, though from Quebec, had made no pledge to oppose conscription and would follow King, while King himself was "profoundly depressed."[19]

Meeting followed meeting in an attempt to avoid the looming catastrophe and the disaster that faced Mackenzie King. Stuart insisted that "the reinforcement pool would be exhausted shortly, unless replenished from Canada."[20] But there was one hopeful disclosure: Ralston revealed that about one hundred and twenty thousand general servicemen in Canada had volunteered for overseas duty; ninety thousand more volunteers were stationed in England. Surely it should be possible to take the fifteen thousand required from this enormous pool? Power and Macdonald were delegated to meet Ralston and the senior military officers to see what could be done.

The meeting took place on the evening of October 26 and achieved nothing. In Power's words:

That evening was spent in . . . an attempt to find out why it was considered impossible to use the 120,000 . . . in Canada, and the 90,000 . . . in England. This immediately brought up the question of the physical and mental capacities required to be an infantry soldier in these days (officially known as PULHEM)† including physique, hearing, mental capacity and stability, among other things. We were told that there was a regulation that no one over the age of thirty-eight would be allowed to serve in the army in the actual theatre of war. This rather surprised me and also Macdonald, who I am sure was able to recall a large number of excellent soldiers in the front line in the previous war who were well over forty. The army officers had an answer to this: conditions were different; it was not a trench war, and

* This was T. A. Crerar, Minister of Mines and Resources, not General Crerar; J. L. Ilsley was Minister of Finance.

† The PULHEMS rating, introduced by Dr. Brock Chisholm.

infantrymen were often called upon to take long marches. Moreover, the weapons were more complicated and required a higher degree of intelligence than the mere handling of a rifle required. . . . I began to suspect that in the minds of the generals the infantryman of today must almost be an honours college graduate.[21]

At any rate, Power and Macdonald abandoned any hope of prying loose fifteen thousand men. Professor R. MacGregor Dawson said of this meeting: "Any layman . . . simply cannot believe the statement that it was impossible to secure 15,000 men from a pool of 210,000. . . . It may be so; but the sceptic will be pardoned if he retains the conviction that a Sir Clifford Sifton or a C. D. Howe would after two weeks' work have extracted the men or the army would have lost a large number of officers."[22]

A careful analysis shows this: of the 120,000 in Canada, 42,000 were suitable for the infantry, 11,000 unsuitable for various reasons, and 67,000 were either medically unfit, too old or too young for overseas service. A total of 48,000 (which included under-age men) had reached certain levels of training, and of these 3,000 would be ready for overseas service before the end of 1944 as well as 1,500 formed from NCOs and tradesmen. Thus only 4,500 could be found in time from Canada and of the 88,626 in England 35,631 were non-effectives (unfit, in hospital, serving sentences and so on), nearly 30,000 were in static units and the remainder were in the training stream. Few of the latter would be available before the end of December. Remustering in the United Kingdom would still produce a couple of thousand infantrymen, but too late.[23] Neither in Canada nor in Britain, then, were there hidden pools that would immediately yield fifteen thousand infantry soldiers. Certainly, by lowering medical standards and sending out partially trained men they could have been found, but the country would not have stood for that at a time when trained NRMA men were standing idly by in Canada.

The events of the next three days were not clearly known by McNaughton—he was not as yet embroiled. We know from King's diary that, on 26 October, the Prime Minister was certain that Ralston would resign unless conscription was introduced; that he believed the Minister "was prepared to precipitate the most appalling domestic situation."

From an entry on the same day we also know that the Prime Minister was becoming more and more disgusted with the Defence Department. As he told the Governor-General, "The truth of the matter is the men on whom we have had to rely have not given us the facts or told us what was to be considered. Rather, they have given us assurances in the opposite direction and now they expect us to clear up the whole situation." The message from Crerar, pointing to the true position with regard to reinforcements, which had been suppressed by Stuart, still rankled: "It is past the middle of October before we knew of the word sent by the Commander-in-Chief of the Army to the Chief of Staff in August." Small wonder that he concluded later that day, "I am convinced that the Department of Defence has made a terrible mess of our whole war effort."[24] And for Ralston himself, "I feel more and more, as McNaughton does, about Ralston: that there is something inhumanly determined about his getting his own way, regardless of what the effects may be on all others. In this thing I believe there is a desire to justify in his own mind his earlier convictions for conscription."[25]

Yet King, who at first regarded Ralston's threatened resignation as a disaster that must be avoided, was moving towards a compromise solution between the extreme positions of the conscriptionists and their opponents. On 1 November the Prime Minister told McNaughton that "Ralston had agreed to an appeal to N.R.M.A. personnel with the proviso that, if it was not immediately effective, conscription would be enforced."[26] But when King said that, he had already changed his mind about Ralston and determined that he must go. His decision came about on the evening of 30 October. He had concluded that there was now a faction, prompted by a leader—Ralston—which would stand for conscription and desert him to bring about the downfall of the Liberal government. His mind flashed back to the conscription issue in 1917 when a group of Liberal conscriptionists had forsaken Laurier to join ranks with Borden and help him form the Union Government. In short, the Prime Minister felt there was "a conspiracy against him." He was disposed to fight but to do so he needed McNaughton: "If it comes to opposing Ralston and his conscriptionist friends, McNaughton will soon find himself at my side strongly fighting for the people's rights in this country."[27] Therefore on 30 October, it is evident, King had reached a firm resolve: to get rid of Ralston if McNaughton could be persuaded to take his place.

On 30 October McNaughton had no idea of this. That day C. D. Howe invited him for luncheon at the Rideau Club to discuss postwar reconstruction. Howe, the newly appointed Minister of Reconstruction, wanted to review research and industry with McNaughton. "His ideas of reconstruction," McNaughton concluded, "seem to be to support private enterprise with research," and he agreed with that. Howe said he was setting up a commission on coal: would McNaughton become a member or a technical adviser? McNaughton, with the governor-generalship in mind, said he would be "glad to help, except that the Prime Minister and I had agreed on other work," but as that was not about to materialize just yet, he "could work on this without publicity for a couple of months." Saying he would get in touch, Howe left the matter there.[28]

Early on 31 October, King sounded out certain members of the cabinet. Ian Mackenzie (then Minister of Agriculture), Gardiner (Minister of Veterans Affairs) and Howe all wanted McNaughton brought in. St. Laurent agreed it would be wise. As King put it, "McNaughton has none of the enmity towards French Canadians which so many conscriptionists entertain." To bring him into the government would give confidence in Quebec. But would McNaughton accept? It seemed to the Prime Minister to be

> . . . above all other questions whether McNaughton or Ralston is right in the way in which our armed forces were disposed of. I feel that, if I offer McNaughton the position of Minister of Defence, he will see in this crisis two things particularly: one, the appalling situation which would be created in Canada if conscription were made an issue at this time, that his aim is like mine to preserve the unity of Canada above all else; and, two, that he will see a complete vindication of his own judgment and an opportunity of making that clear to all concerned.[29]

His assessment was very shrewd.

At 10:35 on the morning of the 31st King telephoned McNaughton. "Wants quiet talk with me," McNaughton's record reads, "would come my house or Laurier House. Agreed 12:20 today Laurier House." He arrived a couple of minutes early, waited a few minutes for King, then the two talked for an hour before luncheon and for an hour afterwards. McNaughton, contrary to his usual custom, kept only the briefest notes, of which these are the most significant: "NRMA"; the name "Ralston" heading five others, "Drew, Hepburn, Bracken,

McTague, Borden"; "Governor General, Citizen, 20 Oct"; "National Emergency, not Political, at PM's disposal if required"; "No request to authorities overseas to lighten task. They should indicate task wanted—If we accept we will discharge—How we do it is strictly our business"; "we must satisfy ourselves and Public that there is a real necessity before invoking conscription"; "Possible savings on L of C to Med—my thoughts we should re-integrate—Med Force is not under control of Canada and might be involved in a Chanak Incident."[30]

Such notes by McNaughton were rarely in chronological sequence and we may assume that this was the case here.* A likely interpretation of them is as follows. The first item to come up, very probably, was a report in the Ottawa *Citizen* of 20 October headed "Lord Linlithgow May Become Next Governor-General," which McNaughton brought to King's attention. This, King assured him, was no more than guesswork on the paper's part designed to scotch his own chances† and could be ignored. He had more serious things to discuss. There was, he was convinced, a challenge to his government based on the emotional question of extending the terms of service of NRMA men so that they could be drafted overseas. All-out conscription would follow the defeat of the government and he feared the outcome in Quebec; national unity—everything he had worked for— might well be destroyed. Ralston stood firmly for conscription and if it were not enforced would be likely to resign.

The leading Conservatives, Bracken, McTague, Drew and Borden had come out openly for conscription in June and Hepburn in Ontario was with them on the issue. This was a national emergency and, for McNaughton, an opportunity to serve. It was bigger than politics. Would McNaughton stand at his side to help save Canada? More particularly, would he, an avowed anti-conscriptionist, accept the Defence portfolio if it should become vacant?

McNaughton did not think about it long. To impose immediate conscription might aggravate the very situation it was designed to cure. Suppose, as seemed likely, there were disorders in Quebec? As he

---

* These notes came to light a few days before the General's death and the writer had no chance to question him closely. From previous conversations, however, he is reasonably certain that the interpretation which follows is correct.

† Governors-general, traditionally, had been British. The idea of a native-born governor-general met with considerable opposition in some quarters when, much later, Massey became the first Canadian to fill the office.

was to say to King seven months later, "We had in November no adequate numbers of troops available for the maintenance of law and order. It was manifestly absurd to count on conscripts to enforce conscription. . . . If the situation had developed adversely we would have had to break in on our training organization and perhaps to bring back troops from Europe or even we might have needed to ask for help from the U.S.A."[31] Thus, if there had been disorders, there would have been fewer–not more–reinforcements for Europe; and immediate conscription might provoke insurrection. That is why McNaughton, with no further thought of the highest office in the land which he had accepted, said that he was at the Prime Minister's disposal if required, as King was certain he would say. Canada came first with McNaughton. And fearing the consequences of any other course, he had said, "We must satisfy ourselves and the public that there is a real necessity before invoking conscription."

McNaughton knew he was being used by King, but he did not mind that if what he did would be of service to his country. King's record of this meeting makes McNaughton's attitude clear. The defeat of King and an election on the conscription issue would do irreparable harm in Canada; more, "it might affect the Presidential elections in a real way and also . . . the Germans would, of course, construe a conscription election as evidence of the breaking up of the Empire and use it with their people for prolonged resistance." This was "not a matter of Party with him but of service to the country at the time of great need." McNaughton therefore counselled King "to stay at all costs"; he would be at the Prime Minister's "disposition" and "prepared to take on the task" if Ralston did resign.[32]

The rest of McNaughton's notes are his first thoughts about what he should do as Minister. Operational tasks, as indicated and accepted, would be discharged, and the Army would be reinforced to that extent; there was no question that it could be otherwise. And his thoughts about the Canadians in Italy are interesting. To re-unite the Army in Northwest Europe obviously could not be done at once, and thus would not help to solve the crisis. But it should be done eventually, not only to bring all the Canadians under the control of Canada but to save manpower. And there would then be no adventures in the Balkans, which this early he clearly foresaw would be suggested and which might lead Canada to another Chanak crisis.

King's account is more specific:

He spoke of having always felt it was a mistake to divide the army and to have some of our men fighting in Italy as well as in northern Europe . . . that a great many men were used up in lengthy communications. What he thought should be done was to end the Italian campaign, so far as our men were concerned, and have them brought back and the army made one in Northern France. . . .He of course disagrees cordially with Ralston on some of the things he has done and says he does not trust Stuart's judgment at all.[33]

Whether McNaughton stipulated that the NRMA soldiers would have to be ordered overseas if voluntary methods failed is not recorded; he himself thought that he had done so at this first meeting, and King's account does not rule it out: "He thought [conscription] would work great harm in the years to come but he had great faith in the common sense of the Canadian people and believed that one had only to tell them the truth to get them to support any policy that was necessary."[34] If not, he certainly did at his next meeting with the Prime Minister on 1 November. That day, at about 10 A.M., King telephoned and asked if it would be convenient for McNaughton to come to Laurier House again. McNaughton arrived twenty minutes later and was shown into King's study. King had brought certain papers with him, in particular a despatch from the British government outlining the war situation ("Much as I would have stated it," said McNaughton, "based on the Press"), and the General was particularly interested in an intercepted message which indicated that the Japanese had learned of German despair about the war—the Germans considered it had been lost. It would be the utmost folly, McNaughton concluded after reading that, to introduce conscription at once, or even to say that it would be introduced should an appeal for volunteers fail. In King's words, he "stressed the fact that between now and a month hence when we would be committing ourselves to introduce conscription the whole thing might greatly change, and the need be met in other ways, or be entirely removed. That, he felt, was the strongest of reasons for not committing oneself to any course in the intervening time."[35]

There was, then, a fundamental difference between Ralston and McNaughton though it has been often represented that what the one wanted, and what the other did, were exactly the same. Ralston, according to King, was "prepared to have an appeal made but asks what is the government's policy should it fail?" He wanted the addition to the statement concerning the appeal that "if it fails, we will then pass an order-in-council extending the N.R.M.A., call Parliament together and pass it if need be by closure." McNaughton said: "What Ralston suggests is psychologically unsound. One should not cross a bridge until one comes to it. If an appeal is made and it does not succeed, the time to decide whether conscription is necessary or not, will be then."[36]

The Prime Minister also brought Stuart's report which had been submitted to the War Committee. Casualties, McNaughton noted from the latter, had been running at the intense rate, in place of alternate periods of intense and low, which indicated to him that divisions had been kept in the line for periods that were far too long. It seemed to him that possibly "6-7000 men" on the line of communications were part of the price that had been paid for having a corps in Italy, and he was quick to note that the Armoured Division in Italy had two infantry brigades, whereas "in Western Europe it had one."[37]

The cabinet had agreed, on the afternoon of 31 October, that there should be an all-out appeal for volunteers. It has been said that McNaughton as Minister was far too optimistic, far too confident that he could persuade the NRMA men to volunteer. He may have appeared confident–clearly a defeatist could never have made the voluntary system work—but, as he told the Prime Minister on 1 November, he had reservations. "I spoke of the possibility of putting the zombies to work," he said, "if they would not fight," and then he made it absolutely clear that, if it should prove necessary, he would insist on conscription to maintain the strength of the units in the field.

If, after proper inquiry, I was satisfied that a new method of raising our force for employment in the field was necessary for the proper discharge of Canada's part in the war I would not hesitate to say so and give reasons. I had not this information now and therefore I could not say. Certainly conscription was not a method to be used unless it were clearly required. . . . Changing from the voluntary system to conscription was a very serious step. We needed to know that it was really necessary before we took it. If it was necessary we should be able to convince the public in whose judgment I thought we should have confidence.[38]

This was clearly put but Mackenzie King still demurred at the definition of "necessary"*[39] to the extent, according to his own diary, that McNaughton agreed there would have to be "a supreme necessity."[40] It must have occurred to King, however, that to bring in conscription on the recommendation of the anti-conscriptionist McNaughton, after efforts to produce the required reinforcements by voluntary methods had failed, would obviously be more palatable to the Province of Quebec than if the measure were to be taken on the advice of Ralston. He said that he proposed to ask the Governor-General that afternoon to accept Ralston's resignation but (somewhat unconstitutionally) if other ministers should tender their resignations, "he would counsel non-acceptance until the situation cleared." King asked McNaughton to be ready between 4 and 6 P.M. to go to the Governor-General to be sworn in; they would then meet the press together. They spoke of Ralston, and McNaughton gave his opinion frankly: he had ". . . over-centralized the Department, paralysed decision at the lower levels and burdened the senior executives with detail."[41] McNaughton spoke of Stuart, "in whom I have no confidence" He would have to go. They discussed the Conservatives: Bracken, McTague and Borden. McNaughton said that he must tell those with whom he had been dealing his reasons for joining King's cabinet. "It was an emergency. I belonged to no Party but I was a servant of Canada." Then, after saying he would wait for the Prime Minister's call at his home, McNaughton left.[42]

He drafted letters to the Conservatives that afternoon, though they were not mailed until 2 November when his appointment had become definite.

> Two days ago [he wrote] I learned that the situation as regards conscription had become acute. I felt intensely that one more effort must be made to raise the men required as volunteers and so I put everything else aside to undertake this task.
>
> I just want you to know that until the emergency I never had the least intention of joining this Government.[43]

This time, unlike the occasion when he told them of his acceptance of

* The Prime Minister quoted McNaughton as follows: "If an appeal is made and it does not succeed, the time to decide whether conscription is necessary or not will be then. My view is that if the 'attendant circumstances' at the time all matters related to the war and our armies in reference thereto should make it necessary, then conscription would have to be adopted."

the "task proposed by the Government" (the governor-generalship) a fortnight earlier, there was no reply. It is perhaps understandable that he was thought to have played the role of a Judas, even though, as we have seen, his motives in joining the Liberal government were of the highest and his integrity beyond question. Conservative hostility against McNaughton, whose role was decisive in ruining the best chance the opposition had during the war of toppling Mackenzie King, was to become very evident when McNaughton tried to obtain a seat in Parliament in future elections.

King went to the Governor-General immediately after leaving McNaughton. In a marked overstatement of the case he said that McNaughton "was quite positive he would be able to meet the situation short of conscription by methods that he would propose." At Mackenzie King's next words Athlone coloured, for he knew the consternation that would be caused in military circles in Britain: "When I ask you to accept Ralston's resignation, I will also recommend General McNaughton as Minister of Defence." This was the man whose name had been proposed such a short time before for a much different kind of office. The Governor-General, however, could not do otherwise than agree to approve McNaughton's appointment when King had explained that "he was the one person who possessed the knowledge that was required to meet the situation successfully."[44]

The cabinet met later that afternoon. Ralston was conciliatory. He was prepared to consider the question of an appeal and he no longer mentioned conscription, should it fail. Mackenzie King "kept waiting, expecting Ralston would repeat what he had said [previously] that he would have to resign unless so and so were done." But Ralston did not, and Mackenzie King determined, as he put it, "to reach a conclusion." A way had to be found to "save the government," to save "a terrible division" and to secure reinforcements if that was possible. The man who could do this was McNaughton, a man highly regarded by the troops overseas, by mothers and fathers at home and by every Canadian citizen. Though he "was not a Liberal, a Conservative, or C.C.F." he was "liberal in his policies."[45]

The evening's events were dramatic. At 5 P.M. King telephoned. "Discussions were taking longer than he had expected. . . . It may be late before the meeting is over. He had just called me to let me know so that I would not be waiting without news." McNaughton replied that his only desire was to get a proper solution, "If it can be done peacefully without a cabinet split all the better." King said that "his particular differences with the Minister (Ralston) will come at the last, *he was keeping it to the end* *. . . he was trying to establish the greatest measure of agreement in the Cabinet beforehand." He would continue, McNaughton said, to hold himself available.[46] Then, an hour and a half later, the Prime Minister telephoned again. "First step taken," McNaughton recorded, "a letter of resignation renewed [revived?]" "Council . . . all told. Know that I will take over. All relieved. . . . Nothing to be said by anyone tonight. Will get in touch with me about noon tomorrow. I expressed my best wishes, my willingness to help solve difficulties in the emergency and my confidence we could win through. My satisfaction at the turn of events which enabled me to serve under him personally. Mr. King spoke his thankfulness to Divine Providence at outcome."[47] Late that night the news was out: "10.20, Montreal *Gazette* carrying story I am to succeed Ralston, nothing to say at all"; "11.25, Ottawa *Journal*, nothing. . . ." and next morning the calls still came in. "Please do not press," McNaughton said, "I will let you know as soon as I have anything to say."[48]

King's efforts to keep his cabinet together were successful, and his remark—"all relieved"—was possibly the case at the time. Those who favoured conscription, notably Ilsley, Crerar and Macdonald, were perhaps persuaded to remain for the moment to see what McNaughton could do. At any rate they did not follow Ralston out of the government as many had supposed they would. As for Ralston, it is obvious from McNaughton's note that King had used the letter of resignation which the Defence Minister had tendered in 1942, and this is an unpleasantly ruthless feature of the business. No evidence has yet come to light that Ralston was behind a conspiracy to overthrow the government as King had supposed, and his loyalty and integrity have not been questioned. Whatever McNaughton's opinion of Ralston as Minister of National Defence, there is no doubt that Ralston was always sincere, if misguided. But in defence of King's action, Major-General Pope, who was the Prime Minister's Military Secretary

* Italics supplied.

at the time, stated that although it was said that King butchered Ralston, "King had two choices–Ralston or Canada–and he put Canada first,"[49] and if we take the long view, the Prime Minister certainly made the right decision.

That evening Mackenzie King went to Government House where he told Athlone what had transpired. "It was apparent from His Excellency's face and manner," the Prime Minister wrote, "that while he was saying nothing he was greatly concerned."

> I think his mind related all the time to the trouble there had been with McNaughton in Britain and the effect this might have on the High Command* there. In this connection I am glad I let Churchill know in advance that there might be serious repercussions. His reply to my wire [concerning the reinforcement crisis] fully justified me in going ahead as he said he would prefer me to act without any reference to himself and do what I could to meet the situation. There would have been a terrible row if I had not got some word to him before taking the present step.[50]

Though there would be consternation in British minds Mackenzie King was delighted that McNaughton was to replace Ralston. He summarized the advantages: no other minister was likely to resign, not even Macdonald and Ilsley, for it would be "a declaration of want of confidence in McNaughton, which would put them in a terrible position"; McNaughton's appointment would be popular with the Army ("I am afraid Ralston does not begin to have anything like the following McNaughton has in that relationship"); it would give McNaughton "the chance to get the Canadian Army together as one"; and he would be "a fine Minister of Defence." Furthermore, McNaughton would no longer be a threat to himself: "I think this step will greatly strengthen my position; it will be a knock-out blow to Bracken and his Tory friends in their endeavour to secure McNaughton. The C.C.F. and Labour have been after him as well. Altogether I think, in truth, only Providence could have brought things together to work as one at this moment."[51]

---

* This is revealing. Why should McNaughton's appointment upset the British "High Command"? Having been removed, he no longer commanded the Army. As Minister of National Defence, however, his strong views on the autonomy and Canadian control of the Army, so troublesome to the British and which had been largely surrendered by Ralston and Mackenzie King, would once more become effective.

Despite the use King made of Ralston's 1942 resignation, Ralston formally resigned again on 1 November. In a letter to King of that date he said, "I consider myself bound by what I have said in the House. Our differences are fundamental on the vital matter of reinforcing our troops, and consequently and as requested by you, I at once tender my resignation as Minister of National Defence. In the stand I have taken I have considered that my first thought should be my duty to our fighting men in our Overseas Army." Then, with marked sarcasm, "I wish every success to the distinguished citizen who I understand will take up the duties of this Department. . . ."[52] In his reply of 3 November, accepting the resignation, King recalled that it was not until 19 October that any report was made to the government that the Army's reinforcement was causing concern; he drew attention to the assurances given by Ralston in August "that additional military commitments then being sought" (the extra brigade in Italy) would not affect adversely the reinforcement position. He also referred pointedly to Canadian troops being in two theatres. He took exception to Ralston's statement that their differences were fundamental on the "vital matter of reinforcing our troops," saying that to him, too, the matter was vital. They differed only in method:

There has not been a time since the war began that it has not been recognized that resort to conscription for service overseas would occasion the most serious controversy that could arise in Canada. I can think of no course of action fraught with greater danger to our war effort—to say nothing of the unity and strength of Canada today and for generations to come—than a general election at this late stage of the war on the conscription issue. I believe that such an issue would almost certainly arise were the House of Commons to be asked to endorse an Order in Council extending the terms of service of N.R.M.A. personnel to include service overseas. Until it is apparent that conscription for overseas service is necessary to the full support of Canada's forces overseas, and that its application would prove effective, the Government would not be justified in taking the risk of widespread national dissension.[53]

As to King's request for Ralston's resignation, "You will recall that on several occasions you stated that if the recommendation contained in the report of the Chief of Staff was formally made to the Cabinet and not accepted, you would feel compelled to resign. . . . Subsequently you stated that you saw clearly that the Cabinet were not disposed to accept your recommendation." Mackenzie King had then, he said, sounded out McNaughton and once that was done had made sure "there should not be further delay in reaching a decision."[54] His tactics were superb; in the light of future events they were fully justified.

On the morning of 2 November the Prime Minister called for McNaughton at his house; together they drove to Government House where McNaughton was sworn in. Then they left for the East Block to meet the members of the cabinet. On the way King, who felt concern that Macdonald and Ilsley might still follow Ralston out of the government, asked McNaughton to make a point of "speaking particularly" to them; to reassure them "as to adequate reinforcements being secured." This, at the meeting of the cabinet, McNaughton did. His purpose above all else, as King described it, was "to see that the war effort was kept up to full strength. That the needs of the army were fully met. That he believed this could be done without any resort to compulsion. He spoke as a man with authority and knowledge, wisdom and understanding. I think he made a profound impression on everyone, and I am sure gave both Macdonald and Ilsley a feeling of confidence, and I could see joy and relief in the faces of all the Ministers."[55] Perhaps McNaughton really did think that the voluntary method would work; but he had not as yet studied the position. He did so, immediately after the cabinet meeting, at the Defence Department.

The news of McNaughton's appointment, released on 2 November, electrified the troops abroad. Press despatches stated that the word was "received with delight by Canada's Overseas Army."[56] In Italy the men shouted, "Andy's back!" as they moved towards Ravenna and it was the same in Northwest Europe where they battled to free the Scheldt. There would be some nail-biting in Whitehall, and Stuart, who was in Ottawa to be with Ralston, wrote out his resignation, which crossed a message from McNaughton relieving him as Chief of Staff. In Canada feelings were mixed. The Tory press, expectedly hostile, came out strongly for immediate conscription and bitterly attacked McNaughton. Other papers deplored this "entirely unfair, ungrateful and unwarranted treatment" and asked that the new Minister be given a "fair chance to implement his plans."[57]

McNaughton saw King on the morning of 3 November to report on what he had found in his new department. The figures prepared were substantially correct; NRMA men would have to be obtained to make up the number of reinforcements required. "He believed," the Prime Minister wrote, "that it should be possible to secure them as the result of an appeal but . . . there would have to be an immediate drastic change. . . . He had found some were working in factories and on farms but were receiving as much as 6 and 7 dollars a day and others were being paid a premium to do extra work while the army pay was only $1.25; that simply meant that men would not think of enlisting for service overseas."[58] Everything depended on the attitude of the NRMA soldiers and the thought sobered King. He realized that it might, after all, come to conscription. Nevertheless, there must be no statement as yet that conscription would follow the failure of an appeal. He reminded McNaughton of what the General had said at Laurier House; that it would not be "psychologically wise to cross any bridge before we came to it." These tactics would be followed. He also warned the Minister of "the impossibility of forcing conscription unless there was an overwhelming opinion. . . ."[59]

A Committee on Recruiting, consisting of certain members of the cabinet under McNaughton's chairmanship, considered how best to make voluntary methods work. The situation McNaughton found with regard to NRMA men was briefly this: the majority were held for home defence, largely on the west coast; some of these, fully or partially trained as soldiers, were being used as a sort of reserve of workers; the remuneration of these workers was "extravagant" and, as McNaughton said, "I do not wonder that those who seek their personal advantage and who have had the chance to make large sums in cash would fail to come forward to volunteer for service overseas."[60] To remedy this McNaughton decided that all physically fit NRMA men would be concentrated in potential fighting units and that those who had not already done so were to complete their military training. NRMA men below the physical standards required for fighting units would go to labour units for work useful to the war effort; and those who were totally unfit would be discharged. The committee recommended that the co-operation of NRMA men, and the public, be sought by stating quite clearly that what was required were reinforcements and unity. The need for home defence was no longer pressing; now reinforcements overseas were wanted, and it was the duty of the men and their families to reappraise their responsibilities in the light of this changed situation. Firm action, also to be clearly announced, was planned to ensure that the first claim on jobs after the war went to men who had served overseas; NRMA men who failed to volunteer would be held back from demobilization to avoid prejudicing the chances of soldiers returning from abroad.

The committee recognized that McNaughton's reputation was the strongest weapon in the government's armoury to induce NRMA men to volunteer. It agreed that he should fulfil longstanding engagements to speak at Arnprior on 5 November and to a branch of the Canadian Legion in Ottawa on the 6th, to provide him with opportunities for stating the position to the people. Messages from the General were also to go to the Army Overseas and in Canada. Then, on the 8th, the Prime Minister would broadcast to the nation, indicating the policy of the government and the action being taken. It was recognized by the committee–in view of the group loyalty of the NRMA, their failure to see the need to volunteer, home pressures, the financial benefits which they had enjoyed and their sense of martyrdom–that the task was by no means easy.

McNaughton's message to the men overseas said in part:

> Through all the long months since I had to leave you I have followed your progress with the deepest interest. Your performance on the field of battle has been magnificent. . . .
>
> Now I have been called again to a position of responsibility in your affairs. It is my concern to see to it that you will have the support you need; and we will maintain our just part in the great Allied effort to bring the enemy to final defeat. It is my concern to watch over your every interest now; and when peace shall again be established.
>
> Then, we will enter a new and changing era of great promise for Canada and my concern and duty will be to ensure that you who have borne the heavy burden in the heat of battle shall have fair play.[61]

He spoke at Arnprior as planned, setting forth the issues and hoping that the "urgent need for support to the Army being known, the public would . . . encourage our young men to take up voluntarily the honourable obligation of service overseas to meet this need."[62] His speech was not well received. Next day, "to give greater emphasis to my appeal,"[63] he addressed the Legion branch in Ottawa. The meeting "developed into a stormy, bitter session."

Interruptions were frequent as mingled boos and cheers broke into his prepared address before an audience of more than 700 tense veterans from both wars.

When the General said that he was firmly convinced that the best hope lay in the maintenance of long traditions of voluntary service, the Legion members all but shouted him off the speaker's platform. . . .

When the defence minister pleaded with the veterans to accept the assurance that "there is no other way to correct the situation in time to be of any use," loud shouts of "Take the Zombies!" . . . stopped the General short. At almost every statement made by General McNaughton the hall burst forth in howling pandemonium. A question period after his address was a hectic, emotion-packed affair finally halted by the chairman as the only way to end what had developed into a bitter debate.

If nothing else, the meeting served to indicate the attitude of the Legion. Observers claimed it ended in a draw—General McNaughton did not break down the unanimous hard-core of the gathering in favour of conscription; on the other hand the meeting did not appear to shake the defence minister in his present faith and adherence to the voluntary system.[64]

These speeches by McNaughton, which launched the appeal and were widely reported in the press, left no doubt regarding the government's policy. Two days later, in his national broadcast, the Prime Minister reinforced McNaughton's words. He pointed to the dangers of insurrection were conscription to be introduced, as was being vociferously demanded, and called instead for full co-operation with the government's efforts to make the voluntary system work. Though it had been suggested by the committee that McNaughton should also broadcast to the nation, and although he drafted an address,*[65] this did not take place; his views were already well known and there was nothing he could add.[66] Nor did McNaughton visit the NRMA camps in the West, as had also been suggested; his place was in Ottawa, at the head of affairs, to direct a co-ordinated campaign in the limited time available for the experiment; fifteen thousand men had to be obtained by the end of the year, and a few weeks were all that remained.

The Prime Minister's appeal aroused as much hostility as

---

* This, significantly, stressed that the men should be obtained voluntarily but warned, "It well may be that later as the situation develops it may be necessary to seek authority for further extension as may be found advisable."

McNaughton's speeches. "What I fear . . ." King reflected, "is that now that both McNaughton and I have spoken, very little else will be done."[67] There had been an air of crisis in the country ever since Ralston's resignation. This deepened when it was known that McNaughton had committed himself to King and to the voluntary system. McNaughton, but a short time before the nation's hero, now found himself the target of invective and abuse. At about this time he told King that he must be the most hated man in Canada but King thought the dubious honour belonged to him.[68] There was apathy in the cabinet and an excited public opinion (exploited by the Conservatives) across the nation, except in Quebec, clamouring for the zombies to be sent. In this atmosphere King's appeal for the co-operation of the public was almost certainly completely wasted.[69]

On 13 November the cabinet approved the summoning of Parliament, as King wished. McNaughton told King after the cabinet meeting that enough reinforcements had been found to meet all expected needs until the end of the year. The information was a great relief to the Prime Minister who had had it from Ralston that "men would have to be conscripted before the end of the month so as to be put aboard ships by November the 30th at the latest." Next day he reflected: "Where would Canada have been today without McNaughton's appointment?"[70] Two days later McNaughton was able to report "a gain of some 3,000 in the figures of those trained men available in Britain," so that "the acute situation might not arise until the latter part of January."[71]

Without public support a successful result now depended on the attitude of the NRMA men themselves. A good response from them would end the crisis and it was to this that McNaughton turned his efforts. His prestige in military circles had always been high and he still counted on inspiring the senior officers with some of his own enthusiasm and zeal. Messages went out to the senior commanders in Canada and on 14 November they assembled in Ottawa to hear what the Minister of National Defence had to say. That morning McNaughton addressed them: he set out the requirement, the policy of the government and the need for a careful explanation of the

With General Smuts in Ottawa, 1945

The Liberal government, June 1945 (NATIONAL FILM BOARD)

The McNaughtons meet their son, Edward, on his return from the war.
With them are their daughters Christina (left) and Leslie

national issues to the NRMA. He hoped that officers who were convinced themselves would be able to persuade the NRMA soldiers to volunteer.

However, his audience was not convinced. The officers commanding districts and camps had encountered the problem in the past and McNaughton had not. Though the removal of additional remuneration for civilian work and the cancellation of farm leave for fit men (which had kept them from volunteering in the past) would help, the feeling of the meeting was that the majority would still wait to be sent and would not volunteer. These senior officers saw no possibility of obtaining fifteen thousand infantrymen under the voluntary system by the end of the year but nevertheless they would go back and do their best to get them. McNaughton, as he announced publicly next day, thought a concerted drive would bring the desired results. His statement was probably no more than an effort to counteract the adverse propaganda and create a more confident atmosphere but it proved most unrealistic.

Yet McNaughton remained convinced that the volunteers could have been obtained. He told King in June of the following year that if it had not been for the apathy, even defeatism, of some officers, combined with attacks by the Conservatives and "their abuse of our proposals," the volunteers would have come forward.[72] During an interview with Professor MacGregor Dawson in 1953 he mentioned "certain Conservative leaders [who] had raised conscription as a means of attacking the government. I indicated . . . that I had been convinced that what concerned these people was the defeat of the King government, not the reinforcement or the good of the Canadian Army or the effectiveness of its operations."[73] In particular he blamed Major-General G. R. Pearkes "and the difficulties he had made"[74] and he appeared to have good grounds for that.

At the meeting of 14 November Pearkes had been particularly prominent in protesting the futility of further efforts. He had returned to his command—Pacific, and the key to the problem in that it housed the bulk of the NRMA men—but it was not before 20 November that he convened a conference on recruiting and, worse than this delay at a most crucial time, he had not prevented his officers from making statements to press representatives who were there. The officers who were interviewed made no bones about the fact that they believed the campaign would fail; some actually said they were convinced that the only way to obtain the men would be to introduce conscription.[75] This situation could only make the position worse; it contained, moreover, ominous undertones of military defiance of the established policy of the government. "It is quite apparent," Mackenzie King wrote, "that there is a conspiracy there [Vancouver]. One after the other has been coming out and saying that the N.R.M.A. men were just waiting for the Government to do its duty and send them overseas. That looks like the Army defying the civil power. These men in uniform have no right to speak in ways which will turn the people against the civil power."[76] From the time of hearing reports of these happenings in Vancouver McNaughton had few illusions. His experience in Britain had been that orders, no matter how unpopular, had been loyally obeyed; he now began to feel that the organization in Canada "was full of rotten stuff on which to lean and if I had pressed unduly at any point it might have broken in my hand. I had . . . to go very carefully and temper my orders to what was realizable."[77] Mackenzie King himself said, "Our men were suspicious of Pearkes. They felt he had never tried to make the voluntary system go. . . ."[78] Three months later Pearkes was relieved of Pacific Command at his own request; he then stood for Parliament as a Progressive Conservative and obtained a seat.

At this time the number of reinforcements obtained since the beginning of November was as follows: NRMA men volunteering for active service from 1-4 November, 96; for the following week, 173; for the week 11-18 November, 280; and general enlistment (including NRMA conversions) for a three-week period ending 21 November, 3,077. Thus a grand total of 3,626 men only had been obtained, which was far from encouraging.[79] But can it be said that the campaign had been given a fair trial? Certainly not in Pacific Command where the most was hoped for. Nevertheless it is not surprising that the CGS (Lieutenant-General J. C. Murchie) and other senior officers in Ottawa, after reviewing these figures, concluded that the campaign would not produce the desired results. Parliament, not due to sit until the end of January, 1945, but which had been recalled by King in this emergency at the cabinet meeting of 13 November, was to reassemble on 22 November. Murchie apparently considered it right to make his views and those of his colleagues known before the parliamentary session began and he telephoned McNaughton on the morning of the 22nd, as we shall see.

It was vital for McNaughton, who was not a Member of Parliament, to appear before the House when it assembled, to make a statement. Power thought King "should immediately appoint McNaughton to the Senate; a joint meeting of the two houses could then be held on the grounds of a national emergency, and McNaughton could tell his story publicly and officially." McNaughton said firmly that such things as dates of sailings, rates of casualties should not be mentioned except in a secret session. King favoured "forming the House into a Committee of the whole and bringing McNaughton before it in a secret session."[80] McNaughton, Power said, objected to being appointed to the Senate because he "did not wish it to be thought that he was accepting a bribe." He told Power that there had to be secrecy, "but he would give in his public statement all the reasons for secrecy . . . and that he would promise in the same address to withhold nothing in the secret session." Power discussed "a senatorship for the period of the coming session only" but McNaughton preferred to "face the electorate just as he wanted to face the House of Commons and any other public body." McNaughton gave his views on the reasons for the crisis. Power quotes him as saying, "We would not be in this mess if the Canadian Army had been kept together" and that "the matter of infantry reinforcements could have been dealt with without any trouble . . . in August" if the trend of infantry casualties had been acted upon then.[81]

Monday, 20 November, was a black day for Mackenzie King. Five ministers—Macdonald, Crerar, Ilsley, Gibson* and Mulock†—insisted that a date should be set for the "experiment in voluntary recruiting" to end and wanted 1 December. If it was unsuccessful, then the NRMA soldiers should be conscripted. This reflected the mood of the country which now seemed unanimous on the issue, outside Quebec. When McNaughton saw the situation in the cabinet he

addressed the Council very emphatically, saying that members must realize what was involved in attempting to enforce conscription, apart altogether from what the effect upon Canadian unity might be. That he had given the matter a lot of thought, and he had to tell them the position of the army which was that the men who would have to

* C. W. G. Gibson, Minister of National Revenue.
† W. P. Mulock, Postmaster General.

maintain law and order in Canada should any opposition arise when conscription was attempted were for the most part N.R.M.A. men. That they would be the men who would have to be responsible for seeing that trained N.R.M.A. men were sent overseas. He feared situations that might arise with the possibility of bloodshed. Once that sort of thing started it might spread like a prairie fire throughout the country.[82]

McNaughton told the Prime Minister that "he was fully convinced that the [voluntary] policy would work if given a chance" that it "was a matter of principle and we should stand on principle." King agreed. McNaughton then shook his hand, saying, "I will stand by you in this."[83]

Power recorded that McNaughton was "confident to a degree"; that he was convinced he could "break down the core of resistance" in the NRMA units and get the men to enlist–but that was on the afternoon of the 20th, before the General had heard of the disastrous events in Vancouver, the news of which came through that night. [84]

The next day was equally dark, King being plunged "in the depths of depression" and thinking he should resign. Macdonald had made up his mind to get out at once. He told Power that in his opinion "the majority of the Liberal Party would stand by King but there would be a sufficient number, added to opposition groups including the C.C.F., to defeat King in the House. He was now convinced that McNaughton's effort for voluntary recruiting would fail and saw no reason why he should wait any longer."[85] It was a day of proposal and counter-proposal as to how the situation should be met but nothing was resolved. King was in the eye of the storm. No matter which way he turned, tempestuous seas, mounting higher, lay ready to engulf him.

And yet the morning of the 22nd brought the first of a series of events that would smooth the waters and roll the clouds away. At about 10:30 Murchie called McNaughton. McNaughton, after asking to have the message confirmed in writing, telephoned King at 10:35 to pass along serious news. His record of the call reads:

"C.G.S./ A.G. Possibility of V system gone. (West Coast incidents.
(Quebec

MacF resgn. Military end disintegrating. Will come to Laurier House as soon as I have this advice in writing."[86] It requires little explanation. The Chief of the General Staff and the Adjutant General were now convinced that voluntary methods of recruiting could no longer be effective. There had been incidents on the west coast and in Quebec. One officer (Brigadier R. A. MacFarlane, commanding Military District Number 10) had resigned over the issue and, if other senior officers did the same, military control would disintegrate and finally disappear. That was the gist of McNaughton's message. The Prime Minister said that McNaughton called the news "a blow in the stomach" and that would not be inconsistent.

The written confirmation that the CGS submitted to McNaughton shortly afterwards was merely a factual statement of the position. The crux of his memorandum read:

4. After a careful review of all the factors, including the latest expression of their views by the District Officers Commanding, I must now advise you that in my considered opinion the Voluntary system of recruiting through Army channels cannot meet the immediate problem.

5. The Military Members concur in this advice.[87]

This was the message that McNaughton took to King. Can it be construed as an ultimatum; as an attempt to overthrow the government; as a revolt of the generals, as it has been described? "It certainly can *not*," McNaughton said, "it was no 'palace revolt.' The C.G.S. tendered his considered advice and that of the other responsible officers in a proper and constitutional manner. It was his duty to do so; he did his duty."[88]

Already, when McNaughton reached Laurier House with the message, King had decided, as a result of McNaughton's telephone call, to impose conscription. His object was to stave off mass resignations from the government and the decision was his. McNaughton had always had a clear understanding with King that conscription would be adopted if McNaughton should advise it, and the Prime Minister, with his sure political acumen, knew that the time had come. "It is apparent to me that to whatever bad management this may have been due," King wrote, "we are faced with a real situation which has to be met and now there is no longer thought as to the nature of the advice tendered, particularly by General McNaughton. And if so tendered by General McNaughton who has come into the government to try to save the situation, it will be my clear duty to agree. . . . A situation of civil war in Canada would be more likely to arise than would even be the case were we to attempt to enforce conscription."[89] McNaughton, with the Vancouver incident of 20 November in mind and having now heard the responsible opinion of the senior military officers, concurred. The army must be reinforced. The voluntary system had been tried and there was less danger in conscription now than three weeks before. An Order in Council was necessary to send the required number of NRMA men overseas.[90] He said much the same to Professor MacGregor Dawson at an interview in 1953 regarding Mackenzie King's biography: "Our efforts to persuade volunteers had been blocked to the point that trained reinforcements would not come forward, in the numbers required, in time. Therefore to grip the situation we had to turn to such measures of compulsion as would be sufficient for the need (and incidentally show . . . the government intended to govern).[91]

The House met that afternoon. Ralston's resignation was discussed, the correspondence read, and the House adjourned. King, who now had a firm course on which to steer, handled the wheel with masterly skill. At the meeting of the Liberal caucus later in the afternoon his new confidence became apparent, though he could say little before calling the cabinet. The first snow of winter began to fall. In cabinet that night King suggested what was in his mind but he left it to McNaughton to review the various factors that had led to a change in policy. Recognizing the inevitable, McNaughton's colleagues listened quietly and made little fuss. The "conscriptionists," who had been about to quit the government, remained; and of the "anti-conscriptionists" the only casualty was Power who remembered the pledge he had given in Quebec that the Liberal government would never introduce overseas conscription. Thus, with an undreamed-of degree of unanimity, King prepared to brief the Liberal rank and file in caucus on the morning of the 23rd before the House should meet that afternoon.[92]

The night of the 22 to 23 November was unbelievably tense for McNaughton. The speech he had prepared for Parliament advocated voluntary methods of raising reinforcements but the day's events had changed all that. He worked far into the night with Brooke Claxton (formerly King's Parliamentary Assistant, and Minister of Health and Welfare since 13 October), J. W. Pickersgill (an External Affairs officer on loan to King) and Major-General Maurice Pope to produce a speech that would fit the new policy. At 2 A.M. McNaughton ran a hand through his tousled hair. "I can no longer think clearly," he said. Claxton observed the deep, tired lines cut into the General's face. "Go to bed," he said kindly, "you have a speech to make later on today." McNaughton did so and the others worked until five o'clock when the speech was finished.[93]

On the afternoon of the 23rd King exploded his surprise packet under the feet of the opposition. He read an Order in Council which had been prepared to give McNaughton authority to dispatch overseas an approved number of sixteen thousand NRMA soldiers. Then McNaughton (who, though not a member, had been permitted by the House to do so) came to the clerk's table on the floor of the House of Commons to make his statement.

His speech was remarkably consistent with what he had set out to do. There was no difference *in purpose*, he said, between his predecessor and himself. Both were equally determined that the Canadian troops in battle would not lack reinforcements:

Where my predecessor and I differed was on the methods which should be used to provide the reinforcements needed. In October last he took his stand on the basis that we had outrun the possibilities of maintaining our forces overseas by voluntary enlistment. At that time I took the stand that our traditional system which we had used throughout this war and long before, was capable of producing the results we sought and of producing them in time; *provided always that it was given the full public support which was requisite.**

I submit that it would have been *very dangerous indeed*† to hazard a change until all measures proper to the existing method had been developed to the full and it had become clear and evident to all that they would not suffice. To do so might well not have increased the rate of flow of reinforcements but might instead have reduced it most seriously. This is what I think would have happened if voluntary service had been abandoned for compulsory service in the late days of last month when the question had first come up. Since then *conditions have altered,* public attention has been riveted on these matters. *The paramount need of reinforcements in adequate supply has been recognized throughout our land from coast to coast.* For myself I frankly state I prefer our present system in its application to the

* Italics supplied.

† Italics here, and in the succeeding passages, have been supplied.

conditions we have in Canada *but I have not excluded the other method in circumstances in which it is absolutely necessary*. . . .

The lives of our men in the fighting lines must be guarded in every way that is dependent upon action which can be taken from here. The numbers required to make up the reserve that is necessary are larger than could be provided in time by the voluntary conversion of trained and fit personnel of our N.R.M.A. men to general service. . . .

I have brought these facts before my colleagues in the Cabinet. An Order-in-Council has been passed extending the service of this number of men to the European theatres of operations. . . .

If we enter this new era united, we will be strong and we have all that is required to carry the happiness and welfare of our people to heights as yet undreamed.[94]

McNaughton delivered his speech firmly and with no embarrassment. Although it would seem to those with no knowledge of the background that he had been a turncoat twice—first when he had deserted the Conservatives to join the Liberal government and now, having been the arch-apostle of voluntary methods, he had stolen the position of the Tories to advocate conscription—such was not the case. But appearances were against him. The opposition concentrated the entire weight of its attack upon McNaughton uninterruptedly until late that night. The derision, scorn and fury which the Tories had felt for McNaughton since he had saved King by joining his government at the beginning of the month poured down on the man they considered their betrayer. The tolerance with which maiden speeches are invariably greeted in Parliament was thrust aside and McNaughton was dealt with savagely. Aware that he had acted in accordance with his principles, McNaughton stayed calm, even under the most bitter insinuations, and answered the questions of his examiners comprehensively, courteously and patiently. "I could see he got very flushed around the temples at times," King remarked, "but he maintained a sweet calm." The Prime Minister himself felt isolated from the uproar, hardly "conscious of the presence of another soul" and feeling "in better shape than I have felt at almost any time."[95] Though physically exhausted at the end, McNaughton kept up this remarkable performance until eleven o'clock that night, when the House adjourned. King, who was considering his own retirement, admiringly told the Defence Minister in the elevator afterwards that he saw in him a future prime minister of Canada and this, from a man who assessed politicians and affairs with almost flawless skill, is a revealing if hollow compliment.

For a fortnight more the bitter debate continued. Ralston (who, like the rest, not unnaturally misunderstood what King and McNaughton were driving at) condemned what he termed the piecemeal method which had been adopted but declared that he would support the government. It was a magnanimous gesture that must have brought over many, who condemned King's treatment of Ralston, onto the side of the Prime Minister; and it effectively set at rest any suspicions that Ralston had plotted the overthrow of the Liberal government. In the early morning of 8 December, after an all-night sitting, King obtained by a margin of 143 to 70 the vote of confidence that he sought. Louis St. Laurent, the French-Canadian Minister of Justice, had also followed King*[96] (in what McNaughton termed "one of the historic decisions of the century") and his attitude, with that of Ralston, had been decisive. There was no general election to disrupt Canada's war effort. And though thirty-four Liberals from the Province of Quebec voted with the Tories, it was no more than a gesture.†[97] Unlike 1917/18, there was little rioting against conscription in French Canada; Canadian unity was not disturbed. It was no doubt realized in Quebec that every effort had been made to avoid conscription and that the limited conscription applied was unavoidable. There were few outcries. Power was surprised that all had passed so smoothly. He had been obsessed, as we have seen, "with the idea that a policy of conscription . . . would have the effect of almost dividing the Canadian

* St. Laurent revealed what he had felt at the time in an address which he delivered at the Chateau Frontenac, Quebec, on 29 November, 1946. "Not for a moment," he said, "was I able to hesitate as to the line of conduct that I was in duty bound to follow. But that line of conduct was indicated to me by facts known to me and that I could not indicate to my friends. I therefore found myself in the painful necessity of following a line of conduct which many of you, my friends, sincerely and honestly felt to be wrong. It was all the more painful because I could not but recognize your good faith and because I was convinced that if I had been able to reveal to you the motives which made me act, you would have approved them unanimously. . . . By [June 1945) . . . happenings which had occurred in the interval constituted a complete demonstration that we had been right."

† On 28 November Mackenzie King, with McNaughton, attended a caucus of the Liberal Members from Quebec. McNaughton, in King's words, "made a fine speech. They gave him a very warm reception. In fact they gave him an ovation when he began and at the conclusion of his speech."

nation in two." The threat to unity had been foreseen and avoided by method and careful timing.

The tactics followed had been McNaughton's: they had been implemented superbly by Mackenzie King as he, with no false modesty, admitted:

> I think the Members must have come to feel it was pretty fortunate that I had steered the whole question as I have, and particularly that the Government and country were most fortunate in having McNaughton as Minister of Defence. He makes an excellent impression, giving a sense of real knowledge of the situation and complete understanding of what is to be done in addition to appealing to French Members by his known desire to have matters carried out on a voluntary basis. . . . On all sides there is a feeling that it was a mighty fortunate thing for Canada that Ralston had been replaced by McNaughton.[98]

The Honourable J. W. de B. Farris, speaking in the Senate on 30 November, 1944, said this of King and McNaughton:

> What has happened as a result of [McNaughton's] effort has made it possible for the Government to carry on without an election. . . .
>
> I make this prediction . . . that when the turmoil and the bitterness of the present occasion is past, when the actions and conduct of Mackenzie King are seen in proper perspective, he will be fully appreciated and generous tributes will be paid to him, and along with Borden and Laurier he will go down in the history of this country as a great statesman who served Canada faithfully and well.[99]

That prediction, concerning Mackenzie King, has in large measure been fulfilled; McNaughton on the other hand is still regarded by many with bitterness. It has been suggested that he was naive, that King played upon this and sacrificed him to preserve his government.* McNaughton disagreed. Certainly King used him to stay in power, but he knew what King was doing and he knew what he himself must do for the good of Canada. Seven months later, when his personal fortunes were at their lowest ebb, he talked of "my purpose in joining the Government." That was "the restoration and maintenance of our Army" and his recognition that this required "the maintenance of unity" in Canada. "The policy we had pursued, despite the abuse and

* See Bruce Hutchison, *The Incredible Canadian*, pp. 380-383.

falsehoods spread by the Progressive Conservative Party, had carried us through. . . . I felt my purpose in joining the Government had been achieved. I have since gone over every step many times and if I had it to do over again I would alter nothing."[100]

The sixteen thousand reinforcements that were obtained through the Order in Council of 23 November meant that the Canadian Army maintained its strength until the end of the war with Germany. McNaughton watched the situation closely. He recalled Lieutenant-General Sansom from retirement leave to keep an eye on the reinforcement situation in Europe so that if any adverse trends set in, he could be informed at once. There was, in fact, little danger of any further crisis, for the decision to order sixteen thousand NRMA men overseas prompted many to volunteer freely. But all this was not achieved without decisive action on McNaughton's part.

The news, dispatched from Ottawa on 23 November, that sixteen thousand NRMA men were to be ordered overseas prompted mutinies in Pacific Command. So much, then, for the views of officers who had said the NRMA soldiers were waiting only to be sent. In Vernon, B.C., NRMA men marched through the streets shouting "Down with conscription" and "Conscript money as well." Pearke's comment was that newspaper reports said a thousand men had been involved, whereas the correct figure was less than two hundred. Two hundred was bad enough. This was child's play compared to Terrace, again in Pearke's command, where control of the camp was completely lost. More than a thousand rioters (1,600 is the figure given) mounted anti-tank guns on the terrace overlooking the town to control movement in and out. Thus they could prevent the dispatch of the reinforcements needed overseas. McNaughton, told of the situation on the 28th, while the acrimonious debate on conscription still continued, passed a note to Mackenzie King who clamped censorship on the press.[101]

Mackenzie King, St. Laurent and McNaughton got together after the evening's sitting to discuss the matter. "I have seldom seen a look of greater concern on anyone's face than that which came over St. Laurent's face as he thought of the situation as it might develop," the Prime Minister wrote:

The gravity of the situation was apparent the moment McNaughton told us what he had said in Council might happen. That if the troops began to resist he had not the soldiers or the men to enforce law and order. He added that what he could do was to call out the Militia in aid of the civil power but its men were for the most part employed in munitions and the like and that would mean cutting down supplies of ammunition and create other difficulties as well. He thought they could be trusted to do their entire duty, but it was a matter of trust. He added that in the Defence Department there had been nothing prepared in the way of plans to meet a possible situation of the kind. They had evidently taken it for granted that all that had to be done was to say the men would be conscripted and they would be. He had been kept in the House all day, unable to work out any plans with his officials and unable therefore to give much attention to this immediate situation, but would begin doing so tonight. *Here was exactly what McNaughton had told the Cabinet would happen if we had to resort to conscription.** The officers had lied by saying that the men were ready to go and were anxious to go but wanted to be ordered by the Government. . . . [McNaughton said] we could only trust to the people themselves; there was really nothing else. He was hopeful that knowing how serious the situation is, wise counsels would prevail. But the facts simply are that we have not the needed protection, either through the army or the police, to keep law and order if the situation gets out of hand. Here indeed would be a state of anarchy, not merely in the philosophical sense of no government existing, but as well, in the actual sense of civil strife. One can only pray to God that we may be spared anything of the kind. I believe we will, but if we are it will be because of the belief that the people of Quebec still have in myself and now also in McNaughton. It makes it very clear that had I not taken the course I did, and the Government passed into the hands of conscriptionists, that a measure of civil war would have been inevitable.[102]

McNaughton considered the situation, especially at Terrace. Were it to be handled in the traditional military way, troops would have to be sent from somewhere to suppress the mutineers. Apart from the militia, there were no troops for this; every available man, as McNaughton fully realized, was in the reinforcement stream and to use them at Terrace would defeat the purpose of reinforcing the overseas Army, which the government was determined to fulfil. Not only would the mutinous NRMA men be unavailable; the employable reinforcements would have to be used to subdue the mutineers, thus rendering the situation overseas doubly critical. McNaughton tackled

---

* Italics supplied.

the dangerous situation swiftly and unconventionally using a blend of firmness and guile.

The mutineers were armed and fully munitioned. Anti-tank guns, as has been noted, covered the road and rail approaches. "Pass the word around," McNaughton advised the Adjutant General (Major-General A. E. Walford) "that the Minister is a soft-hearted old man. He knows that these men are not really mutineers; they merely have a grievance. No matter how urgently they are required, they will be given the chance of going home to tidy up their affairs and kiss their wives and sweethearts. They will be given leave. Nor will they be burdened any more than necessary. The opportunity will be given every man to turn in his rifle before going home and these will be shipped for them to the embarkation point." The bait was swallowed. Unpunished for their actions in the majority of cases, freed from the stigma of being termed mutineers, effectively disarmed, the men went off on leave. The majority who reported at the concentration point received their rifles and embarked. Those who did not were hunted out and apprehended within the limits of the law. The reinforcements, who otherwise would have remained in Canada to be almost certainly involved in bloodshed, sailed as scheduled.[103]

No charges of mutiny were brought, on McNaughton's orders. The ringleaders alone faced disciplinary action and these only the lesser charge of disobeying orders. There were few convictions in Canada, and in England, where some men had arrived before disciplinary action could be taken, the charges were dismissed. Once in the fighting line the NRMA soldiers fought well and bravely.[104]

Chapter 3

Minister of National Defence

The Order in Council of 23 November, 1944, authorizing the dispatch of sixteen thousand NRMA men overseas, really ended the conscription crisis. We know that now but it was not known then. Throughout the country agitation for a larger measure of conscription continued and, as has been mentioned, McNaughton kept a close eye through Sansom on the reinforcement situation in the overseas units, ready to act in good time if the need arose.

The political strategy that had been adopted, so inexplicable at the time, had mollified Quebec and was to bring a united nation to the end of hostilities. Mackenzie King, through his alliance with McNaughton, had been able to abandon an untenable position by making a bold, but limited, retreat; and that was to prove enough. What had been the alternative? As McNaughton saw it, a new government, full-scale conscription and a country torn apart; in all likelihood, bloodshed in the streets of Quebec. McNaughton had done too much for his country to sit passively and watch it be destroyed. But if he had ignored King's overture at the end of October he would have avoided much pain and misery. Only King, who did not forget, knew the part played by his abettor. By practically everybody else McNaughton was regarded as an abject failure.

He was fifty-seven now, trim and austere, with a grey moustache and hair that had whitened at the temples. His eyes still probed and pierced under brows that remained as dark as ever. He wore glasses for reading but he took them off and frequently nibbled their ends in conversation.

The ministry held no terrors for him; administration and organization came as second nature. One of his first acts as Minister was to introduce morning conferences to review the routine tasks at hand and to cut down paperwork; to do the business of the day in the day was what he aimed at. The arts of "man mastership" and persuasion had been familiar to him for more than thirty years, and he was a good and practised speaker. His subordinates in Defence—especially his old assistant Cameron, a colonel now and McNaughton's Military Secretary—took his success for granted and indeed he conducted the Army through the period of its heaviest fighting with little fuss or bother. After an important meeting of the War Committee on 13 December, "one of the most satisfactory they had had," the Prime Minister noted: "I find General McNaughton, as all members do, exceedingly helpful. He thinks so clearly and understands so well the problems he presents."[1]

But this was behind the scenes; he had no seat in the House, should there be another session of Parliament, from which to report and keep the country abreast of day-to-day developments. "Could anyone be chosen better fitted," one newspaper asked reasonably, "to be minister of national defence during the most critical period of the war? Yet because this great Canadian scientist and organizer, as a result of his experience and observation during the last great war, believes that a volunteer army makes a better fighting force than a conscript army—although he adopted conscription when volunteering failed to provide the necessary reinforcements—a number of politicians and newspapers have done their utmost to blacken his reputation and undermine his prestige. . . ."[2] But such favourable notices were extremely rare.

The theme of McNaughton's tenure as Minister is best given in his own words to a friend: "I am convinced in the rightness of what I have set out to do, which is to see to it that our Army has full reinforcements by whatever method is necessary; to see to it that they have guns and ammunition and other equipment in . . . abundance; and to get things set to give our men every care and opportunity on their return to Canada."[3] His first concern was with the B.C. mutinies and their aftermath, as we have seen, and this lasted until after the turn of the year. In January six thousand NRMA men, who had been warned for posting overseas, were still missing from their units; "very close," McNaughton said, "to the proportion forecast as likely last November, when I had to decide how I should set out to dissipate the menace of this wrongly disciplined body of men."

Now we have sifted out from the mass over 8,000 of the better disposed and got them overseas. They will give no more trouble and I think will eventually make good soldiers. Some thousands of others have now returned and they are, I think, repentant and chastened and they too should give no more trouble.

The 6,000 now A.W.L. [absent without leave] have no arms in their hands. They are away from the influence of the leaders of resistance to authority and are scattered throughout the country as fugitives from justice and without any real public sympathy so they are ripe to be dealt with individually.

We have many difficulties yet to overcome but, with patience, I am certain a final solution will be reached which will give assurance both that our Army will be kept fully reinforced and, equally important, that unity will be restored in our own country.[4]

By the middle of January McNaughton was able to report in a statement to the press that the reinforcing programme was up to date; in fact, somewhat ahead of schedule.[5] Early in February the position stood at "well over the maximum number which had been stated by Staffs to be required at this date" and he was "profoundly thankful that this result has been brought about before the major battles in Europe have been rejoined."[6] A month later the position was so satisfactory that McNaughton was able to decline the offer of shipping space for ten thousand men from Canada during April and May and thereafter he had little anxiety.

On the afternoon of 8 December McNaughton heard from Mackenzie King that he was to contest the Ontario constituency of Grey North, Ontario, in a by-election that would be held early in February, and Qu'Appelle (which contained his birthplace, Moosomin) in the general election.[7] The Prime Minister had been against the latter as "an impossible riding to contest" but finally agreed that McNaughton "had a moral obligation to accept nomination in Qu'Appelle when it had been offered."[8]

Mackenzie King confidently expected an acclamation for McNaughton in Grey North, a courtesy traditionally accorded Ministers in search of a seat. But, as it turned out, the by-election campaign, fought in the snow and cold of midwinter, was to be sternly contested and one of the most publicized political battles in Canadian history. The Conservatives at first seemed willing to concede the seat;[9] then there was some suggestion that John Bracken, who still had not sought a seat two years after his selection as Tory leader, should run against McNaughton, but the stakes were considered too high;[10] finally Garfield Case, a York County farm boy who was chosen mayor of Owen Sound in 1942 (and who had been an unsuccessful candidate at Grey North in 1940) ran. The Tories, who were determined to beat McNaughton, rallied solidly behind Case. A third candidate also contested the rural riding for the Co-operative Commonwealth Federation. This was Earl Godfrey, a retired RCAF Air Vice Marshal and wealthy manufacturer. When due for retirement Godfrey had approached McNaughton to ask him to intercede on his behalf for an extension of service, and the Minister, who knew Godfrey's capabilities, would have gladly done so had this not savoured of favouritism. He told him so frankly and Godfrey, annoyed, opposed McNaughton at Grey North.

McNaughton proved a poor politician, and it did not take him long to demonstrate that. For his nomination he travelled to the Grey North Liberal convention–a convention that had in reality been called merely to ratify the decision of the Liberal hierarchy–on Saturday, 16 December, in a private, radio-equipped railway car, despite the warnings of Arthur W. Roebuck, the Ontario campaign chairman of the Liberal Party.[11] McNaughton would not compromise; he must, he insisted, remain in touch with National Defence Headquarters while the war continued. Press reports pointed to the "luxuriousness" of the car, the *Dominion of Canada No. 102* painted on its flank, and all its "magnificent appointments." The "contrast between the King government candidate's surroundings "and those of "the fighting men" in Europe was "a subject of amazed comment by the astonished people." One reporter had teenage boys speculating why the car was there. "Do you know why they have that special coach on?" said one. "Why?" asked the other. "Because the Governor-General of Canada is on it, that's why."[12] (Ironically, the boy, or the reporter as the case may be, would have been on the mark but for the reinforcement crisis.) Roebuck, on reading the hostile press, began to "shudder at the black eye involved in a loss."[13]

Mackenzie King conceded that the by-election was important: "Were McNaughton to be defeated it would be a considerable blow to the prestige of the government and [a general] election would be necessary immediately, and we would be starting into the campaign with a very great handicap."[14] He was alarmed when it seemed likely that the opposition would field candidates and had considered cancelling the by-election by Order in Council, should they do so. There were also two other possible courses of action: to dissolve Parliament, and discontinue the contest in Grey North; or to instruct McNaughton not to file nomination papers and then to announce that there would be no further session of Parliament.[15] But Mulock, the Postmaster General (who was McNaughton's campaign supervisor) reported that the General's chances were good. McNaughton himself felt, in King's words, "that on no account should we dissolve until the by-election was over. He himself would not like it if we did. He would be credited with pulling out. He said the best informed members of the Committee were prepared to concede a majority of 900 . . . he was making good headway wherever he went."[16]

King then said that the by-election would go on. He began to believe that McNaughton would win, "and win handsomely . . . He is an excellent candidate. Already the tide in North Grey [*sic*] has begun to swing strongly in our favour." And the effect of that on the result of a general election, about which King was pessimistic, would be great: "The one hope on the horizon is McNaughton. If he wins North Grey, it will galvanize life into the whole party. He has the personality which would help in binding the party together"[17]

There was very little that was not used by the Conservatives against McNaughton. On 5 January Dr. T. T. Shields of Toronto, speaking in the strongly Protestant Grey North constituency under the auspices of the Canadian Protestant League, assailed the Minister as being under Roman Catholic influence. Shields had heard that McNaughton was a Roman Catholic and had wired him to find out. McNaughton replied that he was, and always had been, an Anglican, while his wife, as always, was a Catholic.

The Roman Catholic Church, the speaker asserted, wields much of its influence through women. "It is a singular fact that General McNaughton has adopted the theory [teachings] promulgated by the church to which his wife belongs. You can draw your own conclusions!" And the Roman Catholic hierarchy, he claimed, "is determined to keep its followers at home, to breed their kind, while our men are allowed to go overseas to be killed or wounded, so that the balance of the population will be more speedily adjusted."[18]

Case consistently referred to the Minister as "Mr. McNaughton," an honourable form of address, certainly, but not in the sense he intended. He hinted that McNaughton had been relieved of his rank, though he was promoted full general after his return from England. He charged that McNaughton had built the finest army in the world and then abandoned it, whereas he, Case, wanted the men to get the reinforcements and the help they needed. This was a strange argument for Case to use since, as McNaughton's aides discovered, he had waited to be conscripted before joining the army in the First World War. Had McNaughton brought this up it would have done Case great harm but the General would not stoop to it. George Hees, a future Tory minister but then a major on leave from Europe, was brought in, "his wounded arm strapped grotesquely under an army walking-coat," to make an impassioned plea for men.[19] McNaughton was an outsider, Case said as he hoisted banners bearing the words, SEND A GREY NORTH MAN TO OTTAWA, NOT AN OTTAWA MAN TO GREY NORTH![20]

But it was Bracken who did the most damage. He came into the constituency at the end of January and remained for several days. Speaking at Owen Sound on 31 January he declared that "reinforcements are the only issue." Though he had been with the troops overseas that month he described the situation as it bad been *in October* before the reinforcements had started to arrive. He had talked to "hundreds of our young men, in England, in France, in Belgium, in Holland, and in Italy. . . . I have talked with men who lived through the storm of bullets and shells in Normandy and Belgium and to those who are now fighting in the mud and snow of Holland and Italy." He had talked about the reinforcement situation. "If Colonel Ralston was wrong, I wanted to know it. But they told me he was not wrong. They told me his was an understatement rather than an overstatement." As for McNaughton, the delight with which his appointment as Minister had been received by the troops, Bracken said, had soon faded when he had denied them reinforcements. "About one thing the men overseas are in tight agreement. General McNaughton left them in [*sic*] a cloud but he continued to hold their affection and confidence which he had enjoyed while he led them as a soldier; but for his action in supplanting Ralston, on this issue so vital to them, most of them hold him in utter contempt (*applause*)."

Let there be no doubt about this—there is a very general feeling that, by reason of the King-McNaughton plan, the men have been let down by their Government. . . . No, those men will find no comfort in the news if General McNaughton should be chosen to represent Grey North. . . . On the contrary, the best news that could go overseas to the troops who have still to fight their crucial battles would be a crushing rebuke . . . to the man who, having had their confidence, bartered that high trust for an office that Ralston had ably and honourably filled, and lent his support to a politically expedient but totally inadequate policy. . . . The man who stood out for reinforcements was pushed aside and one who accepted the policy that failed was taken in.

He referred to the aftermath of the mutinies and said that thousands of potential reinforcements were "deserters." The "Minister was using censorship orders to conceal from the public the fact that 6,300 men had failed to return to their units after embarkation leave."

The question I submit is whether or not you would endorse as Minister of National Defence a man who manipulated censorship to hide from you facts that might embarrass him.

As for Mackenzie King's contention that McNaughton was indispensable, "where in his pathetic plea as to the indispensability of General McNaughton does he say a word to justify that man being in office, much less being worthy of your confidence and support?" McNaughton was far from being indispensable: "If you defeat McNaughton, Prime Minister King can bring in Mr. Ralston as his Defence Minister (*applause*)," and Ralston was "the one great Liberal of his generation who had the courage to state the issue as he saw it and stick to his guns. . . ."

To the troops overseas, you men and women of Grey North can be, in a very real sense, the voice of Canada. . . . Electors of Grey North . . . it is falling to you . . . to give light and leadership in the greatest human issue that ever faced this country. It is my confident belief that you will meet that challenge unafraid, that your vote will be for Canada, united, indivisible and free, with one standard of rights for all—a land that our boys in khaki, grey and blue will be proud to call their home (*applause*).[21]

McNaughton spoke well, as he always did. He presented the issues plainly, his resonant voice ringing with conviction. He had sent more than eight thousand draftees overseas and he was pledged to send more as they were required. "North Grey can depend upon it," one newspaper reported, "he will not let the army down. Abundant reinforcements are overseas at present and under McNaughton that condition will be maintained."[22] "You have conducted your campaign with deep sincerity and dignity," wrote one of McNaughton's correspondents. "I cannot believe that the voters in Grey North will fail you."[23] But in the clamorous, emotion-packed halls of Grey North few listened to McNaughton's reasoned statements. "None of the professional politicians had the sort of magnetism McNaughton had," C. P. Stacey, one of Canada's foremost historians, has written. "That was a misfortune for Canada; and it was a further misfortune that when McNaughton himself was led into politics it was under conditions that neutralized his strong personal magic."[24]

It was Bracken's charges that got the headlines, not the rebuttals by McNaughton, and at Meaford on 1 February Bracken attacked again. He now admitted that reinforcements had been sent from Canada but he contended that there had been a wholesale throwing away of arms:

> Let General McNaughton tell you about those who were sent over. Let him tell you what happened. Let him tell you the whole sordid story. Let him tell you what happened on troopships carrying draftees overseas. Let him tell you why some of these men arrived in Britain without rifles which they were expected to have. Let him tell you how rifles were thrown overboard and how ammunition was thrown overboard.[25]

Had McNaughton been an experienced politician or had he had a flair for politics, he would have swiftly turned that one against his attacker. It is difficult to see how Bracken's charge, even if true (which it was not), could have been discreditable to McNaughton. All the soldiers had been trained in Ralston's time; and Bracken himself had insisted with passion that all NRMA soldiers should be sent overseas. He had blamed McNaughton for not doing so. If the conscripted men really had thrown away their rifles it might have been used as an argument against conscription. The General was hopelessly unfamiliar with these strange—and, as he saw it, unscrupulous—weapons; he missed the opening which Bracken had presented and once again his denial did not catch up with the charge before the votes were counted.

Polling day on 5 February brought a triumph for Garfield Case; McNaughton was decisively rejected, while the third candidate, Godfrey, lost his deposit.* Case, five-foot seven, referred to himself as "The Mighty Atom" after toppling McNaughton and he arrived in Ottawa a celebrity. But the press, which had watched the election closely, pointed to Bracken as the real architect of the Tory victory which, patently dishonest though his charges were, he obviously was.

Paradoxically enough, it was McNaughton, not Bracken, who was accused of dishonesty. The Tory press, gloating over his defeat, said as much:

> Grey North electors knew the reinforcement story. . . . They knew how General McNaughton had joined Mr. King to stop conscription and get rid of Ralston. They knew also how, when overnight Mr. King and General McNaughton were compelled to change their minds by a few determined Ministers, General McNaughton–now forced to do the thing that Ralston wanted in the first place, and still insisting that he disbelieved in it–held on to his portfolio to do it.
>
> They did not believe that from such men could come a conscription policy that was honest.[26]

But the Prime Minister had not lost confidence in McNaughton. He thought he had done wonderfully well. "To have been faced in the course of the campaign with the NRMA absentee problem, to have had so little time in the constituency and to have come to it a complete stranger, shows how truly remarkable is the impression which his personality made upon the electorate."[27]

McNaughton was out, at any rate, and a gleeful Tory caricature depicted him as Napoleon, hunched on a mule, leaving Grey North in driving snow for his own special retreat–oblivion.

On the contrary, McNaughton was in his office next morning as usual.[28] But he was not yet through with Grey North. While he had not seized on the "rifle incident" as reported by Bracken for his own advantage, he now saw the damage it would do to Canada. He had a draft prepared by the Judge Advocate General for the consideration of the Privy Council. Bracken's Meaford speech was quoted, and it was pointed out that these words, coming from a man who had recently

---

* The result was:	Case	7,333
	McNaughton	6,097
	Godfrey	3,118.

returned from the Army Overseas, had been given wide publicity both in Canada and in other countries. The general effect had been deplorable, especially in the United States, where Senator Burton K. Wheeler had claimed that between sixteen and eighteen thousand Canadian soldiers had thrown away their weapons. The effect on Canadian soldiers in the field, and on their allies, could not be calculated. The officer in charge of administration at CMHQ, London, had investigated the allegations and had found that one soldier (an unstable man who had attempted desertion) did throw his rifle and his kit overboard at the point of embarkation; the kit had been recovered but not the rifle. This was the only untoward incident in what had otherwise been an orderly movement. It was suggested that Bracken be asked if he accepted the findings from this investigation as a true representation of what had occurred or if he would prefer that a judicial inquiry should be made into the charges he had brought.[29]

Though McNaughton was bent on action against Bracken, the draft was considered in Council on 9 February when it was decided that nothing should be done. It was noted that to carry the charges so carelessly made by Bracken into the courts or before a judicial inquiry would, as McNaughton recorded, "inevitably lead the public to believe that the courts were being made use of to pursue a policy of political vengeance." King, instinctive politician, noted that McNaughton failed to see that there are some things in politics that cannot be dealt with by legal action, or other forms of discipline, as they are in the army.[30] McNaughton accepted the decision without demur. "We must concern ourselves with the future," he went on, and perhaps this could best be done "by my making a simple and frank statement . . . that the Canadian Army has full reinforcements;" that "these expressions on its discipline are unfounded"; that "the Army is heavily engaged in decisive battles and needs the united support of the people of Canada"; that though "some may think this should be given in one way and some in another . . . we cannot go all ways at once." The course that had been charted would be "pursued to a final conclusion in Victory" and then "those that wish to do so may make any conclusion they wish."[31] Though he tried to make his message clear*[32]

* "That Army of ours is equipped with the best there is; it has been denied nothing that is required. It is properly reinforced to carry through our just part in the battles

he was not widely reported in the increasingly hostile press. The "whole galaxy of political skullduggery," of which he had complained in the by-election, continued to be used against him, and in the newspapers there was "a flow of misquotations, misrepresentations and abuse both in their articles and in their editorials."[33]

After Grey North, McNaughton was in low spirits and he gladly turned to the affairs of his department. His actions there had been foreshadowed in a conversation with Ralston at the time of his retirement from the Army, when he had given Ralston his views on a Canadian occupation force in Europe after the defeat of Germany:

I had great anxiety about future relations U.S. and U.K. with Russia. I thought our commitment to maintain troops in occupied Europe should be cut down as soon as possible. They were hostages for our compliance with British schemes. Canada should co-operate in the British Empire *as a nation.* I believed in the British Empire, but not on the basis of providing raw materials and detachments.[34]

He had discussed the Canadian Corps in Italy:

In insisting on sending 1st Corps out . . . the Government of Canada had run counter to agreed plans (COSSAC) and had played into the hands of [those] who wished to break up the Canadian Army for political reasons. We were too independent for them. . . . There was a school of thought in the W.O. [War Office] and F.O. [Foreign Office] who sought to divide us in detachments so that we might not have autonomy. We had lost the potential position we might have had for our government when, if we had a real army, their view would have carried weight. As it was, our government's opinions were brushed aside.[35]

And finally, he had considered Japan:

As regards the future in the Pacific I thought the [Americans] were logical in insisting on the maintenance of their supreme authority. They knew how to wage mechanized war and to exact casualties from

---

still ahead. It is highly-trained and skilled in all the arts of war. There is youth in the ranks and youth in command. It is under wise, determined leaders. It will be fully maintained until victory is achieved."

the Japs in the ratio of perhaps 20 to 1. The British did not. They treated war as a game and would go plunging men against men in the jungles where our men were perhaps not even the equals of our enemy man to man, having regard to the fact that the Japs could maintain themselves with very little in the way of supplies. . . .[36]

Taking Japan first, McNaughton, on becoming Minister, found that the part Canada should play in the Far East after the defeat of Germany had been under consideration since early 1944. By the middle of the year the British had said that their manpower would be strained by the war against Japan, the occupation of Germany and other parts of Europe, and defence of the lines of communication to the Far East. Large Canadian forces were suggested, perhaps two divisions against Japan and another to help in the occupation of Europe. In August the CGS (Murchie)—possibly with the reinforcement situation in mind—proposed a contribution of only two divisions, one against Japan and one for occupation duties. In September the Canadian Chiefs of Staff thought that the division for the Far East would be best employed against the Japanese homeland though this would obviously mean that it would come under American command. Churchill stated at Quebec, during September, that Canada need supply no more than a token land force[37] and noted that she was anxious to be assured that her forces would operate in the northern parts of the Pacific, as Canadian troops were unused to tropical conditions. This was sensible; it meant that time spent in acclimitization and jungle training could be dispensed with.[38] On 20 November, soon after McNaughton's appointment as Minister, the Canadian cabinet approved the commitment of one division, plus ancillary troops, against Japan.

In January, 1945, McNaughton, faced with a request from the British, agreed to contribute forces for the occupation of Germany, but advised the Canadian government that this should be for only a limited time. Not more than one division should be made available and he suggested that the matter be reviewed before the end of the next fiscal year (31 March, 1946). The Department of External Affairs informed the British accordingly, pointing out that the plans for the control of Germany which had been drawn up gave Canada no voice in policy. The outcome can be anticipated: a reconstituted 3rd Canadian Infantry Division, Canadian Army Occupation Force, consisting of volunteers and men with a low priority for return to Canada and release, occupied an area in Northwest Germany under British command. In January, 1946, Clement Attlee, who had succeeded Churchill as Britain's Prime Minister, asked Canada to reconsider its decision concerning the early return of the force but Mackenzie King remained firm. Withdrawal began at the end of March and was completed in June.

At the ministry McNaughton also found that Ralston had done something about having the corps returned from Italy. Ralston, eager to have the last word as Mackenzie King had shrewdly prophesied, had sent the corps off to the Mediterranean; he had also got rid of McNaughton. While he probably never regretted having replaced McNaughton by the more compliant Crerar, he certainly had second thoughts about the divided Army. He could never really have realized the consequences. Though the move to Italy had been difficult to arrange as well as fantastically expensive in terms of time and shipping, even before the corps had fought in any major action Ralston wanted it back. McNaughton was astounded when he read the file.

In May, 1944, Ralston, obviously uneasy about the split, had suggested in a message to London that the Canadians in Italy should be returned to be reunited with the Army "as soon as military conditions permit." In words strongly reminiscent of McNaughton, Ralston had said the move was "desirable from a national point of view and in order to make the most effective contribution." The British, with tight shipping programmes and more to do on the eve of the invasion of Northwest Europe than ferry thousands of men back from the Mediterranean, not unnaturally ignored the Minister's appeal. Against their better judgment they had complied with Ralston's original request; he had got what he asked for and his vacillation later was just ignored. As it had turned out, it suited the British very well to have a less independent Canadian force to deal with. They, therefore, did nothing; on the contrary, in October, 1944, the War Office informed Alexander that it did not intend to withdraw any part of the force at his disposal.[39]

This was in direct contrast to the American attitude which, Brooke recorded, "was unfortunately one of–'We have already wasted far too much time in the Mediterranean doing nothing, let us now lose no more time in this secondary theatre. Let us transfer and allot all available resources to the main theatre and finish the war quickly in

Germany.' "[40] But Churchill strongly supported a Mediterranean strategy: Brooke tells us that there lay in the back of his mind the desire to "form a purely British theatre, when the laurels would be all ours."[41] In June, 1944, Alexander, according to Brooke, was hoping to break through the Ljubljana Gap, in Jugoslavia, into the Hungarian Plain; in July he was talking of "an advance on Vienna."[42] This would entail an amphibious operation based on Italy against Istria, on the eastern shore of the Adriatic. Once through the Ljubljana Gap alternative plans could be developed, to strike south against Greece or northwards through Hungary to Austria.

In December, 1944, Brigadier E. G. Weeks, whom McNaughton had known in England and who had just returned from Italy, called at National Defence Headquarters and the General heard from him that the corps in Italy might be used in an amphibious operation against the Dalmatian coast. McNaughton told King, reminding him of what he had said to the Prime Minister in October about the danger of involving Canadian troops in another Chanak Incident; he knew that "Chanak" would strike a responsive note in King. The Prime Minister feared the incursion of a Canadian force into an area where a brush with communism might be expected; and McNaughton thought of Salonika, where Canadian hospitals had been deployed in the First World War, and the prevalence of fever and malaria. What happened is best told in McNaughton's words:*

By my instructions, after I had consulted the War Committee of the Cabinet, Alexander was informed that 1 Canadian Corps would not be so employed but that if it remained in Italy it was not to be directed to the east of Austria. At this time Alexander was again told that it was desired that 1 Canadian Corps should rejoin 1 Canadian Army at the earliest practicable date.

An angry protest followed from Churchill which was replied to in very firm words by the Prime Minister. I also sent a directive through C.M.H.Q. to the Commander 1 Canadian Corps instructing that he was not to accept orders for operations out of Italy without approval by the Canadian Government.†[43]

---

* Though written from memory, his account is substantially correct. For details see the official history of the Canadian Army, *The Canadians in Italy*, pp. 658-60.

† Commenting on this to Colonel G. W. L. Nicholson (author of *The Canadians in Italy*), McNaughton wrote: "These telegrams are important because they represent the first occasion on which the Government of Canada insisted upon its rights to consider and sanction a plan of campaign in which Canadian troops were to be

A few days elapsed and then came the proposal from the War Office for the move of 1 Canadian Corps to France and my instructions sanctioning it and restoring the Canadian units in Italy to the Northwest Europe establishments; also restoring armoured and artillery units which had been transferred [converted, because of the shortage of infantry reinforcements] into infantry . . . and liquidating commitments for lines of communication units and base establishments in Italy.[44]

Ottawa heard during the first week in February, 1945, that the Canadians were to be withdrawn from Italy. "I am very glad to learn from you," Mackenzie King cabled Churchill, "that operational considerations now make it possible for the Canadian Army to be united again." On 15 March the corps, made responsible for the left wing of the 21st Army Group, assumed an operational role in Northwest Europe.[45]

The best summation of Canada's Italian role is provided by Colonel Stacey.* Quoting McNaughton as saying, "The acid test of sovereignty is the control of the armed forces," he continued:

. . . I don't think there can be much doubt that on the question of dividing the Army McNaughton was absolutely right and the Government absolutely wrong. In a military way what was done was absurd. At the request of the Canadian Government Mr. Churchill forced on the Allied Mediterranean command military formations which they hadn't asked for and didn't want. In terms of Canadian national policy the thing was still sillier. The only way a country of Canada's stature can hope to exercise any influence in a coalition war or to maintain effective control of its own forces is to keep those forces as concentrated as possible, not to scatter them around the world. The best commentary is the fact that the Canadian Government asked for its Corps back from Italy before it had even fought a major battle. But it didn't get it back until a few weeks before the Germans surrendered.

In sharp contrast is Brooke's version. "Thank heaven," he wrote, "I have at last got the whole of the Canadian Army now assembled in France." It had been "a difficult task":

---

employed. There was a violent reaction to this from Churchill to which a completely firm reply was made by Prime Minister King confirming the position I had taken."

* Cited in *The Canadian Forum*, August, 1966.

In the early stages McNaughton's opposition to the splitting of the Canadian forces had to be overcome. Then a portion had to be sent out to Italy when I had difficulty with their lack of trained commanders. Finally they had to be withdrawn again from Italy and assembled as a whole Army for the final stages of the operations against Germany.[46]

This might be thought a tortuous method of going about getting the Army assembled in France when, as late as in the middle of 1943, the Canadians *had* been poised in England, *as an army*, for that very task.

McNaughton was very pleased that the Army was at last united. He "shuddered to think what would have happened if we had been drawn off into this subsidiary theatre [the Balkans]. . . . The unification of the Canadian Army in the decisive theatre in Northwest Europe was a strategical necessity of the first importance."[47] Though, as he admitted to Alexander in Ottawa the following year, "*amour propre*" had been involved, that was not the reason for his insistence; "it was necessary to strengthen the striking left wing in Northwest Europe and I had verified this need with Eisenhower on his recent visit."[48] McNaughton feared the effect of flying bombs and rockets on the British people. It was a very good thing, he contended, that the invasion had not been left any later. As things were, it was a close call. "Hitler's great mistake in 1944," he said in 1947, "was that he did not put sufficient effort into the trans-Channel means of bombardment."[49] The bombardment by flying bombs began in June. "Casualties were as heavy as in September 1941 and damage too was very severe. This new bombardment imposed a severe strain on people living in the area of attack."[50] In July Brooke reported, "We are at present devoting nearly 50% of our air effort in [*sic*] trying to stop these beastly bombs, added to which about 25% of London's production is lost . . ."[51] Then on 8 September long-range rockets arrived and this campaign was to last seven months. It was a strategic necessity of the first importance, McNaughton considered, to remove this scourge from England in the shortest possible time. "I doubt that the people of the UK would have stood the V2 attack if it had been much intensified or if it had been much prolonged. . . ."[52]

We came through this war with a very narrow margin indeed. We saw in the last phase of the war the development of . . . a newer and better weapon which very nearly . . . altered the whole trend of world history. . . .

It so happened that the bridgehead that we were holding and on which we were depending for the subjugation of Europe—the British Isles—was within effective range of these new weapons and I hate to think what might have happened if we had not, for any reason, been able to make the landings on the Normandy beaches . . . and carry them through. . . .

If that had been postponed by months, if those operations in Normandy had been less successful, if we had been drawn off to another theatre of operations, I think we would have gone through a period of great anxiety as to whether in fact we could have brought the war in Europe to a satisfactory conclusion.[53]

That is why McNaughton had wanted all the Canadians in the vital theatre. He had also, as has been recorded, been "glad to check Churchill's ideas" of "scattering Canadian" into the Balkans.[54] But even the reunification of the Army, in the climate that prevailed in Canada, was used against McNaughton. It was alleged by the press that the corps had been withdrawn from Italy to be broken up and used for reinforcements in Northwest Europe.

McNaughton had other preoccupations as Minister. Planning the shape that Canada's postwar force would assume began under his leadership. And demobilization and rehabilitation plans drawn up in Ralston's time were revised and polished until, as was to be generally admitted, they compared favourably in generosity and fairness with any in the world. General Smuts, who visited Ottawa during June, 1945, was especially interested in the scheme and went over it carefully with McNaughton.[55] The discontent that followed the earlier war was thus avoided.

We have seen that Canada had agreed in November to commit only one division against Japan. On 4 April, 1945, King spoke of this commitment, which he modified, in the House of Commons. It was not intended, he said, to detail men for the Pacific; the men to be employed would be chosen from volunteers. With that, the Prime Minister abandoned even his limited form of conscription. McNaughton, who had been consulted, agreed; King's statement, in fact, had been based on a draft that he had received from McNaughton the day before and of which the Prime Minister said:

When I began to read it, it was again as though an Unseen Hand had placed it before me just at the moment that it was most opportune to

With the Spanish ambassador at the inauguration of President Dutra of Brazil, 1945

Canada's representative at the United Nations, 1946-1950

have it. It really set out the whole background as I felt it should be set out. Instead of narrowing the picture to the Japanese war, it read naturally out of the present situation in Europe into the possible activities of the future. By very little revision, I was able to incorporate practically all that he had written. This with what had been prepared before re Japan . . . I was able to put into a very clear, correct statement, concise, yet comprehensive, covering all the really difficult points. First: no conscription, for any of the forces; second: Army force being made up in Canada, and men having a month's vacation here. 3rd: No detailing from general service but all to elect for service in Japan. The size of the contribution, indefinite and all still to be determined on further consultation with the U.K. and the U.S. Implied no compulsion at all re men serving in Europe. No compulsion to send N.R.M.A. men before others. Finally a pretty well implied possibility of no army force going at all because of the scarcity of ships and the great need if men abroad are to get home, that they shall have transportation for the purpose. Were the troops needed against Japan or even wanted, the situation would be different but there is no justification for incurring, for reasons of prestige or above all to suit the army's pleasure and desire to increase its own importance, further sacrifices of life and enormous unnecessary expenditures. . . .

This means that those who want no conscription will certainly be drawn in our direction and I am sure they are, so far as Japan is concerned, by far the greater majority. Secondly, it means that our opponents will have to bring forth conscription themselves as a policy and fight on that ground, if that issue is to be introduced. It is one thing to say that we ought to have conscription for reinforcements and another to advocate it for the war against Japan.[56]

But Mackenzie King's statement opened up the old conscription issue, with all its rancour, even wider. What about the zombies? "Surely," one newspaper demanded, "the Government should say: 'We are going to send to the Pacific first of all whatever Home Defence personnel we have available, with selection after that on the lines of the shortest service in Europe.' "[57] Otherwise the zombies would not be touched; they had refused to volunteer for service in Europe and, it could be considered certain, would not volunteer for service against Japan. This, inevitably, became a major campaign issue in the general election of 1945.

Early in May McNaughton began to prepare in earnest for the election which would be held on 11 June. The war in Europe being nearly over, he devoted himself to this campaign. He received advice from a former Member of Parliament: ". . . a large smooth-working organization, the great majority of which are women and volunteers, contributed more than all else towards electing me. Speeches count in the country as a whole. But *real* workers count even more in the riding one contests."[58] McNaughton studied Arthur W. Roebuck's *Campaign Management and How to Win Elections* and found himself considering another sort of "army":

The outside workers are the Candidate's shock troops, and on their number and devotion, intelligence and knowledge, much, if not all, will depend. Accordingly as large an army as possible should be recruited. . . .

The field force should consist firstly of a goodly number of drivers in motor cars; secondly, of sufficient men on foot to cover all the streets, calling at the houses of supporters; thirdly, women operators for all phones available; and, fourthly, of messengers sufficient to make the rounds of the polls and bring in memos from the Inside Scrutineers. . . .

The fateful morning has arrived.

Drivers, footmen, telephone operators and scrutineers have each been handed their instructions. Now the telephones are ringing and the pleasant voices of the women operators are politely urging the favorable voters to come to the polls at once. The foot workers are hurrying down the various streets calling at the houses of the friendly voters in consecutive order, with their cheerful persuasion to remember without delay the first duty of citizenship, and the drivers of the cars are darting away for supporters who require to be conveyed to the polls. . . .

In every Polling Division, the Poll Captains and their committees of outside workers are shepherding, coaxing, dragging the favorable voters to the polls. . . .

"Shepherding, coaxing, dragging . . ." Kissing babies and smiling at indulgent mothers. McNaughton surveyed the prospect with a sinking heart. He knew quite clearly that if this was what the people expected he could never be the sort of politician they wanted.

Walter Dawson, an ex-private from the First World War and then the Liberal organizer in Saskatchewan, took over McNaughton. From his office in Regina, Dawson, recognizing that "the General is no politician," issued orders: where to go, whom to meet, and what to say. The programme was intensive but from that point of view McNaughton, despite his fatigue, was an "ideal candidate, working like no other in this province." Dawson promised during the final week really to put on the pressure: "the General will be holding meetings every afternoon as well as evenings." At first McNaughton spent the afternoons being taken "around the shops and some of the homes to shake hands" and this he did dutifully. He tried to meet Dawson's wishes, going as far in one speech as to say that he was no longer a man of "bombs and bullets"; he was one of "baby bottles and bonuses." But all this was ludicrously foreign to his nature and he could not suppress himself for long.[59]

Qu'Appelle was a sprawling constituency that stretched from the eastern outskirts of Regina all the way to the Manitoba border. Moosomin was McNaughton's campaign headquarters. But he had left his birthplace more than forty-five years ago, so he was conducting his campaign as a man from Ottawa rather than as a native son. The constituency had sent the same man, E. E. Perley, a Conservative, back to Ottawa for the past five terms. He lived and farmed at Grenfell and thus was a local man, greatly respected, who could stand on his record. Mrs. Gladys Strum, the CCF candidate, also lived in the constituency; she had done extremely well in provincial politics and was bound to be strongly backed at the federal level. Of the three, McNaughton would have the hardest fight.

There was one hotel at Moosomin, the Queen's, and there McNaughton, his wife and daughter, Leslie, stayed. "You check in," Richard Jackson of the Ottawa *Journal* wrote, "are given a key on a big iron ring, and carrying your own baggage, you go upstairs to find your own room. There is no running water, and if you want a warm sponge bath, you leave word the last thing at night with a desk clerk to bring up a pitcher of hot water for the crockery basin in your room when he comes with your morning call. . . . It's a bit rugged."[60]

In spite of changes in the town, some of the landmarks were familiar to McNaughton. His father's store, facing the Canadian Pacific line, was still there. Now it had a modem front, painted yellow, but, though owned and run by local businessmen, it perpetuated the old name, THE R. D. MCNAUGHTON COMPANY, LIMITED.

From Moosomin, McNaughton drove out to campaign meetings in other parts of the prairie constituency "in a big, cream-colored car," the Tory press reported, "owned by one of the proprietors of his father's store."[61] But it was at Moosomin that McNaughton broke loose from the trammels of Dawson's good advice; there, for the first time, he departed from the orthodox electioneering guidelines. On Saturday, 12 May, he spoke of the war against Japan. Canada did not intend, he said, to send huge numbers of men to the Pacific "to slug it out hand-to-hand with a foe perhaps better fitted than we to survive in the jungle war." These words, it will be remembered, were almost the same as he had used to Ralston the previous September but, no matter how sensible, it was one thing to use them in private conversation and quite another to announce them from a political platform at a time when the government's policy of sending only volunteers to the Pacific was under strong attack. Nor did he explain them, apart from saying, "We will assist to smash the Jap forever with a maximum of machines and explosives . . . and our share will be just and fair."[62]

This was no more than Currie's old maxim, with which McNaughton was so strongly imbued, to pay the price of victory in shells and not in the lives of men. And it is interesting to consider the success or otherwise of the policy of fighting the war against Japan with volunteers. In all, 1,180 officers and 17,819 men were required, and a campaign to raise them began on 11 May. By the middle of July 2,796 officers and 36,386 men (out of almost double this number who had volunteered) had been found fit to go. Thus the requirement, as well as enough reinforcements for a three-month period, had been easily obtained by voluntary methods.[63] The provision of further reinforcements, as would have been necessary had heavy fighting been encountered, was never tested (indeed, the force itself was never sent) for hostilities ended on 14 August. A factor in both Mackenzie King's and McNaughton's speculations might well have been the atomic bomb, though neither could talk of that. McNaughton, as president of NRC, had known of atomic plans and progress since early in 1940. Because of his duties with the Army he had taken no part other than to ensure that Canadian scientific liaison officers had been accorded the facilities they needed.[64] And on 16 June, 1944, at a private meeting with Mackenzie King, McNaughton recorded that the atomic bomb had been discussed: "We spoke of the —— and state of readiness and the date R—— had said it might be used. I explained the nature of this device and the continuing danger in the area. . . ."[65]

In May, however, McNaughton's words brought nothing but adverse comment. "It is regrettable," commented one newspaper, "that the Japanese are such unorthodox fighters, prone to indulge in such ungentlemanly activity as jungle warfare, but the General is quite willing to let others do the dirty work. . . . While Britain maintains conscription, universal service and an all-out effort, General McNaughton and his chief [King] choose this moment to announce suspension of the draft, abandon all pretense of equality of service, and cap it all with this astounding declaration of distaste for jungle warfare."[66] This was not the voice of Canada, someone wrote to the Ottawa *Journal*; it was "but the mutterings of an old soldier stumbling around in a maze of politics into which he has but recently been inveigled."[67] In the place of Ralston, "with his vigor and his honest courage" was the "inept"[68] McNaughton who, "a desperately ambitious man, behind Mr. Ralston's back, made a deal with Mr. King to prevent the Home Defence draftees being sent overseas as reinforcements." This was another attempt to win favour with Quebec: "What he really means is, the Government is not going to send the well-fed, well-sheltered, healthy, trained draftees. . . . It is going to send the spirited, unselfish, gallant but battle-weary youth who fought our battles in Europe . . . in whatever numbers, and by whatever means, it can compel them to volunteer again."[69]

The background of staccato, adverse comment in the Tory press remained. Premier George A. Drew of Ontario described McNaughton as "the most pitiful figure in public life." He made quite clear what the Tories had against him. "Once the commander-in-chief of Canada's army, having the respect of the men under him and at the time deservedly so, he came back to Canada and put his great prestige at the disposal of Mr. King."[70] But, as in Grey North, Bracken, who swept into the constituency of Qu'Appelle on 23 May to speak at Regina, Indian Head and Moose Jaw, hurt the most. He attacked McNaughton's statement. The people of Qu'Appelle, he said, would know "how to deal with a man who talks like that," giving the impression that Canada was "walking out of the war."[71]

King, according to the Ottawa *Journal*, had abandoned his Defence Minister to his fate, which, like so many other press reports, was untrue:

Most pathetic spectacle of this campaign so far is that of "Andy" McNaughton, fighting all by himself out in the Qu'Appelle constituency of Saskatchewan, shunned by his colleagues, seemingly deserted by the Prime Minister who used him last November.

Mr. King was in Saskatchewan this week, but though he had four days in his own riding of Prince Albert, he passed by "Andy's" door without as much dropping in for a cup of tea, or to pass the time of day with him. . . . Last Fall, when he thought "Andy" handy, he called on him for help; if now he thinks him no longer an asset, but in fact a terrible liability, that's no reason for his dropping him like a hot potato. . . .

Poor "Andy"! As he drives over the lonely prairie in his army car, with not a colleague within a thousand miles of him, what passes through his mind? Does he ask himself why his erstwhile friend and political sponsor, MACKENZIE KING, should now speed past his riding as though it were a leper colony, abandoning him to fight a sort of private war on a forgotten front?[72]

Nevertheless, despite the presence of his wife and daughter (both of whom, Dawson admitted, would have made better politicians than McNaughton), it was indeed a lonely battle. He looked "very tired, and must be," Richard Jackson wrote, "for his campaign schedule keeps him hustling from morning until early the next morning. There are deep, black circles around his eyes, and that lock of grey hair has fallen even further down over his forehead." When speaking he stood "stiffly erect, clutching his notes tightly, but seldom looking at them. He is deeply tanned from Prairie wind and sun. There is no animation; just a plain, straightforward recital in a voice which seldom changes tone, and then only in anger."[73]

There was little to be animated about. Though "he would far rather talk about the future," as he said at Fleming on 30 May, he would have to talk about the reinforcement issue because there had been so much controversy and misinterpretation. And the reinforcement crisis, to McNaughton's mind, had been satisfactorily disposed of. It was in the past—done with—and he was never interested in the past nor in discussing dead issues. People speculated about why McNaughton had left the Army, but he could not say much about that: "In view of the wish which Mr. King had expressed, which I had taken as a direction," he noted later in 1945, "I had not developed this matter at all in the election campaign."[74]

But in speeches such as the one at Fleming, on the dry dusty boards of a dance hall in this small prairie town of two hundred or so people, he gave a perfectly truthful account of his leaving the Army and joining King's government which, in its very simplicity, failed to convince his hearers, who had read too many inaccurate press accounts which pointed to different motives. He had left the Army, he said, because "of a very serious difference in views on policy" between the former Defence Minister and himself. He had returned to Canada with the idea of entering politics and had looked at the parties and their programmes. As for the Liberals, he decided that he could not support a certain group then in power—an obvious reference to Ralston—and so made up his mind to stay out of it. Then, he was "just about to do something else" (here again McNaughton was handicapped, for he could say nothing about the governor-generalship) when, on 31 October, the Prime Minister had called him and asked for advice on the reinforcement problem. A day later King had asked him "to put into operation the advice I had given him" and he had done so with the "one and single purpose" of "immediately and completely correcting the reinforcement situation," in order to "preserve the unity of the country. "He had been "disgusted and dismayed" by the terrible campaign of misrepresentation that had followed.[75]

That, it was reported, was "the end of the story of his troubles," and once again, as might be expected, he got a bad press.[76] His audience was genuinely puzzled. McNaughton's manner was convincing enough, but his hearers asked themselves why he was in the race at all. Did he have political ambitions? Why bother, when he could have rested on his record as a scientist and a soldier? Was he vindictive, anxious to get back at Ralston? It didn't seem to be that; there had been no signs of it during his last few months as Minister. His purpose in joining King, he had said, was to see the war through to a proper conclusion; but then they thought of his obvious reluctance to take all the zombies, who were obviously shirkers, and somehow things did not add up. And yet when he spoke they felt that he was so sincere in his beliefs that, whether they agreed or not, they wondered how anyone could doubt him. On the other hand, there was a mystery somewhere and straightforward prairie folk don't like mysteries, especially in the man who seeks to represent them. Then—and this was damning—his father, known and trusted, had been a Tory. What was his boy doing standing as a Liberal? And Andrew—as they always called him—had married a Roman Catholic.

Yet, in spite of every thing, "he was given close and respectful attention, and applause when he finished his speech." The General, it was stated, was making friends.

He is drawing good crowds—better than his opponents—although some of the following are those people curious to see and hear a national figure. . . . "If the campaign went on long enough," said Editor McKay [of the Moosomin *World-Spectator*] who knows the people of Qu'Appelle, "the General would win easily. It is instinctive to like him when you meet him."[77]

McNaughton's wife thought otherwise. "Andy was not a politician," she said. "There's no doubt about it. My argument is that he was far too straight."[78] She went with him everywhere, and "driving through the gumbo at that season of the year is something unbelievable if you've never done it." Once, approaching Regina in the middle of the night, dreadfully tired, she looked out and saw the lights of the city like a great star on the plains. "Thank God!" she said, "We'll soon be there." "Cheer up," said the driver, "it's still forty miles. You're on the prairies now." It was during this visit to Regina that the war in Europe ended. "I shall never forget," Mrs. McNaughton said, "the sound of cheering outside. Leslie and I went out and bought red and white flowers, and as near blue as we could get. She and I went to the Catholic church that afternoon while Andy went to an Anglican service. It somehow seemed unbelievable that the war had ended."[79]

Whether at his campaign headquarters in Moosomin or elsewhere, the arguments over his candidacy continued to rage in McNaughton's birth place. "In the hotel lobby, in the town's one tavern, on the streets and in the stores" it was the main topic of the day. "There seems to be no half-way neutral opinion publicly expressed. They are either violently for him or dead against him. . . . They are asking but not getting a satisfactory answer to one all-important question—why did the General leave the Army? '. . . The General must answer that question if he hopes to represent us,' declared [an] old farmer. 'Seems funny to me that he won't. We can't understand it.' 'It's my guess,' said another, 'that he'll keep his silence even at the price of losing the election.' "[80]

McNaughton said no more than he had said at Fleming and on election day, 11 June, he was defeated. The final result, after the soldiers' returns, was: Mrs. Gladys Strum, 6,146; Perley, 5,415; McNaughton, 4,871. The soldiers' vote went heavily against

McNaughton. That night, while the results were coming in, McNaughton was called to the telephone at the hall in Moosomin. It was Mackenzie King, speaking from Ottawa: "Well, Andy, how are you getting on?" "I'm not," returned McNaughton shortly. "Nor am I," said King. Soon after he got up on 12 June, King telephoned again. Though McNaughton had been defeated, it appeared that King had won Prince Albert and McNaughton congratulated him warmly. The General himself was still hopeful that "the soldiers' vote might yet change things."*[81] King told him not to be concerned: "We would find him a seat somewhere else."[82] In the end Mackenzie King was defeated by the servicemen's vote (his unpopularity with the services was well known). He lost his seat but the Liberals won the election. In a House of 245 seats they carried 127 against 68 for the Conservatives, 29 for the CCF, and for the others, 21. Thus, in a House of 245 seats, the Liberals had a narrow majority and King, despite his own defeat, counted the result a vindication of his policies.

McNaughton's short political career was nearly over, but King refused to admit as yet that he might have to part cabinet company with him.[83] It was generally assumed "that Mr. McNaughton, after being decisively rejected by the votes of two far-separated constituencies within six months, will not persevere with his ill-starred adventure into politics."[84] "The swan song of McNaughton," the *Globe and Mail* hinted, "will soon be sung."[85] It was his own fault, some said. He had been "very anxious to sit in parliament no matter what side was in." Had he stayed a Conservative and been "selected to lead the Conservative party instead of Mr. Bracken he would have been our new Prime Minister."[86] "Many an old soldier, Mr. Editor, thought their 'soldier Andy' might become Canada's first Field Marshal." Instead, he had entered politics "to suit the immediate political convenience of a shaky Prime Minister," and now his "short but hectic political fling is over."[87] His was a "melancholy case," rejected for the second time "and this time by the votes of soldiers he once commanded. Truly, General McNaughton's star failed him tragically on the day that he

* Referring to this conversation four days later, King wrote: "My recollection is that in 'phoning he rather kept believing that he could win. I am inclined to think that his Scotch pride and a certain touch of vanity make it difficult for him to look either defeat or the possibility of defeat in the face. He is wholly inexperienced in politics and feels that because in his own mind he ought to win he will win. Unfortunately that is not true, as one comes to know in politics."

yielded to Mr. King's blandishments and consented to take the place in the cabinet of a soldier comrade."[88] "By the grace of Mr. King only" he continued to administer Defence, "one of the most important departments at Ottawa. . . . His unsatisfactory record both as a military commander and as a politician was responsible for his emphatic rejection by two sets of voters and is striking evidence that he has forfeited the confidence of the Canadian people which he had once enjoyed."[89]

McNaughton had, in fact, lost more than a political career, had he wanted that, at Qu'Appelle. More had been riding on the issue than his ministry. With two political defeats behind him he could no longer be the next Governor-General of Canada. That was impossible for a man who had "forfeited the confidence of the Canadian people "and it was no surprise to him, therefore, when King broached the matter at Kingsmere, his country home, on the afternoon of Saturday, 16 June. It was raining. After looking at the view for a few minutes from under a tree they went inside. King talked of the governor-generalship and who, *in the United Kingdom,* might be considered. This was the reason behind King's invitation, "so that he could indicate to me that with the march of events he had now no intention of proceeding with the arrangements we had made about mid-October last."[90] All that the General ever asked was that King should place on record that he, McNaughton, had been "named to the Office of Governor General of Canada; that this had been approved by H.M. The King; and that he had given up this post in order to serve the greater need in the crisis of November, 1945"[91]. Certain British persons,* including military leaders, were then discussed and McNaughton, from his knowledge of them, gave his views on their suitability or otherwise.

In an effort, perhaps, to soften the blow, King spoke of a "Canada Medal" which had been authorized and which he hoped would be awarded; he said he thought the first should be given to McNaughton. Then, naturally, the talk turned to politics. When speaking by telephone on the morning after the election King had said that he wished McNaughton to remain in the cabinet; the situation would clear and he would then try to open a seat where McNaughton might

---

* Among them George Macaulay Trevelyan, the historian.

expect an acclamation. At Kingsmere, however, McNaughton said that if it were more expedient for him to drop out, King could feel that his portfolio was at his disposal. While King "thanked me for this offer," he "made no comment by which I could judge his attitude."[92] King commented in his diary: "It is a very baffling situation for me. I would like McNaughton in the Government and feel that he is a real strength. I can see, however, that some of my colleagues may feel it would be inadvisable to risk a third election with him at least at present."[93]

Lord Alexander* became the next Governor-General. He had first heard the news from Churchill at the Potsdam Conference. "Canada has asked you to be their next Governor-General. What do you think of that?" Alexander, his biographer says, had doubts. Almost all his salary would have to go on entertaining and there was Government House, with its servants, to maintain; "it would all be so different from his modest little Georgian home in Windsor Forest." And the thought of McNaughton haunted him. "There were still elements of the Canadian Army and community–mostly those who had no immediate contact with the matter–who remembered that it was Alexander who had refused to permit General A. G. L. McNaughten [*sic*], who had commanded the Canadian Forces in England, to visit his own 1st Division during the invasion of Sicily."[94] Nevertheless it was a great honour, and Alexander accepted. One of his first acts as Governor-General was to make peace with McNaughton at Government House.[95]

Brooke, now Lord Alanbrooke of Brookborough, heard a rumour, according to his biographer, that *he* might become Governor-General. "Though he could not believe it–and in view of the part he had so reluctantly had to play in the reorganization of the Canadian Forces it would have been clearly out of the question–he could not repress a wild hope that it might be true." When he heard that Alexander was to go, however, he bore the news nobly: "I remain with a few heartburns which I think Kipling's *If* has taught me by now to overcome."[96]

As for Montgomery, McNaughton never did become reconciled with him though Montgomery proffered an olive branch. At the end of the war in Europe the Field-Marshal said: "For the first time in history, Canada has put into the field an army of her own. I first met

* Raised to the peerage in the New Year's Honours List of 1946.

and got to know that army in 1941 when General McNaughton was forging the weapon—and right well he did his job. . . . It is my earnest desire that I may one day visit your great country—if you ask me."[97] Montgomery came to Canada the following year where he announced: "The former Canadian son of two good bloods, French and British, has grown to full manhood; a worthy son has now become a full partner, whose advice and counsel is eagerly sought in the old home and whose strong right hand is ever ready to be raised in the cause of freedom."[98] (In realizing that, McNaughton contended, Montgomery was a quarter of a century too late; that conclusion could easily have been drawn at the end of the First World War.) Montgomery expressed a wish to meet McNaughton whom Mackenzie King asked to come to Ottawa from New York, where he was then representing Canada at the United Nations, to shake Montgomery's hand in public. But McNaughton, who was extremely busy, refused. He was chairman of the Atomic Energy Commission that month (August, 1946) and felt that it would not be right to leave even for a day. But, as he pointed out, he was not making that an excuse for not meeting Montgomery; he just did not want to meet him. "I felt that I had received grievous injury from him, that he had stabbed me in the back. That despite this and to satisfy a public interest at the time, I had journeyed from England to Sicily, gone to see him and shown no sign of my feelings. I really did not wish to see him again. The matter was far too painful and I would only do so if it were a matter of public interest." McNaughton was not pressed to return to Ottawa.[99]

This was characteristic of McNaughton; he had the blood of the Highland Scot and did not forgive easily. It was a weakness, especially in politics, and Mackenzie King—a master of compromise—remarked on it during August, 1945, when McNaughton was still Minister of National Defence. He was "much too strong in his suspicions and dislikes and hatreds . . . McNaughton was not a good man in politics for that reason. Sir Wilfrid [Laurier] was right when he said that it does not do to cherish resentments in public life."[100]

Four days after his meeting with King at Kingsmere, McNaughton was called by the Prime Minister again. The older members of the cabinet, he was told, were considering where he should run in a by-election and would reach a conclusion in a day or so. King mentioned the personal attacks upon McNaughton, which were still continuing, but thought it logical that they should; the opposition was seeking to weaken McNaughton "as one of the *strong men* in the government" in the eyes of the public. McNaughton reiterated that his portfolio was still at King's disposal but that he had "no intention of quitting"; he would continue to fight the Conservatives. King did not request it. "He recognized that without the action of myself and St. Laurent," McNaughton wrote, "there would now be no Liberal government and the opposition knew this as well."[101] St. Laurent had assured the Prime Minister that McNaughton could be elected in any constituency in Quebec[102] but McNaughton's record does not make clear what his answer was. He said afterwards that to have been successful there would have made him "a prisoner of Quebec" and he had no wish for that.[103] Then, on 29 June, King "spoke of Nova Scotia and being able to get a seat there but he [McNaughton] thought it would open at once the contention between Ralston* and himself. . . .[104] By 17 July King asked himself whether, in McNaughton's own interests, "a hiatus at least for a while might not be all to the good."[105]

A week later King took McNaughton to dinner at the Country Club, where the Prime Minister told him "of our all feeling it would be taking almost too great a risk to open a seat." McNaughton, King recorded, "had evidently felt that would be the decision we would reach. He said at once that just now it did not matter so much. He had got things in shape in the Defence Department where anyone else could carry on all right. Had accomplished most of what he had set out to do. . . ." King "tried to have him see that once he is out of the arena there may be a considerable change of opinion and feeling that he has not had fair treatment by the public."[106] On 26 July Howe and St. Laurent echoed this view: "Both . . . were quick to see that a reaction would follow the minute he left the Ministry. The public will begin to say we had not treated McNaughton properly. That his genius ought to be kept for public service in Canada."[107]

The press attacks still continued. Towards the end of July the Montreal *Daily Star* drew attention to a dubious "record" that McNaughton, by this time, had achieved. He had been a cabinet minister for eight months without a seat. This was much longer than Laurier, his closest runner-up, who in 1877 had held office without

---

* He was a Nova Scotian.

being an elected member for one month and twenty days.[108] "Mr. McNaughton," the *Globe and Mail* announced on 6 August, "must not stay in office any longer if responsible government has any meaning."

That day McNaughton wrote to King:

In confirmation of the view which I have expressed to you in our discussions, I write to say that I believe the time has now come when I should leave your Government.

. . . the purpose which I sought to serve in undertaking the duties of Minister of National Defence for the Army has been accomplished. . . .

I would have been very happy indeed to have continued to serve in the Cabinet under your leadership, at least until the war with Japan had been brought to a conclusion, our men and women brought home from overseas, cared for and re-established, and our post-war reorganization set up for defence; but now a new Parliament is shortly to assemble, of which I am not a Member. . . .

Consequently, not having the privilege of membership in the House of Commons, I think that the public interest requires that I should stand aside. Therefore I place my resignation in your hands. . . .[109]

It was more than two weeks before King replied: "holding strongly to the view that the war with Japan might end even before the assembling of Parliament," he had felt it in the public interest to retain McNaughton's services. But "the war with Japan is now, too, happily at an end," he wrote, so that "if your resignation is to be accepted, this would appear to be the appropriate time. . . ."

The course adopted with regard to conscription had been "fully vindicated." In addition, King acknowledged

what your membership in the government has meant to Canada in helping to meet the problems which have arisen in the concluding months of the war, in the work of demobilization, and in post-war organization. Your wisdom and far-sighted vision in dealing with these problems has been of in estimable value.[110]

By this "exchange of mutual compliments," the Toronto *Evening Telegram* reported in a final thrust, "both agree that what they did was for the best and that to have followed any other course would have been to court disaster. That is the happy ending the collaborating authors have written for the chapter, and the Canadian public is now expected to roll over and go to sleep."[111] But another newspaper, which had been consistently hostile, warmed to McNaughton again now that he was no longer in politics:

An unhappy episode thus comes to a close. Most Canadians, we think, looked with distaste on the spectacle of General McNaughton's . . . political interlude, because we much prefer to think of him as the distinguished soldier, the eminent scientist, to whom Canada stands in debt for many splendid achievements. He is the man who built the Canadian Army which others were to use, and victoriously. He gave this country notable service in the First World War and between the wars. Canada has few men of his quality, few with as good a brain, few with as great a zeal for their country or as great a capacity for disinterested endeavour in the public interest.

These are the things we propose now to have in mind. For McNaughton, soldier and scientist, the country has respect, admiration and gratitude, and he reverts now to the role he never should have quit.[112]

More spontaneous was the estimation of an unknown Canadian manufacturer whom an American writer met in the dining car of a train between Los Angeles and Chicago at the time McNaughton was under severe attack in the spring of 1945. "The conversation turned to the war, the difficulties over conscription, and the . . . election campaign in Grey North. The manufacturer did not know of the special interest of his questioner when asked about the General's standing with his countrymen, and the place he held in public regard. His answer was quick and brief: 'The first gentleman of Canada!' "[113]

Perspective, as the Kingston *Whig-Standard* rightly pointed out, has shown that Grey North and Qu'Appelle voted better than they knew. They denied McNaughton a seat in Parliament–and, indirectly, the office of Governor-General– and he felt it deeply. But McNaughton was a statesman, as later events were to prove. He was not–and never would be–a party politician, because compromise and expediency were not in him. Had he been Governor-General, what then? After his term of office, retirement; and he would have chafed in that. As it was, the voters of 1945 opened the way for him to be of even greater service to his country than he ever could have been in the offices that they denied him.[114]

Chapter 4

At the United Nations

General McNaughton, as is obvious from the results of the election campaigns that he fought in 1944, had temporarily lost the confidence of the Canadian people. Nevertheless, he had not lost the confidence of Mackenzie King who knew what McNaughton had been about and who, victorious in a by-election, remained Prime Minister.

On 21 August, 1945–the day he accepted McNaughton's resignation as Minister of National Defence–King named him chairman of the Canadian section of the Canada-United States Permanent Joint Board on Defence, a highly important body. Early in January, 1946, the Prime Minister invited McNaughton to represent Canada at the inauguration of President Dutra of Brazil, to be held in Rio de Janeiro on 30 January. On McNaughton accepting for his wife and himself, King said he was glad to know that the General was once more "in active association" with him and the government of Canada.[1] In April, 1946, McNaughton was appointed Canadian representative on the United Nations Atomic Energy Commission. In the Birthday Honours List of June, 1946, he became a Companion of Honour and in August Belgium named him *Grand Officier de l' Ordre de Leopold.* Towards the end of September the Prime Minister appointed him to the presidency of the Atomic Energy Control Board of Canada and in January, 1948, he selected McNaughton to be permanent delegate of Canada to the United Nations and the representative of Canada on the Security Council. Even after November, 1948, when King ceased to be Prime Minister, he still watched over the General and he was consulted about the appointment of McNaughton to the International Joint Commission on 1 January, 1950.

In April, 1960, the General was able to look back with equanimity at Grey North, his first election battle, however painful it had been at the time. "I cannot say I regret the result," he wrote. "If we had won that contest I would not have had the remarkable sequence of experiences which fortunately have come my way in the United Nations Atomic Energy Commission, the Security Council, the Canada-U.S. Permanent Joint Board on Defence, and more recently the I.J.C., which I value most of all."[2] It seems appropriate, therefore, to discuss McNaughton in his many roles in the order in which he placed them.

The greatest problem facing the world when hostilities ended against Japan was the threat posed by atomic weapons. The atomic bomb, as had been terrifyingly demonstrated at Hiroshima and Nagasaki, was a weapon whose use in a future war might well destroy civilization; and yet the peaceful uses of atomic energy would undoubtedly be of great benefit to mankind.

Canada had been associated with the United States and Britain in the development of the atomic bomb. Atomic energy research in Canada had its origin in 1899 when Rutherford came to this country and began his investigations of the nature of radioactivity; he made, both in Canada and in England, discoveries of great importance in the development of atomic energy. These, however—as McNaughton, who had attended Rutherford's lectures at McGill, said—were "concerned with driving back the frontiers of human knowledge rather than with any so-called 'practical' application of the knowledge thus gained."[3] In 1939 a German scientist, Otto Hahn, made the fundamental discovery of the nuclear fission of uranium, induced by fast neutrons, which *had* practical implications; these in time led to the successful production of atomic bombs. Fortunately for the Allies, the bomb was produced first in the United States and not in Germany.

Shortly after Hahn's discovery the Second World War broke out. Then, as McNaughton said, the possibility of producing a decisive weapon based on Hahn's principle of fission in a chain reaction became of immense importance to all governments. If this could be done, obviously that nation which succeeded first might well gain thereby the power to eliminate all its enemies. The enormous potential value of the result was quickly recognized by the war leaders of all countries. Nowhere was this more so than in the United States, where support grew in crescendo on a tremendous scale, in a carefully planned and directed programme of mass effort that swept all obstacles before it. Never has there been any comparable undertaking or achievement in research and development in world history. We can rightly claim for scientists from the United Kingdom, from Canada and from elsewhere in the British Commonwealth a share in the inception of these undertakings and in the evolution of the scientific basis on which progress was made. We can claim a share too in the working out of the actual methods which were used. But we must pay our generous tribute primarily to the United States, for without their effective participation in production, not even the best ideas could have been translated into accomplishment.[4]

In 1940 a Canadian scientist, Dr. G. C. Laurence, was already experimenting with petroleum coke and uranium oxide, studying the neutrons in this material to determine whether a great quantity of energy could be released. In short, he was attempting to produce a reactor. That year information on atomic energy was exchanged by Britain, the United States and Canada, and in 1942 a British research team moved to a laboratory which was established at the University of Montreal under the direction of the National Research Council. Research in this laboratory at first was closely co-ordinated with that taking place in the other two countries. In 1943 it was decided that studies should continue on a pilot-plant scale, and as such a plant in Britain would be exposed to the hazards of air attack, it was further decided jointly by Britain and Canada that the plant should be built in Canada. Dr. J. D. Cockcroft,* a British scientist, became a director of the project, which was under the administration of the National Research Council. The design was developed in the Montreal laboratories, while the plant itself was built at Chalk River, Ontario. There, on 5 September, 1945, more than two years after the Americans had built their first reactor, the experimental reactor went into operation, the first to do so outside the United States. Since both Britain and Canada had worked on the production of the atomic bomb with the United States, they knew its secret and for that reason

* Later Sir John Cockcroft.

they were intimately concerned about the atomic threat that faced the world after V-J Day in 1945.

The control of atomic energy was vital if it were to be directed towards peaceful ends and the world protected from atomic bombs; and for this an international agreement was required. The three nations that had been associated in the wartime project took the first step with a declaration issued from Washington on 15 November, 1945, by President Truman, Prime Minister Attlee and Prime Minister King. These leaders, said McNaughton, "recognized the need for an international agreement and proposed as a matter of great urgency the setting up of a Commission under the United Nations to study the problem and to make recommendations for its control."[5] It was vital to obtain the support of the Soviet Union, and accordingly a meeting of the Foreign Ministers of the U.S.S.R., the United Kingdom and the United States followed in Moscow at which, during December, 1945, the Washington proposals were endorsed. Finally, at the meeting of the General Assembly in London on 24 January, 1946, the United Nations Atomic Energy Commission was set up by unanimous resolution—the first resolution, incidentally, passed by that assembly.

This commission, McNaughton continued:

> composed of delegates from each country represented on the Security Council [the "big five" nations: the United Kingdom, the United States, the U.S.S.R., France and China], as well as Canada when Canada is not a member of the Council, together with six non-permanent members, was charged with making specific proposals:
>
> (a) For extending between all nations an exchange of basic scientific information on peaceful ends;
>
> (b) For the control of atomic energy to the extent necessary to ensure its use only for peaceful purposes;
>
> (c) For the elimination from national armaments of atomic weapons and of all other major weapons adaptable to mass destruction;
>
> (d) For effective safeguards by way of inspection and other means to protect complying states against the hazards of violations and evasions.[6]

"Thus the other nations of the world," in McNaughton's words, "recognizing the importance of the contribution which Canada had made, appointed this country as one of the original and continuing members of that Commission."[7]

The commission met for the first time in June, 1946. McNaughton was "an obvious choice" as Canada's representative, wrote George Ignatieff who accompanied the General to New York as his adviser,* because of his proved leadership in many fields, including science. "What was not known was whether he would also prove to be a diplomat." The Prime Minister had no doubts whatsoever on that score, for, as Ignatieff continued: "Mr. King, when making his choice, had inquired what kind of representatives the other countries were sending to the United Nations Atomic Energy Commission. He was told that some were appointing diplomats, others scientists, and some were sending politicians. Mr. King had commented: 'McNaughton—all three.' "[8] McNaughton was fifty-nine. At the United Nations temporary home at Hunter College in the Bronx and at the Henry Hudson Hotel where later sittings took place, he met other representatives: the American, towering Bernard M. Baruch, adviser to presidents, seventy-six years old, "a millionaire at 30, a celebrated national figure at 50 and . . . a legend by the time he was 70";[9] handsome Andrei A. Gromyko of the U.S.S.R., only thirty-six years of age and little-known as yet; Alexandre Parodi of France; Dr. Quo Tai-chi of China; and Britain's "perfect diplomat," Sir Alexander Cadogan, who had been Permanent Under-Secretary of State for Foreign Affairs before going to the United Nations.

McNaughton flew to New York and Ignatieff has described the trip:

> I remember that we left Ottawa on a sultry day in June 1946. Because of storm warnings, civil aircraft had been grounded. As he already had an appointment that day in New York, the General would not brook any delay; so we flew off in an R.C.A.F. DC.3 aircraft. Over Albany we ran into an electrical storm. The lightning seemed to glance off the wings . . . and the aircraft danced on the air as we struggled through the turbulence of the storm. Suddenly we found ourselves literally falling through space. We had hit an air pocket. The cases containing official papers lying on the floor of the aircraft hit the roof; my neck seemed to be wrenched out of its socket, as if I was being hanged. I

---

* And who was subsequently Canada's permanent representative and ambassador to the United Nations.

frankly felt frightened; but when I looked at the General, only his glasses had fallen off and he was still holding his papers. All he said was: "It seems to be a little bumpy."

It was this quality of being cool and yet exhilarated in face of crisis which I came to admire in the General when we plunged together into the recurring crises of those early days in the United Nations.[10]

The object of the commission's first meeting on 14 June was to negotiate a treaty. "It would," wrote Frederick Osborn, who joined the commission early in 1947 as deputy representative of the United States, "have been hard to pick an international group more highly qualified for their task. They were competent and ready to negotiate. The atmosphere was one of expectation."[11] Baruch immediately made proposals on behalf of the United States. The American plan

generally resembled that outlined in the [Acheson-] Lilienthal Report, which had been released in the United States a few months previously. It called for the formation of an International Atomic Development Authority, which would foster beneficial uses of atomic energy and would control atomic activities in all nations either by direct ownership, management, or supervision, in the case of activities potentially dangerous to world security, or by a licensing and inspection system in the case of other activities. This system of control would be set up by stages and after it was in operation the manufacture of atomic bombs would stop. Existing bombs would be disposed of, and the world authority would be given information regarding the production of atomic energy. In addition, the United States proposal emphasized that the veto of the Great Powers in the Security Council should not apply in the event that any nation was charged with having violated the international agreement not to develop or use atomic energy for destructive purposes.[12]

Baruch, then, while accepting the basic principles of the Acheson-Lilienthal Report, had added a rider of his own which concerned the dropping of the veto. Why had he done this? An authoritative article which appeared in the American *New Republic* of 3 April, 1950, points to the reason:

Baruch . . . had an essential distaste for the policies of the Acheson-Lilienthal Report. It was not merely that the report reduced him to the role of spokesman rather than statesman; Baruch apparently doubted that US security would be strengthened by the control of atomic weapons—the only effective counter-agent to the Red Army in Europe and Asia short of our maintaining a large standing ground and air force and a continuing state of near mobilization. But instead of flatly proposing that the plan be broadened to include other forms of disarmament, which he was perhaps at first tempted to do, he undertook to strengthen the "limited control" of the Acheson-Lilienthal plan by adding his own provision for punitive sanctions. . . . As anyone with an elementary understanding of the dynamics of the UN would have understood, a proposal to abolish the veto even in atomic matters was doomed to failure.

Not only did Baruch want to abolish the veto, which had been accepted by all the Great Powers as the basis for collaboration within an international organization and which was part of the Charter of the United Nations: he planned to impose sanctions–"condign punishment"–for "violations of the rules of control which are to be stigmatized as international crimes."[13] Instant and condign punishment for contravention of the control arrangements, which could only mean atomic retaliation, would be decided upon in the Security Council where Russia was in the minority and where veto rights, under Baruch's proposals, would have been abolished. "But," wrote an American student of international politics, "the Soviet Union has good reasons for believing that the veto constitutes an essential element of her security. It makes it impossible for the rest of the world to conspire and 'gang up' against her in a coalition disguised as a world organization."[14] Thus the proposal that the veto should be abolished was a controversial issue that Russia would certainly oppose. As one Soviet writer put it, Baruch "stubbornly tried to impose upon the Commission proposals which he knew could not serve as a basis for agreement among the members of UNO, since they contradicted its Charter and its principles."[15]

Five days later Gromyko put forward a Soviet plan which differed fundamentally from that of the United States. In McNaughton's words, "It proposed the immediate outlawing of the atomic bombs and the destruction of all existing stocks of atomic weapons within a three-month period. To this end the Soviet delegate tabled a draft convention which, he said, should be negotiated forthwith as the first step towards the establishment of a system of international control. The Soviet delegate was prepared to discuss methods of control and inspection but he maintained that this should not hold up the immediate prohibition of atom bombs."[16]

To McNaughton this plan seemed "very unreal." The idea that atomic bombs could be outlawed simply by signing an international agreement prohibiting them simply would not do:

> The experiences of the last twenty-five years have shown that international agreements without accompanying safeguards are not enough to guarantee the peace. The prohibition of the use and manufacture of the atomic bomb at the present time would merely seriously reduce the military strength of the United States, the only nation now in possession of atomic bombs. It would be an act of unilateral disarmament which would give no assurance that any country engaged in atomic energy activities would not or could not make and use the bomb in the future. Fissionable material, the essential material for such peaceful applications of atomic energy as the development of industrial power, is also the explosive element of the bomb, and in the absence of effective inspection and control could readily be diverted from peaceful to military uses by a nation secretly preparing for atomic war.[17]

The Baruch Plan, ostensibly, was realistic, even generous, and the Canadian government supported it. McNaughton, however, thought that the United States should not sacrifice its enormously strong position except in favour of the international authority that would ensure control, as had been suggested. But it would not be easy to persuade the Russians to accept the plan because of the element of international ownership of atomic energy plants and the proposed international inspection which would force the U.S.S.R. to lift the iron curtain. For that reason he had reservations about the abolition of the veto, as Baruch proposed, because he thought it would antagonize the Russians unnecessarily and promote entirely the wrong atmosphere for concessions. As McNaughton saw it, agreement was essential. Even though the ideological war between the free nations and the Communist states continued, with an effective control plan the risk of atomic war would be negligible and atomic energy could be put to the service of men—"to lift the people from an environment of fear into one of plenty." "That may seem naive today," he said in 1966, "but it seemed worth working for at the time."[18]

In McNaughton's view, then, a Soviet pledge not to use atomic weapons was not enough. The United States would be wrong to reduce its atomic arsenal by even one small bomb simply because the Soviets, as they were willing to do, gave such a pledge. A pledge was not a plan. Arms control could exist only through the acquiescence of the great military powers. And Russia, which at first would be a controller not so strong as the controlled, would have reason to acquiesce in a plan for the control of atomic energy; and with that plan marking the bounds of authority the veto was not important. The General, therefore, assigned no great significance to the veto question. "If the majority proposals were followed in other details," he said, "it would be possible to pass information to all nations, upon whom, in the last analysis, any deterrent to the breaking of agreements rested. There would, therefore, be no way of suppressing information of violations or seeming violations, and the veto lost its significance."[19]

At the next meeting of the commission, on 19 June, McNaughton was the first to speak. He announced that the Canadian government welcomed the American proposals and supported the principles on which they were based. But he saw the veto as a "potentially divisive issue" and suggested trying "to promote an atmosphere of mutual confidence by moving promptly to exchange scientific information." Gromyko, who spoke on the same day, stated flatly that the Soviet Union "rejected any tampering with the veto."[20]

Most members of the commission were in general agreement with the principles of the American proposals, diametrically opposed though they were to those of Russia. There being as yet no basis for agreement on the two plans, the commission put aside the political aspects of control and turned instead to a study of whether the control of atomic energy was practicable from a technical and scientific point of view. A commission made up of scientists and technical advisers examined these aspects and in September of 1946 reported that *"we do not find any basis in the available scientific facts for supposing that effective control is not technologically feasible."* And though the U.S.S.R. delayed, its representative did sign the report, making it unanimous.[21]

Then, reported Sir George Thomson, a Nobel Prize winner and Britain's scientific adviser, "the next step was to continue to consider methods of control. Here the Commission was greatly helped by a programme of work put forward by General McNaughton, the chief Canadian delegate." This work involved "a systematic survey of the degree of control which an international body would have to exercise over the various establishments concerned in order to make reasonably certain that no diversion of potentially dangerous material to improper use could occur." As an indication of the difficulties, the "establishments" Sir George referred to included everything from the mines where the ore was dug to the piles where the energy was finally produced; and the controls required would vary from simple inspection to actual management by the international authority. Measures to detect secret mines and factories were also considered. Meanwhile, Sir George continued, "an important discussion took place on disarmament in the Assembly of the United Nations and the Russians accepted, in principle the idea of an international inspectorate . . . This greatly eased the situation and led us all to a general feeling of optimism."[22]

McNaughton has described these events in the General Assembly. They show that there was a good deal of acquiescence from Soviet Russia, at any rate up to the middle of December, 1946:

> Together with the control of atomic energy the General Assembly was faced with the problem of the control of other armaments. This matter became one of the main issues at the Second Part of the First Session of the General Assembly held in New York in October 1946. The discussions were initiated when the Soviet Union . . . introduced a resolution, couched in general terms, calling for a general reduction in all categories of armaments. . . . Finally, two resolutions were passed by the General Assembly on December 14, 1946, one on the general regulation and reduction of armaments, and the other calling for the Security Council to determine the information that the Member nations should furnish in regard to their armed forces in order to implement the resolution of armaments. No one had expected that it would be possible for the general membership of the United Nations to agree so early in the life of the United Nations to take such an important first step toward general disarmament. The adoption of the resolution on the general regulation and reduction of armaments by unanimous vote therefore did much to restore confidence in the United Nations as an instrument of world peace and led to a feeling of optimism. . . .[23]

Frederick Osborn has described the Soviet attitude towards atomic energy control in those early days as "passive and acquiescent" and he is undoubtedly right.[24]

The U.S.S.R. would, however, require careful handling if any agreement on control was to be reached. The United States was the only nation in the world to possess the atomic bomb. Never before, in all history, had one country taken such a lead in weapon development that every other country was at its mercy. Russian cities, in Russian eyes, were under the threat of destruction at the time; the threat to American cities lay some years in the future. It was all very well for Western nations to say that the United States was peace-loving, moral, and would never subject the Russians to nuclear attack. The Russians had seen the Americans, no matter how humane they considered themselves, turn two cities of an enemy already tottering on the verge of surrender into flaming ruin. Group Captain G. L. Cheshire, who had been a British observer on the plane that dropped the atomic bomb on Nagasaki, when speculating on whether the weapon would ever be used again, said, "I know that if the weapon is powerful enough, the stakes sufficiently high, and the attainment of the goal is within grasp, human nature can lose its sense of values and not even know that it has done so."[25] Soviet Russia must have been very sensitive about her position and torn between accepting international control of atomic energy—and accepting American supremacy until effective controls were set up—or taking the risk of American supremacy for a longer period while she strove to develop a bomb herself.

The paramount ascendancy enjoyed by the United States in the nuclear field, as McNaughton pointed out, was no more than a fleeting moment in history.[26] It would not last forever: "Within a few years," he said in a radio address, "it may be expected that other countries . . . will be in a position . . . to make atomic bombs themselves" and consequently, he continued, "if we are to forestall an atomic weapons race, it is of the utmost importance that we make every possible effort to work out a practical plan for the international control of atomic energy."[27] It could do no harm, and possibly a great deal of good, for the United States to make every possible concession to the Soviet Union without damaging her own position. But that was not the case.

With the General Assembly's resolution on disarmament approved on 14 December, the climate was favourable for progress in the control of atomic energy. A middle group, including the Canadian delegation, began to move towards a compromise plan that would "water down or eliminate Baruch's proposal on the veto."[28] Whether, in fact, even without the veto question, a compromise could have been reached is doubtful; but it was worth trying at this most propitious time. The AEC met shortly afterwards on Monday, 17 December,* when Baruch attempted to hustle the delegates into the adoption of his plan. But Gromyko would not be rushed. He raised two objections: the American proposals did not agree with the resolution of the General Assembly, and to abolish the veto would violate the charter of the United Nations. "Offering the prospect of a unanimous decision, Gromyko called for additional study and discussion" and that seemed reasonable to the other members. Baruch, although very anxious to put the American proposals–which he had submitted in the form of a resolution on 5 December–to a vote, recognized that a postponement was inevitable; but he would agree to no more than three days. Gromyko considered the time too short. Nevertheless Baruch insisted, and a deadline was set for 20 December. "This peremptory demand," the *New Republic* commented on 3 April, 1950,

> raised a storm–small, but significant none the less as a clue to latent unrest and uneasiness. The opposition was led by the Canadian delegate, General McNaughton, a military man of some independence who had no taste for submitting to pressure from any source.

"McNaughton of Canada," we find in *The New World,* "while supporting the American plan, thought Baruch was driving too hard." Not only that: the General agreed with Gromyko that Baruch's resolution of 5 December should conform in wording to that of the Assembly. Baruch, who saw cracks appearing in a front he thought was solid, sent a member of his staff to Canadian headquarters, who conferred with McNaughton throughout the Thursday night, finally getting away at three o'clock next morning. McNaughton let him understand that a Canadian amendment would be offered to Baruch's resolution to the effect that the wording of the principles, the resolution of the General Assembly concerning all categories of armaments, and the relevant parts of the Working Committee's report, be made to conform in a draft report to be submitted to the Security Council.

---

* These December meetings, which were important, can only be dealt with summarily here. An excellent and detailed description of them is contained in *The New World, 1939-1945, A History of the United States Atomic Energy Commission,* Vol. I, on which this brief account is based. All quotations, unless otherwise referenced, are from this.

McNaughton has given the reasons for his stand:

There was a most *active* effort made in the U.N. from the inception of the U.N. Atomic Energy Commission, but always in the background of this effort I felt there was, in the U.S. at least, an ever-present purpose, more especially when Baruch presided over the U.S. Delegation, to concede nothing which might compromise the technological advantage thought to be held by the U.S.A.; and which was considered to be important to balance the Soviet preponderance in numbers of men and in the amounts of conventional armaments they had available. This condition, among other aspects, was the reason for our advocacy of combining the discussions of atomic and conventional armaments in one commission.[29]

The Atomic Energy Commission again met, in accordance with Baruch's deadline, on Friday morning, 20 December. Gromyko said bluntly that he was not ready to discuss Baruch's proposals: their wording should first be checked against that of the resolution of the General Assembly. McNaughton then brought out his amendment, which, he said, "endorsed only the principles" on which the resolution was based. "It did not bind the Commission to adopt Baruch's proposals as they stood." Gromyko still wanted time for study and held out for another postponement but Baruch accepted the Canadian amendment and insisted on a vote. The commission rejected Gromyko's appeal; by a vote of 10-0 (neither Russia nor Poland voted), it approved the American proposals as amended. This meant no more than that the Working Committee would take the three documents which have been mentioned and from them fashion a consistent report for the Security Council.

On 27 December, when the Working Committee met "behind closed doors," McNaughton proposed that Baruch's draft of 5 December should be revised. He produced a document that reflected a behind-the-scenes effort "to assuage Russian sensitivities," which he considered important. He had obtained some concessions from the American delegation but he was unable to move them on the veto question; for here the Americans and Canadians could not agree. McNaughton's proposal dropped Baruch's assertion that "a violator of the terms of the treaty should not be protected from the consequences of his wrongdoing by the exercise of any power of veto"; he substituted "there shall be no legal right whereby a wilful violator of the terms of the treaty or convention shall be protected from the consequence of violation of its terms." The Americans insisted on the insertion of four words, "by veto or otherwise," so the clause would read, ". . . no legal right, by veto or otherwise, whereby a wilful violator . . ." and they also objected to McNaughton's modification of "condign punishment" which excepted the right of self-defence from wilful violation of the treaty's terms.

The committee discussed McNaughton's documents clause by clause, especially the words "by veto or otherwise," on which the Americans insisted. Parodi of France wanted them deleted; so did Cadogan "if omission would aid in achieving unanimous acceptance of the principles." But Baruch insisted that they should remain:

> The American people believed that the United Nations stood for the sanctity of treaties. They would—and should—*withdraw their support from an organization** which became "no more than a debating society and a place to exchange pious-sounding documents." To refuse to take the course the United States delegation had advocated "would shock the moral, common sense judgment of the world. . . . If the violators of a treaty can legally and with impunity escape the consequences of a violation of an agreement voluntarily entered in to, then every treaty executed under the auspices of the United Nations contains the fatal defect that it is binding only so long as the major nations want it to be binding." Baruch said he could neither recommend to the American people nor advocate before the Senate that the United States surrender the atomic weapon under any system open to nullification of punishment by subterfuge. His peroration was brief and uncompromising. "Gentlemen, it is either—or. Either you agree that a criminal should have this right by voting against our position (or you fail to take a stand on the question by refraining from voting), or you vote for this sound and basic principle of enduring justice and plain common sense."

Baruch, however, did not have his way—at least, not yet. McNaughton and Cadogan had misgivings, as they made clear. Parodi of France, supported by Juliusz Katz-Suchy of Poland, wanted no part of "by veto or other wise." Only Brazil and Egypt supported Baruch. It was agreed that the chairman of the Working Committee should submit a report to the AEC together with a letter stating that the passage was in dispute. Alexandrov, a Russian delegate "who had sat

---

* Italics supplied.

With Philip Jessup of the U.S. delegation, UN Security Council, April, 1948

A discussion with Arthur Creech Jones of Britain and Warren Austin of the United States

The president of the Security Council

silent throughout the long, gruelling session, asked that the letter note that he had not participated in the discussions."

At about this time Gromyko was being cold-shouldered by the Western group, but not by McNaughton. At a banquet, where the Russian sat next to Mrs. McNaughton, he told her: "It's so easy to work with the General; *he* is never rude."[30] And McNaughton later described how, in the commission, he was under instructions from his government to support a particularly American point of view. He sought to change those instructions, in the interests of a settlement, at first without success. Then he was told by telephone in the middle of the night that he was at liberty to attempt to change Baruch's stand. "When I got to the meeting the next day," he said, "Gromyko, as shrewd as a pet fox, read in my face that I was going for Baruch. He could see that my papers were not ready and so he took the floor and talked nonsense, really nonsense, until he saw that I had sorted the notes. Then he said, 'I now yield the floor to the representative of Canada,' and sat down."[31] Gromyko, while he may have found McNaughton's support hard to understand, was more than willing to let the Canadian go ahead. As Frederick Osborn wrote:

> Gromyko appeared to recognize McNaughton's position on the Commission. But at times he appeared deeply puzzled by it. The fact that the members of the majority faction of the Commission were friends who enjoyed each other's confidence did not mean that they always agreed. Once when McNaughton and the American delegate were engaged in quite a violent argument over some point in the plan, I remember glancing at Gromyko and seeing an extraordinary expression of bewilderment on his face, as though he were trying to figure out whether or not the whole thing was being staged for his benefit. . . . Certainly no Soviet delegate other than Gromyko ever thought for a moment that the majority of delegates did anything except on orders from the United States.[32]

But McNaughton was sincere in his efforts to obtain an agreement that could be accepted by the U.S.S.R. He was considered by some to be the only type of man who could deal successfully with the Russians. As the Montreal *Star* put it on 18 December, 1947, "He knows his own mind and is quite willing to speak it but he has the patience of Job. He will need it to get the Soviets to agree to anything, but if the McNaughton approach fails nothing can succeed." And he was always anxious to deal with the Russians as he would wish to be dealt with himself. An indiscretion on the part of George Ignatieff, which occurred at a later period, may be cited as an example. As Ignatieff tells it, this is what happened:

I . . . learned that although the General was a stern and completely selfless and dedicated man, he had an unconscious humour which made personal relations with him extremely pleasant. . . . There was . . . the famous Cold War anecdote which, at the time, got me into trouble; only very slightly, I must admit. At one of the monthly Security Council dinners which were held in those days, General McNaughton was talking with Secretary Wallace, at that time Secretary of Commerce in the Truman Cabinet, and with Mr. Gromyko. . . . The latter was in a critical and rather gloomy mood, complaining of life in the United States. Gromyko said something to the effect that American food was mostly tasteless and not interesting like the food in the Soviet Union. He complained that he had been unable to find any apples in New York that tasted like apples. Secretary Wallace . . . turning to General McNaughton . . . said: "I am sure that the General can suggest some Canadian apples which Mr. Gromyko might find more acceptable. . . . What are some of your favourite varieties in Canada, General?" Crisply, the General replied: "MacIntosh Reds and Northern Spies." The conversation, which took place when the revelation of Gouzenko's spy activities . . . was very much in the public eye, dropped like a lump of lead.

The next day the General was most concerned that he might have offended Mr. Gromyko. I assured him Gromyko was probably laughing at the joke. When the General learned I had passed the anecdote to Mr. Pearson by telephone, who had mentioned it at a press conference, the General looked sternly at me and said, "Ignatieff, I don't like humour."[33]

As Walter O'Hearn put it in the Saskatoon *Star Phoenix* on 21 August, 1948: "When the story got wide publicity he was distressed, not only because the incident made him appear rude but because it made him appear a wit. He is not. He is a very serious minded man in a job which calls for high seriousness."

The Atomic Energy Commission next assembled on Monday morning, 30 December. Gromyko, who spoke first, at once attacked the American position with regard to the veto question, which, he alleged, despite Baruch's denials "defied the charter of the United Nations." Without commenting on the Russian's objections, Baruch moved that the report, which had been forwarded by the Working Committee, should be adopted. What was hanging over the heads of the delegates, as Baruch had made clear on the 27th, was the threat that the United States might withdraw from the United Nations unless his wishes with regard to the veto were accepted. As the London *Economist* observed on 4 January, 1947, "The hesitations of several members of the Atomic Energy Commission were in the end overridden by Mr. Baruch's uncompromising attitude, which led him to state that the United States 'would–and should–withdraw their support from this organization' if his plans were to be rejected." This would be a greater disaster than the antagonism of Russia. One by one the delegates, including McNaughton, rose to support Baruch despite their reservations. "Only Lange of Poland spoke in opposition. Deploring Baruch's threat . . . he suggested submitting the report without taking a vote but giving due notice of disagreements. If that proved unacceptable, he favoured amendments to reduce the area of disagreement." The commission adjourned for three hours to deliberate.

On the return of the delegates Baruch swept aside the Polish suggestion. He demanded "a recorded vote." Oscar Lange again suggested "that the chairman indicate in his letter of transmittal that certain passages were not approved by all the delegations." But it was the entire report that was voted on, and it received ten votes, with Poland and the U.S.S.R. abstaining. From then on Russian intransigence became apparent. Some writers have suggested that American atomic bomb tests at Bikini, which started on 1 July, 1946, were tactless and might have been the reason; another says that Baruch's presentation of the United States position, as a whole, on the first day of negotiations might have made the Soviet Union suspicious from the beginning ("negotiations are not usually started that way");[34] but Professor P. M. S. Blackett, the British scientist and Nobel Prize winner, was probably closer to the truth when he asserted that "the question of whether the veto should operate in relation to matters concerning atomic energy was probably decisive in producing the final deadlock."[35]

Why was the veto so important to the U.S.S.R.? The attempt to apply "sanctions against a Great Power," said Blackett, "could only lead to a major war" and thus the "unanimity rule is an essential safeguard against the possibility that the United Nations Organization will be used to 'legalize' a third world war."[36] That is why it was a part of the charter. Now let us look at the veto in the context of the imposition of sanctions—"condign punishment"—against a violator of an atomic energy agreement. Blackett has analysed the subject thoroughly. The Baruch plan, containing clauses for the abolition of the veto and the implementation of sanctions, could quite easily have affected the U.S.S.R. adversely. Under the plan atomic energy would have been used for peaceful purposes under an international Atomic Development Authority (ADA). Would this authority have had the power to slow down, or even stop, peaceful atomic projects in the U.S.S.R.? Apparently so, for the production of power on a large scale was classed as a dangerous operation, easily adapted to bomb production. And the ADA, for the next ten years as far as could be foreseen, would contain a majority drawn from the "Western" nations that would vote with the United States if the U.S.A. considered it important to apply the brake. "The Soviet leaders might well consider that the American Government, out of fear of the economic and political consequences of a possible economic advance, would be most unwilling to permit the Soviet Union to develop atomic energy on a large scale, even if this was done without danger of diversion to bomb production."[37] Suppose the Russians went ahead anyway?

On what would happen then, Baruch is quite explicit. In the list of crimes which would merit immediate and condign punishment is not only the "illegal possession of an atomic bomb," but "creation or operation of dangerous projects in a manner contrary to, or in the absence of, a license granted by the international control body." [Thus] a country such as the U.S.S.R. which attempted to develop atomic energy for power production—even with the most peaceful of intentions, but against a majority decision of A.D.A.—might find itself convicted of the crime of "wilful interference with the activities of the Authority," and so become liable to have instant and condign punishment imposed on her.[38]

A letter to the editor of the Ottawa *Citizen* of 12 June, 1954, is not without interest:

I have [confidence] in the wisdom of the American people as long as they are told the truth. . . .

But how much of the truth have Americans or, for that matter, Canadians been allowed to hear? How many Canadians know that Mr. Baruch killed the McNaughton proposals for atomic energy control, the most workable compromise produced in the United Nations Atomic Energy Committee and forced, yes forced, Canada, Britain, France . . . to vote against their better judgment for the Baruch plan? How many Canadians and Americans have been told the truth that the Russians were willing in November 1946 (and maybe still are) to accept adequate inspection of atomic operations if Mr. Baruch had only given up his insistence on having a Western majority control the International Atomic Authority through abolishing the veto? . . . . Can the Russians as a minority group in the U.N. afford to see themselves swamped by a Western majority any more than French Canadians could see themselves swamped by an English-speaking majority in 1841? Much as we may dislike the Communists we must learn to live with them just as they must learn to live with us if all of us are to survive.[39]

Professor Blackett, in another book, deplored that part of the Baruch plan which called for the abolition of the veto and the conception of "instant and condign punishment." He goes so far as to say that these clauses were inserted deliberately to make the proposals, which on the face of it seemed so generous, unacceptable to the Russians:

The story is told that a well-known general who, with Mr. Baruch had been "putting teeth" into the . . . Plan, said when the final draft was finished, "Now we have made it so stiff that even the Russians won't be fool enough to fall for it." Whether true in detail or not, there was certainly a strong military and political group . . . [that] was opposed to any international control of atomic energy and preferred an old-fashioned national arms race–which is eventually what came about, since the Atomic Energy Commission plan was rejected by the Soviet Union.[40]

McNaughton went almost as far as that. He said the insertion of the veto question and the punishment clause made the Baruch plan "insincerity from beginning to end." It was no more than "a sop to public opinion" and a device to ensure that the American monopoly would continue.[41] Commenting on a statement made by Dr. Omond M. Solandt in 1958 ("In the early years of the atomic era, the West was willing to forego the use of atomic weapons merely in return for an agreement that they would not be used, provided that the agreement was supported by an effective system of inspection"), McNaughton wrote:

This was what was said, it is true, but I have never been convinced that it was a reality. Certainly there would have been no relinquishment by the U.S. of the superiority they thought they possessed until a system of inspection was in effect and proved to the point that they would have felt assured that the Soviet was *not* capable of atomic war.

In reality, this was a one-sided proposition but could it ever have been, or could ever be, otherwise "except as a part of general disarmament" which is a rationalization of human endeavour of which, it seems, there is not now any practical possibility in sight.[42]

The veto, and the bringing of the control of atomic weapons into an agreement on conventional armaments, was the crux. McNaughton had recognized that from the outset and it is obvious that the young Gromyko had done the same. The new year brought the start of the cold war and the beginning of the nuclear arms race. In 1949 the U.S.S.R., determined to seek parity, exploded its first atomic bomb; the United States, whose atomic programme went into high gear after the Russian success, tested a hydrogen bomb in 1952 and a year later the U.S.S.R. did the same. From 1953 each side, as far as can be seen, had the nuclear capability of destroying the other. "Had the United States been able to foresee the speed with which the Soviet Union would overtake it in the field of nuclear armament," Dr. Solandt stated in the Shepherd Memorial Lecture which he delivered at Montreal General Hospital in October, 1958, "compromise solutions would undoubtedly have been more fully explored and these negotiations might possibly have succeeded."

The Hearst Press attacked McNaughton. In an editorial of 3 January, 1947, the New York *Daily Mirror* charged him with "running interference for the Slavic group." His conduct within the commission had been "curious." "Does he not realize," the paper asked, "that the United States has the bomb and can prefer to keep it without further discussion?" The editorial went on to deal in a general way with United States relations with Britain and Canada and contended that "should another war occur—and we hope not—Canada would be largely dependent upon the United States for its defences in the Atlantic and the Pacific and even in the Arctic." But that was not the point at issue; McNaughton, characteristically, had "no comment" to make about the references to himself.

The commission did not meet again until 19 March, 1947, but Bernard Baruch by then was no longer a member. He had resigned soon after his "empty victory"–the adoption of the first report containing "general propositions which . . . had no chance of acceptance."[43] He was a man who, according to his biographer, was "unwilling to face defeat"[44] and the prospect of his winning the next round was none too bright, for he had antagonized many of the other delegates. Frederick Osborn, who succeeded him, in an oblique compliment made the best case he could for Baruch. "His strong and able leadership," he wrote, "had given some of the delegates from other countries a feeling of pressure and insufficient time for the consideration of all possible alternatives."[45] Thereafter, he went on, "'In the conferences which were held before the Commission reconvened,"

the French and Canadian delegates, who combined an intimate experience of the previous negotiations with a knowledge of the field and a high degree of personal competence, took a leading part. As the work proceeded, General A. G. L. McNaughton, the representative of Canada, gradually and by common consent, became the leader of the group. Of transparent integrity, great powers of self-control, and a tenacious grip on the fundamentals, he was ideally suited to the role. As head of the Canadian delegation and with an intimate knowledge of the Canadian atomic energy programme, as well as from his own personal qualities, he carried more weight than anyone else on the Commission. François de Rose, of the French delegation, spoke of him as the conscience of the Commission. It was a correct designation.[46]

The New York *Times* and *Herald Tribune*, as well as the London *Times*, "printed columns of his speeches and reports of interviews with him, setting a pace which the Canadian newspapers matched. Correspondents for the home press added personality stories and sidelight accounts."[47] He was an international figure whose views on this vital question were important. McNaughton, Douglas How wrote for the Canadian Press in May 1948, was greatly respected both at home and abroad:

A shaggy, magnetic man stood on an Ottawa platform this week, every inch a big leaguer. He was General A. G. L. McNaughton, Canadian delegate to the United Nations, talking about international control of atomic energy. He was blunt, vigorous, majestic in his command of topic, young with a timeless spring of spirit. At 61, the soldiers' Andy had been a long way up, a long way down, and now he'd come a long way back. Perspective had smoothed away the bitter days when he left his army command and two defeats at the polls. . . . Instead of retreating into melancholy and retirement, he plunged into atomic energy and made himself an expert. . . . In the offices and corridors of Ottawa, you hear only good things about him. The Canadians who have been at UN say he "ranks ace-high" down there. Even foreigners will tell you the man has a personal prestige, a sincerity that often dominates a room. In the assembly, he is demonstrating a political know-how some didn't think he had. . . .[48]

And Ross Munro wrote much the same for the Southam chain of newspapers. McNaughton was "back on top again as a Canadian leader. . . . In the lobbies and conference rooms at Lake Success . . . he presents the Canadian attitude with an earnest persuasiveness and clarity that stand out like a beacon light in an assembly where diplomatic doubletalk is too often practised. . . ."[49]

The later negotiations within the commission are beyond the scope of this account. McNaughton himself has elaborated on these,* with special attention to the position adopted by the Soviet Union which hardened as time went by. It was McNaughton who, in the summer of 1948, after discussions had virtually collapsed, pointed out that it was useless for the commission, or the Security Council, to argue further with Russia over the control of atomic energy. In March the U.S.S.R. had intensified the cold war by a coup in Czechoslovakia. This, step by step, McNaughton had denounced, leading to the conclusion: "The Communist party, by direct intimidation, silenced their political opponents and Czechoslovakia was enslaved. . . . Having in mind the intimate associations between the Communist party in Czechoslovakia, as in other countries, and the Soviet Union, it is difficult to avoid the conclusion that the Communist party gained

* Notably in the *Annual Review of United Nations Affairs*, 1949.

control of Czechoslovakia with the knowledge, approval and some help from the Soviet Union."[50] McNaughton proposed that the issue of atomic energy control should be referred to the General Assembly and this was done "as a matter of special concern." A year later, on 4 November 1948, the General Assembly approved the Atomic Energy Commission's majority report, which modified the Baruch plan considerably, "as constituting the necessary basis" of an effective system of international control of atomic energy and noted with concern the impasse which had been reached. It called upon the AEC to renew its efforts in the interests of peace. No progress was ever made. The best McNaughton could say of the commission's work was that it "did initiate the preparation of some useful documentation" which might prove "invaluable" if atomic parity should some day result in a resumption of serious negotiations for the international control of atomic energy.[51]

McNaughton remained a member of the AEC until the summer of 1950, when he was appointed chairman of the Canadian section of the International Joint Commission. He had been a member the IJC since the beginning of the year, and was still on the Permanent Joint Board on Defence. His reasons for leaving the United Nations are best given in his letter to Mackenzie King who had written to congratulate him on his IJC chairmanship:

> I am continuing also as Chairman of the Canada-United States Permanent Joint Board on Defence which, in a sense, as no one knows better than yourself, has a task which is complementary to the International Joint Commission; but I have now suggested to the Government that while the meetings of P.J.B.D. and I.J.C. can be co-ordinated without difficulty, the same is not true of the United Nations Atomic Energy Commission and the consultations of its six sponsors in New York. And so I am giving up that particular task.
>
> It is now over four years since I went to New York at your suggestion for the opening meeting of the A.E.C. It has certainly been a most remarkable experience and I do not regret in the least the great effort which had to be given because I believe that in the end the world will re-start the effort to secure an effective international control of atomic energy from the basis which was arrived at in the majority reports and which to a very considerable extent represents the ideas supported and in many cases initiated by Canada.[52]

It was his last letter to the old leader. King died five days later, on

23 July, 1950, and McNaughton was an honorary pall bearer at his funeral.

The General's conception of his own role—at the United Nations [wrote George Ignatieff] is contained in the following statement . . . . "I think the members of the United Nations Atomic Energy Commission now clearly recognize the validity of the suggestion first given in the Acheson-Lilienthal Report, that in a war of long duration it would probably not be possible to prevent the use of atomic weapons; that the worst danger to be feared is the surprise use of these terrible contrivances and, in consequence, that the real objective to be sought is to free the world from secrecy in atomic matters and to allay suspicions by giving a certainty of warning to the world; if nations should start to prepare for atomic war." A more succinct and prescient statement of the problem cannot to my knowledge be matched by any other participants in those important discussions which took place at the dawn of the atomic age. Had the nuclear powers chosen the path of diplomacy to cope with the fear of surprise use "of these terrible contrivances," . . . instead of relying upon mechanical devices of surveillance and early warning, we would now not be in such a costly and risky arms race.[53]

Concurrently with the early discussions in the United Nations, the question of the control of atomic energy in Canada came before Parliament. As a result an act was passed which, said McNaughton, "established the Atomic Energy Control Board with the duty of controlling and supervising the development and application of atomic energy in the interest of the people of Canada and generally to prepare to carry out the obligations which it would be necessary to assume under an international agreement of the character and scope which had been indicated in the discussions taking place in the United Nations."[54] As the hopes of an international agreement were never realized, this latter part of the work of the Atomic Energy Control Board lost its significance though that, of course, was not known at first.

McNaughton was appointed to the presidency of this board on 26 September, 1946; his opposite member in the United States was David Lilienthal, chairman of the United States Atomic Energy Commission. The General's early work was mostly organizational: drawing up rules so that the proceedings of the board and the performance of its functions could be regulated; appointing staff; taking over the control and direction of the Chalk River project; and drawing up recommendations for uranium supply and production. Security was a first concern in view of the dangerous possibilities of fissionable materials. "The Board," McNaughton said, "has issued regulations controlling dealings in these substances to ensure that they do not fall into improper hands; similarly some of the information obtained in research has a bearing on national security and naturally the Board is concerned that these matters should be properly safeguarded."[55]

Canada was a large producer of uranium. We have seen that the first important uranium discovery came in 1930 with the finding of pitchblende on the shore of Great Bear Lake. This deposit was brought into production by Eldorado Mines and we have mentioned "the first ounce of radium" of the early thirties. In 1940 Eldorado was forced to shut down because of the wartime dislocation of radium markets but in 1942, when the race for the atomic bomb became intensive, the company was quietly asked to reopen the mine to provide uranium. Then, in 1944, the Canadian government bought the company and that year an extensive search for uranium was begun by Eldorado, with the assistance of the Geological Survey of Canada; and in 1947, shortly after McNaughton took over the board, government legislation encouraged independent prospectors and private companies to join the hunt. The result can be anticipated. Ten years later the production of uranium was twelve times as great as it had been at the end of the Second World War; the annual gross income from that production was $100,000,000, putting uranium in the fourth place (gross dollar value) of metal production in Canada.

At Chalk River the nation also had the reactor, though only an experimental one, which had come into operation in 1945. Two years later on 22 July, 1947, a larger one began work there; it was then the most powerful research reactor in the world and remained so for several years. The Canadian reactors, unlike those of the United States and Great Britain,* used a combination of natural uranium fuel and heavy water coolant and moderator. There was certainly no necessity for Canadian policy in scientific developments to be

---

* The United Kingdom had embarked on its own nuclear energy programme in 1946.

formulated by other countries and McNaughton was convinced that the Canadian type was the best reactor for Canada. The first large-scale use of atomic energy was likely to be in the field of electricity. The combination of natural uranium heavy water would generate lower cost power than any other existing design for Canadian conditions of large base-load stations in a utility system with low capital charge rates. And that was what was required. Long after leaving the board McNaughton kept a sharp eye on the cost of generating power by nuclear means, because of the bearing this would have on his negotiations with the American members of the IJC over the development of hydroelectric power sites.*[56] In addition, Canada pioneered in the production of radioactive isotopes of high specific activity, particularly cobalt-60 for use in cancer treatment units. More than a hundred different isotopes were eventually produced for use in medicine, agriculture and industry and were distributed to many countries.

In 1952 Atomic Energy of Canada, Limited, was established to operate the Chalk River project on behalf of the board but McNaughton had left before then to take up his appointment as permanent delegate of Canada to the United Nations and the representative of Canada on the Security Council. "It is not the sort of work I had planned to do," he wrote. "I had hoped, as President of the Atomic Energy Control Board, to get back into a more direct contact with engineering problems and, particularly, to have a close say in the direction of the far-reaching experiments we have in hand with Atomic Energy [at Chalk River]. However, I am sure that the work at the United Nations will be interesting and, I hope, useful and under the circumstances I am very glad indeed to be able to undertake it."[57]

He accepted this new work at the United Nations on 2 January, 1948:

---

* As an example of this, Dr. W. B. Lewis, vice-president, research and development, Atomic Energy of Canada, Ltd., wrote to McNaughton in 1957: "I heard that . . . you had expressed surprise on hearing that we anticipate achieving 5½ mills per kilowatt hour power for Ontario in about 1968. This made me realize that perhaps we have not kept you informed of developments over the last year. I do not know in what detail you would like to be informed, so I am enclosing copies of four papers which go into the matter in different degrees."

I said that I would place myself at the Government's disposal. . . . I had heard of this proposal from Pearson on 31 December and had had a word about it with the Prime Minister yesterday I said I thought I could retain chairmanship of P.J.B.D. Canadian Section but the Presidency of A.E.C.B. I thought I should give up to Dr. Mackenzie.[58]

So McNaughton's old friend, Dr. C. J. Mackenzie, took over the Atomic Energy Control Board while McNaughton prepared himself for his new duties on the Security Council which would be concerned with reducing tension and lessening the risk of war. He was familiar with the struggle to have the atomic bomb banned from the arsenals of the world but it still existed. Thus it was essential to lessen the likelihood of war, and the possibility of the bomb's use, by removing the grounds for tension wherever they were found. "The only complete guarantee," the General said,

that we shall not have atomic war inflicted upon us is to do everything possible to ensure that all international disputes which might lead to war should be settled by peaceful means, as all signatories of the United Nations Charter have obligated themselves to do. We must therefore strive to make it possible for all nations to live in peace with one another under international law and to eradicate the causes of war.[59]

He would be "no figurehead," the press announced, but "an active participant in a broad field of UN affairs." He "will also give his scientific knowledge to the U.N. Atomic Commission on which he is probably the most informed member."[60]

McNaughton took his seat as a council member on 15 January; his first term as president of the Security Council for the month of February, 1948, began two weeks later. There were three major questions before the council that month: India-Pakistan (over Kashmir); Palestine; and Indonesia. For the first week the council's time was completely absorbed with Kashmir.*

The background of the problem was roughly this: in the nineteenth century the Kashmir territory (Jammu and Kashmir), with its Moslem people, had been sold by the British for a small sum to a Hindu ruler

* This account is based on the United Nations *Verbatim Record* of Security Council Meetings, February, 1948. All succeeding quotations, unless otherwise indicated, have been taken from that record.

as a reward for services rendered in the Punjab War. Though the ruler and his people had nothing in common, no attempt had been made to ascertain the Kashmiris' wishes in the matter. At the time of the transfer of power by Britain to the new Dominions (as they then were) of India and Pakistan, Kashmir had acceded to India but the accession was to be the subject of a plebiscite. Mountbatten, the Viceroy, made that clear. In a letter to the Maharajah of Kashmir he said: ". . . my Government have decided to accept the accession of Kashmir State to the Dominion of India in consistence with their policy that, in the case of any state where the issue of accession has been the subject of dispute, the question of accession should be settled by a reference to the people." Pakistan had objected that it would be impossible to have an impartial and fair plebiscite under the Kashmir authorities. "Very good," returned Mountbatten, "then let it be conducted under the auspices of the United Nations."

Fears that the state would accede to India prompted a rebellion by the people of Kashmir against their ruler. Fighting broke out in October, 1947; in January of 1948 India laid before the Security Council a formal complaint against Pakistan's alleged complicity in the tribal invasion, and the matter was discussed. These discussions "between the representatives of India and Pakistan, under the auspices of the President of the Security Council," McNaughton said when describing the agenda, "will continue so that a basis of agreement may be reached to terminate the fighting, to afford security to the peoples of Jammu and Kashmir under some authority which will be recognized by everyone concerned as strictly impartial, and, most important, to provide for a plebiscite of the people in which all of them will be permitted to express without fear or favour their wishes as to the future government of the State."

There were two rival authorities in Kashmir. First, that of the maharajah* (which was supported by Indian troops), although it had been promised that a more democratic administration was to be set up under Sheikh Abdullah, who would become prime minister. The second was the (Moslem) Azad Kashmir Provisional Government which, reputedly supported by tribesmen from Pakistan, had seized large areas of the state. In this situation lay the seeds of an all-out war between India and Pakistan.

---

* The maharajah actually moved out of Kashmir, but his influence remained.

Gopalaswami Ayyangar, the Indian representative–dark-faced, intense–stated that Pakistani tribesmen, armed and supplied by Pakistan, were riding or marching in as volunteers. They proclaimed the fighting in Kashmir a *jihad* ("holy war") and crossed the cold, rain-soaked hills of Jammu to see, many for the first time in their lives, well-irrigated fields and shops stacked with goods in the Vale of Kashmir. The state was in turmoil through their search for loot and women. Pakistan must compel its raiders to withdraw, so that law and order could be restored. A free plebiscite would then be held through the existing administrative machinery of the state, for to ignore that machinery would be an encroachment on the normal sovereign powers of the state.

Sir Mohamed Zafrullah Khan, the representative of Pakistan, stroked his grizzled hair and rounded beard. He alleged that raids were coming from Kashmir territory against Pakistan. Indian artillery in Jammu and Kashmir State had shelled Pakistani villages, and Indian aircraft had machine-gunned workers in the fields. The ruler of Kashmir, he contended, was an alien oppressor over a people with whom he had nothing in common and that was the reason for the fighting in Kashmir. His influence should be abolished and the Azad Kashmir government, which the people had established, recognized. Why leave the question of a plebiscite to this alien maharajah? There could be no cessation of hostilities if his authority remained; nor if he handed over to Sheikh Abdullah, who was "irrevocably committed to accession with the Dominion of India," and a friend of Nehru. However solemnly given, promises to the people that a responsible government would be set up and a plebiscite held–once they had been subdued and their resistance at an end–would never persuade them to lay down their arms. India, Zafrullah went on, claimed that Kashmir had acceded to her. Pakistan repudiated that. "Until the will of the people is ascertained the question of accession is . . . in suspense."

It is clear from the foregoing that both India and Pakistan agreed about the need for a plebiscite; but they differed on how, and in what circumstances, it should be held. The Indian representative thought a proposal which he put forward, would meet the case:

Whereas India and Pakistan agree that the question of the accession of Kashmir shall be determined by plebiscite and that the plebiscite be held under international auspices.

The Security Council recommends that the plebiscite be taken under the advice and subject to the observation of persons appointed by the Council.

The representative of Pakistan disagreed, for India still insisted on the plebiscite being arranged by the pro-India Kashmir authorities, even though she had no objection to the presence of United Nations representatives. "The whole object," he said, "is to set up an administration with Sheikh Abdullah at its head; to enable that administration, by various steps, to consolidate itself in Kashmir; to have time to do so; and then to hold the plebiscite under that administration although United Nations observers might come at the moment of taking the plebiscite itself. . . ." India had insisted, as a first essential step, that the Pakistani tribesmen must withdraw. But Indian troops would not withdraw, even if the tribal incursions ceased. They would remain, ostensibly to maintain law and order, and only if the plebiscite went against India were they prepared to withdraw. Fighting in Kashmir was bound to continue, with a plebiscite held under the authority of Sheikh Abdullah's administration and menaced by the bayonets of the Indian Army, while the only part that the United Nations could take would be to observe or, possibly, supervise. The people of Kashmir were almost totally illiterate and were very vulnerable to pressure. Pakistan had to be satisfied that the question of the accession of Kashmir to Pakistan or to India, or to be independent, would be fairly and impartially determined. That done, Pakistan intended to exert every means to convince the tribesmen that they, as co-religionists of the majority of people in Kashmir, need entertain no fears for their safety, since the United Nations would see that everything was conducted in a fair and impartial manner.

During the course of these charges and counter-charges by the representatives of the rival dominions the members of the council had been provided with maps of Kashmir and its surrounding territories. "Most delegates," I. Norman Smith wrote in the Ottawa *Journal*, "looked at them indifferently, but the General sprawled his out over his desk and went at it as though he were that night going to invade the place. . . . Brows knit, pencil tracing on the map the speaker's line of argument, General McNaughton earnestly made his own notes on a little pad. You got the feeling that in the private negotiations that were to follow there would be a surprised set of Indians upon realization that this outsider knew their terrain."[61]

At the end of the first week McNaughton advised "the representatives of India and of Pakistan [to] make a further effort to resolve their differences and reach an agreement by direct negotiations under the auspices of the Security Council." Such an agreement could take place, he said, only by "mutual concession." He himself would endeavour "to advance the case by way of agreement between the two parties." That evening McNaughton met the representatives of India and Pakistan and Fernand van Langenhove of Belgium (the *rapporteur* between the two parties). The General set down the trends of the discussions on paper and then drafted what was afterwards referred to as "the President's resolution." This was read by the representatives of both India and Pakistan.

McNaughton, at the next meeting of the Security Council on 10 February, stated in very simple language what steps he thought should first be taken to resolve the difficulties–which related more to "implementation" rather than to "a statement of principles." First, "acts of violence and hostility" must end; then "the withdrawal and continued exclusion of all irregular forces and armed individuals who have entered Jammu and Kashmir from outside "must be brought about, each party using all its influence for that purpose."

That, of course, would mean the withdrawal of the regular Indian troops as well as the irregular Pakistani tribesmen, and it was immediately satisfactory to Zafrullah Khan. How far did these steps go, he asked.

"They go to this extent: that there must be a cessation of violence and hostility, a withdrawal of armed people who have come in from the outside, and measures taken to keep them out. It is obvious that, for the restoration of peaceful conditions, India must withdraw the troops that have been fighting with the population. . . . So long as they are present, it is useless to expect the restoration of normal peaceful relations [and] it also has a vital bearing on the freedom of the plebiscite."

For the preservation of law and order until a plebiscite could be arranged, Zafrullah suggested an international force. "It will be said, 'Well, there is no international force in existence. . . .' Very good, then, let us have British forces, Dominion forces, let us say Commonwealth forces or any force that will be neutral."

The Indian delegate, on the other hand, felt bound to discuss these "trends of opinion" in the Security Council with his government and wished to leave. What he desired "as a trend in the Security Council," said Warren Austin of the United States, was that the council "should take up a position which would amount to that of an ally in war, and should pull off Pakistan and allow India to finish the job by force against the tribesmen." Philip Noel-Baker, the British representative, thought the Indian delegation should be asked to remain: "I think it would be a tragedy if the representatives of India were to go back to their country without having, in their own words, 'thrashed out the difficulties, the real differences that remain . . . .' There is nothing irrevocable about an aircraft ticket. There may be something irrevocable in the departure of the Indian delegation."

And José Antonio Villegas Lopez of Colombia asked, "Are we going to stop discussing the Jammu and Kashmir question until the Indian delegation finds it advisable to return to the Security Council? The position [the delegation] takes is, 'We wish the Security Council to instruct Pakistan to do this and that and get the tribesmen out of Jammu and Kashmir. Next, we want a national assembly convened; then we must have a Government, and, after all this, we will have the plebiscite.' It is because we have not agreed to that order of business that the Indian delegation wishes to return to India. I believe it is the duty of the Security Council to be very clear as to what this adjournment means and as to where we shall stand if we agree to the adjournment."

The Security Council was split on whether the Indians should be allowed to depart or not; if they should go there was talk of imposing a time limit on their absence. Ting-Fu Tsiang of China drafted a resolution that would settle the question by vote but it did not come to that.

"On the evening the crisis came to a head on the question of adjournment," wrote George Ignatieff, "there was an ice storm that covered all the roads between Lake Success and New York, making them like skating rinks. The cars of the delegates slithered and skidded trying to get back from the meeting of the Security Council. And yet this was the moment in which it was essential to find some formula to 'save face' of the two important parties in the dispute. As McNaughton and I found ourselves skidding down a hill on the Grand Central Parkway leading back to New York, we were trapped until a sand-truck could get us out. There we sat in our impasse while the General reviewed every possible way out of the impasse in the Council. It seemed an interminable time before we were released from this strange dialogue which had been forced upon the two of us by the elements.

The General's mind had been made up in that hour or so that we waited for the sand-truck. The next morning he settled the controversy. . . ."[62]

On the morning of 12 February the president took the matter into his own hands. If any formal action were taken, he said—to vote on whether the Indian delegation should be permitted to go or not and for how long—it would form a precedent though it might not have been so intended. McNaughton described for the council his thoughts of the previous evening and his solution:

[I] had a particularly grave responsibility to ensure that no precedent whatever could be created while I had the duty, which goes with the office of President, of maintaining the Security Council's prestige and authority. . . . I felt that the solution of the question which was troubling us was not to be found in any resolution put to the vote. . . .

The Security Council must, at all times, remain the master of its affairs . . . the Security Council should not take any action whatever which would derogate from that position; and that within this principle it should be the duty of its President, officers and members, to ensure the prompt and most effective conduct of the business which it is the responsibility of the Security Council to transact.

I asked myself what this conclusion of principle led to in the particular case now before the Security Council, and it is this: that further progress in the settlement of the case between India and Pakistan requires that the representative of India should now go to his country; and that he should inform the Government in the fullest detail of the views, opinions and consensus of thought which have been expressed here; that he should take the views, advice and decision of his Government and return to the Security Council as expeditiously as possible so that we may resume the consideration of these important matters—matters which we have found cannot be advanced without this very information which we now lack.

Our knowledge of the time required for the representative of India to journey to his country for consultation with his Government and to return to the Security Council is not of such precision that we can specify it to a day or an hour. Indeed, I do not think it would be right for the Security Council to attempt to legislate this detail.

. . . unless I hear objection I shall take it that the Security Council is in agreement with the disposition of the procedural question.

McNaughton's firm handling of the matter brought no demur. Tsiang, who had withdrawn his resolution, felt that he had been right to do so. "Having heard the President's statement this morning, I am more than ever certain that I was right. . . . His statement was both thoughtful and statesmanlike, and it has my support." Noel-Baker agreed. "I am entirely of the view that the President of the Security Council was right in proposing that we should have no resolution but should accept his statement.

There, in February, the Kashmir question rested.

With regard to Palestine, Britain, the mandatory power, had given notice of her resolve to terminate the mandate. The problem of what administration should succeed the British had been before the United Nations as a matter of special concern since April, 1947. In November the General Assembly recommended the adoption and implementation of a plan, the "Plan of Partition with Economic Union," which would divide the country between Jews and Arabs. A UN Palestine Commission had been set up which, in February, reported to the Security Council.

The Arab nations violently opposed the partition plan. The British mandate, they contended, defined the Jewish concession only as "a national home in Palestine," whereas the General Assembly, by its resolution of November of 1947 recommending partition, would create a sovereign Jewish state containing a large Arab population; furthermore, this permitted unlimited Jewish immigration, so that in time the Arabs would be driven out. The statement of the Arab case by Faris el-Khouri Bey of Syria, who presented it with vigour and considerable passion, caused McNaughton to intervene:

> The problem of Palestine, which is before us and before the nations of the world, is how to bring peace to that Holy Land which is now torn and terrified by strife, how to reconcile the anxieties of those who are now its inhabitants with the aspirations of those who feel its claims.
>
> The consideration of this problem provokes deep and conflicting loyalties. The Security Council will agree, I know, that in our consideration of the question we should strive to free ourselves of every feeling except the determination to be constructive,

unprejudiced and realistic, and to avoid saying or doing anything which will aggravate the situation or incite to violence those whom it is our earnest desire to help. We ourselves have the right to urge, indeed, I think, to insist that nothing should be done in Palestine itself, or in other countries whose feelings are deeply engaged in this matter, which will complicate or embitter a situation already tense and explosive. Those who ignore this advice and act against it are not merely reducing the chances of peace in Palestine; they are trifling with the far wider issue of peace in the world and are prejudicing the whole future of the United Nations as an organization capable of maintaining security by collective action.

These wise words affected the discussions that followed. The council went on to a realistic study of the problems of partition. The security aspect dominated all the rest, for the General Assembly plan proposed an Arab state, a Jewish state and international territory incorporating both Jerusalem and Bethlehem. The international territory, demilitarized and neutral, would be under the administration of the United Nations. What were the facts? At the time the city of Jerusalem alone had a force of nine hundred British and three hundred and fifty Palestinian policemen backed up by a brigade of British troops, and even so there was daily killing, bombing and shooting. At least a thousand policemen would have to be found to ensure the preservation of law and order. And what about the troops who would obviously also be required? The British mandate was due to expire in the middle of May. Even the police force could not be recruited by then. And Jerusalem depended upon the surrounding Arab areas for its necessities of life, including oil, most of its electricity, and water. Without troops, such as the British had in the rest of Palestine, Jerusalem could easily be isolated and starved. "Jerusalem," the chairman of the UN Palestine Commission said, "would be doomed even with a United Nations Governor at its head." And was economic union, with mutual freedom of transit between all the entities of portioned Palestine, realistic? The Arabs had already vowed to resist partition by force of arms.

Villegas Lopez of Colombia introduced a resolution calling for a non-Palestinian force to maintain order and security in Palestine in place of British troops. Meanwhile the British should postpone the date fixed for the termination of the mandate until 15 July, 1948. Arthur Creech Jones of the United Kingdom replied firmly that the present administration would terminate its control by 15 May; the United Nations had had ample warning. The military withdrawal and the winding up of the civil administration were both being carried out against a background of increasing violence. In fact the general security position had degenerated very seriously since the resolution on partition had been passed in the General Assembly. "This deterioration, due to outrages and lack of restraint by sections of both communities inside Palestine, has also been aggravated by groups beyond the border as, for example, the intrusion of armed Arab bands and, on the Jewish side, the continuance of illegal immigration."

The discussion, then, was useful in that it focused attention on the urgent needs of the moment. As yet the Security Council had come to no decision even about the General Assembly's resolution that there should be partition. On 25 February, to bring matters to a head, the United States representative submitted a draft resolution "to accept . . . the requests addressed by the General Assembly . . . in . . . Resolution of 29 November, 1947," that is, the partition plan. The Belgian representative, two days later, submitted an amendment to this draft resolution. It called for the committee of five–which would be created by the American proposal–to investigate the possibilities of what could best be done in Palestine, in view of Arab intransigence, before the question of partition was actually debated. But the detailed fate of these matters is no part of our present story.

The Indonesian question first came up during McNaughton's presidency on 17 February. At these meetings the attention of the council shifted with startling suddenness, as important developments occurred, from mountainous Kashmir to the Holy Land and then halfway around the world to the spice islands of the Indies. On 3 December, 1947, a draft truce plan had been offered to both parties in the dispute–the Indonesians, who were seeking independence, and the Dutch, their colonial masters–and a first agreement on an armistice (the "Renville Agreement") had been signed aboard USS *Renville* on 17 January, 1948. As a result of this, hostilities had stopped and a cease-fire line, known as the van Moek Line, had been established. The line, incidentally, left the richest oil and rubber regions of Java and Sumatra under Netherlands control. It was important that a political settlement should follow the military truce and therefore the United Nations Committee of Good Offices, which had done useful work thus far, had to continue its activities. McNaughton, on 18 February, introduced a Canadian resolution, which concerned the continuation of the work of this committee in a way that would most facilitate agreement between the governments of The Netherlands and the Republic of Indonesia. "In drafting the resolution," he said, "I have sought to express what is necessary–and in the least controversial terms. . . . I fully recognize that [the resolution] does not completely satisfy either party. Nevertheless, it is submitted in the hope of bringing about a meeting of minds so that, when negotiations are resumed, they may proceed from an agreed basis." A vote was taken on 28 February. Argentina, Belgium, Canada, China, France, the United Kingdom and the United States favoured the resolution. Colombia, Syria, the Ukrainian S.S.R. and the U.S.S.R. abstained. No one voted against it and the resolution was adopted.

That day McNaughton took issue with the representative of Colombia on a procedural point:

> I would remind the representative of Colombia that rules of procedure are made for the convenience of the Security Council and for the proper discussion of its affairs, and that it behoves the representatives on the Security Council to pay attention not only to the strict letter of the rules but also to their spirit.
>
> Having made these remarks, I am sure that we can rely on the courtesy of the representative of Colombia.

He could indeed. Villegas Lopez made amends with complete generosity. "I need not assure the President," he said, "as to how loath I am to do anything that he may properly call out of order, and to which he may call my attention, the more so when I feel that there is no doubt in his mind not only of my highest respect but also of the fact that we have had an extraordinary example of efficiency in the conduct of our business while he has been in the Chair. I am extremely glad to pay my tribute to the ability with which the President has presided over our deliberations during the past month, a month which has been one of the most fruitful in the life of the Security Council. . . ." McNaughton stepped down from the chair to make room for the representative of China who would preside in March.

Not all the previous chairmen had taken their work so much to heart. One writer drew attention to previous sessions when he had felt that "the chairman wasn't paying much attention to the speakers. He was busy talking to his aides or writing letters or reading documents, or off in the visitors' section paying a social visit to his wife or an old friend for whom he had got a pass."

Not so General McNaughton. Seemingly he has decided the place to do the Council's business is during the Council's meetings. His notes trace the main developments in the argument, the differences, the weaknesses . . . maybe this will take a toll of him. At 61 he is not an old man, but he's taking terribly seriously Canada's first chairmanship of the world's principal legislature.[63]

Another, Walter O'Hearn, also described McNaughton during that first presidency:

A quiet, spare man, grey and drawn in the artificial light, brought down his gavel, sharply and decisively, twenty-one times in February, to open a session of the United Nations' most powerful agency, the Security Council. No light tap of the symbol of the Chair's authority sufficed for General A. G. L. McNaughton. It was a firm bang, leaving no doubt that the eleven members of the Council had opened another business meeting. . . . As decisive too was his handling of the greatest number of Security Council meetings ever held in one month since the peace-preserving U.N. body was established in London two years ago. But never was a speaker unable to present his case fully; he was assured of quiet, polite attention with never an untactful turning away of the Chair.[64]

He had, it was reported in the press, drawn "a tough month" in the Council's life to serve his turn as chairman. And yet he seemed to have been "altogether fascinated" by the challenge. Not for one moment had he been seen in other than "rapt attention."

. . . the picturesque and rather hard old head of the General remains turned towards whoever has the floor. Forward he leans in his chair, his head bent slightly to one side the better to hear. And if a speaker sees him lean even more forward he may take it as a danger sign that his argument is becoming involved. But however tough the going, the General stays in pursuit.[65]

But, the writer asked, could he maintain the pace—this "almost fierce energy in behalf of Council and international peace?" His Security Council work, coupled with his membership of the Atomic Energy Commission might prove to be too much. "It is, surely, more than one man should be asked to bear" and, though this was not mentioned, he had an additional burden as co-chairman of the Permanent Joint Board on Defence, which, as we shall see, was no sinecure.

To some, his attitude was puzzling. "He appears to approach every argument in the belief that only misunderstanding and bitterness prevent a settlement and that if he sufficiently ferrets out the truth everyone will join hands and live happily ever after. This is a welcome touch of good will and good faith in an assemblage not unsoured by cynicism—but whether it is not more naïve than realistic is a fair question." Still, the same writer conceded, McNaughton was a practical soldier:

> Being a military man he is all too aware of the fact that UN just now has no force, no troops, no real agreement to put up troops. So he is fixing on using the Council not as a justice obliged to give decisions and make decrees but as an intermediary anxious to give assistance and create understanding. The words the General has in mind are conciliation, negotiation, mediation, enquiry, arbitration.
>
> These words are . . . the Charter's first weapons; and since the General knows the Charter has virtually no other weapons just now, perhaps, after all, it is he and not those who want to use force, who is the realist.[66]

This last proved to be an extraordinarily accurate prophecy of how McNaughton would conduct himself. Was he a realist, then, or was he naive? We are assured by one who worked with him very closely during Canada's first full term on the Security Council between the years 1948-1950 that McNaughton was a realist who "set the hallmark of patience, pragmatism and mediation on Canadian diplomacy." In doing this he engendered no ill will, decisive though he was: "His forceful and colourful personality, coupled with a natural courtesy and attentiveness which was part of a born gentleman, quickly made him one of the most respected, and even beloved, characters among the diplomats representing their countries at the United Nations."[67]

It was essential, McNaughton thought, that the world should understand what the United Nations was and what it was not. As

George Ignatieff continued, "General McNaughton, in applying this pragmatic approach, took the view that a realistic understanding of precisely what the world organization was not intended to do, as well as of what it was intended to do, was essential for coping with the many contentious and difficult problems thrust upon it in those early years. He abhorred both the defeatist attitude which resulted from an over-critical view of the United Nations, and the excessively idealistic view which he considered equally unrealistic and unhelpful to the world organization."[68] Here are McNaughton's own words:

There seemed to be a feeling that the United Nations is a sort of supernational body instead of an international one. The truth is, of course, that the United Nations has no personality beyond that of the representatives from the 59 governments which make up its membership. The world organization takes no decisions, other than those taken by the majority of those delegates on instructions from their governments. Therefore it stands or falls as an instrument for preserving peace, depending entirely on the will to peace which these 59 governments possess and, ultimately, on the reaction of public opinion, however expressed in their 59 governments.[69]

McNaughton understood the realities of Canada's position as a middle power. The guidelines for her foreign policy had been established by Louis St. Laurent, then Secretary of State for External Affairs, in a speech at the inauguration of the Gray Foundation Lectureship at the University of Toronto early in 1947. McNaughton followed the principles set out by St. Laurent very closely. Canada, as a middle power, must recognize the limitations placed upon her by that status. There was little point in urging international action on the larger nations—who, wielding the world's military and economic power as they did, would to a large extent have to implement the recommendation—if they did not advocate it; Canada must first obtain support for the principles she believed in if there was to be any good result. Then, of course, they could be pressed with vigour. Secondly, Canada's external policy must have the backing of the nation; national unity must be a first consideration. And Canada must be governed by the rule of law in international affairs.

Following these principles . . . McNaughton drew some interesting conclusions which illuminate his conception of Canada's diplomacy as a mediatory power. "In each of these disputes," he said, "the Security Council has played an important role in preventing the spread of local hostilities into a general war; and in each case the Council has succeeded, to a greater or less extent, in inducing the parties to seek a settlement by negotiations, rather than by fighting."

How this mediatory diplomacy is carried out was described in some detail by the General . . . and still serves as a blueprint for the diplomacy which has moulded the Canadian position in world affairs.

The General drew three important conclusions: "*First, the Security Council is not in a position to embark on armed intervention and there is no present likelihood that it will be.** In consequence it can usually do little more in the initial stage than to call on the parties engaged in the dispute to stop fighting and to start talking, and to offer them the means by which they can work out a settlement through negotiations rather than by conflict."

In the second place: "It is, I believe, most important that when the Council calls upon the parties to cease hostilities, it must make such a call both universal and impartial–the Council should therefore make it plain that in calling upon the parties to end hostilities, it is not prejudging the ultimate political solution which may be achieved through its good offices."

The General's third conclusion was: "That to the greatest possible extent the responsibility of solving a political problem should be left primarily with the people who are immediately affected by it. It is of course the case that the people who live in a particular area are most likely to be satisfied with a settlement in which they themselves have directly participated. There is also a great advantage in stability through having an agreed rather than an imposed settlement and this procedure has the useful effect of strengthening the sense of responsibility at a point where this is essential to a healthy political life."[70]

"I have given these quotations from McNaughton *in extenso*," remarked George Ignatieff, "as I believe that there is no better evidence of the degree to which he was a pathfinder in the mediatory posture in Canadian diplomacy at the United Nations in its early days."[71]

These, then, were the lines on which McNaughton worked and we may now return to the three major disputes that were before the

---

* Italics supplied.

Security Council of the United Nations. For the next two years while Canada remained a council member the negotiations continued, both at the third session of the UN General Assembly in Paris during the autumn of 1948 as well as at Lake Success. A detailed analysis is beyond the scope of this book and an account can be given only in general terms.

Throughout the Palestine dispute, it has been stated, the Canadian policy was "consistently pragmatic, searching for a practical and workable basis for resolving the conflict and arriving at a stable settlement."[72] Britain relinquished the mandate, whereupon the Jews proclaimed the State of Israel; the Arabs attacked the new state and in the fighting which followed the Israelis seized territory that was beyond the borders prescribed by the General Assembly in November, 1947. It was with efforts to stop the fighting and arrange an armistice that McNaughton was most concerned. He voted for the council's strong resolution of July, 1948, which, declaring the Palestine situation to be a threat to peace, ordered the rival parties to stop hostilities and to issue cease-fire orders that were to take effect within three days—"the first time that such peremptory demands had ever issued from the council."[73] McNaughton was co-sponsor of a resolution, which followed in November, calling upon the parties to negotiate an armistice.*[74] But the Palestine question was one where the United Nations succeeded to a lesser extent, for though the local hostilities did not spread into a general war, and though an uneasy armistice was achieved, efforts to negotiate a permanent peace were not successful.

Indonesia, however, was a different story and here McNaughton played a decisive part. The truce that followed the Renville Agreement of January, 1948, did not result in a political settlement and at the end of the year the Dutch moved forward in a new attack. Calls issued by

---

* McNaughton's account is as follows: "On 15 November, when the Security Council was meeting in Paris, Canada submitted a draft resolution which won the co-sponsorship of Belgium and France and which was adopted by the Security Council on 16 November. The Security Council at the time called upon the parties directly involved in the conflict in Palestine to seek agreement forthwith by negotiations conducted either directly or through the Acting Mediator on Palestine [Dr. Ralph Bunche] with a view to the immediate establishment of the armistice, including the delineation of permanent demarcation lines and the withdrawal and reduction of armed forces so as to ensure the maintenance of the armistice during the transition to permanent peace in Palestine. Since then armistice agreements have been signed by Israel and all of her Arab neighbours. The final such agreement was signed by Israel and Syria on 20 July, 1949. Thus fighting has now officially ended."

the Security Council for a cease-fire and the release of prisoners had no effect on the Dutch. McNaughton supported a resolution of 28 January, 1949, calling for peace and the reopening of negotiations on the basis of previous agreements, but that too was ignored. In February, however, the Dutch proposed that a Round Table Conference should be held to discuss the transfer of sovereignty, which was considered by many to be no more than a device to offset the Security Council's January proposals. McNaughton did not think so. Though other members of the council pointed to previous resolutions which the Dutch had not complied with, the General still persisted. He thought that common ground could be found between the council's resolution of 28 January and this new proposal by the Dutch, and he sought a compromise.

The climate seemed unfavourable. Representatives of the United States had spoken in strong condemnation of The Netherlands. The U.S.S.R. and the Ukraine had branded the Dutch as colonial oppressors and even Britain, sensitive in matters concerning colonial policy, seemed willing to back a condemnatory resolution. But McNaughton worked untiringly behind the scenes to obtain support for a conciliatory suggestion. He faced strong criticism but remained unmoved. By 11 March he had obtained some support, enough to suggest that both sides might be brought together to sound them out. With the republican capital of Djokjakarta in Dutch hands and most of the republican leaders in Dutch custody there seemed little hope of that, and McNaughton's original supporters began to waver. Let the suggestion go forward as "an idea," said McNaughton, "not the final text of a proposal," and that, with the strong support of the Chinese delegation, seemed to have convinced the waverers.

The outcome was the adoption of the "Canadian proposal" on March 23. It deliberately avoided any mandatory phrasing and merely expressed "the sense of the Security Council" that the Commission for Indonesia [which had succeeded the Committee of Good Offices] should help the parties to reach agreement on implementing the resolution of January 28 and on the timing and conditions of the Round Table Conference in which the Commission might participate. The proposal was not entirely free from ambiguity as to its implications, and in various quarters it met with suspicion and skepticism.[75]

Yet the proposal was enough. "The Security Council's directive of 23 March, 1949," writes J. Foster Collins in *The United Nations and Indonesia*, "marked a turning point in the solution of the Indonesian question." Effective negotiations, pursuant to the Canadian proposal, were at last entered into and these proved fruitful. Lester B. Pearson, who had succeeded Louis St. Laurent as Secretary of State on 10 September 1948, said as much in the House of Commons on 16 November, 1949:

In this case I think we have some reason to be proud in Canada. To a very considerable extent it was as the result of a Canadian resolution, which was attacked from both sides, that the machinery was finally set up by the Security Council which now, happily, in a conference at The Hague has brought about a peaceful and satisfactory solution, one which I think will stick, in respect of this complicated and dangerous Indonesian problem.[76]

In a Christmas message on 23 December, 1949, the Queen of The Netherlands prepared her people for Indonesian independence. "We all have to accept the resolutions which have been attained in a democratic manner," she said. "Whether you are of the opinion that Indonesia has been ready for a long time, or is not yet mature enough, for independence, and however full of cares and fear your hearts may be, I would like to say to you in holy earnest–appealing to your deep goodwill towards Indonesia–that, provided the Dutch play this game honestly and according to the new rules, they will be welcome guests in Indonesia. . . . He who wants to preserve will lose, and he who is prepared to lose will preserve. . . ." In a later speech she described the assumption of sovereignty by the Republic of the United States of Indonesia and the conclusion of a union with The Netherlands, as "deeply moving."[77]

In the general congratulations McNaughton was largely overlooked. But he had his memories. After the Security Council meeting of 23 March, 1949, to his consternation, the representatives of The Netherlands and of Indonesia had embraced him simultaneously;[78] and the first person to be singled out by Dutch reporters for comment on The Hague talks had been McNaughton. He was reticent about his own part and full of praise of the council. "Here we have an example," he said, "of the useful and far-reaching results which can be obtained by the objective, patient and persevering application to grave political

McNaughton broadcasts to Great Britain on the international control of atomic energy, November 1947
(NATIONAL FILM BOARD)

Honorary pallbearer at the funeral of Mackenzie King, 1950

McNaughton, General Henry and Mayor Fiorella LaGuardia
ponder a problem of defence, 1947

problems of the peaceful procedures developed in the Security Council." It was a precedent "which carries forward an example and a message of hope and confidence in the future of the United Nations."[79] Dr. L. N. Palar, the representative of Indonesia at the United Nations, had called on McNaughton at his office in New York towards the middle of December. He extended to the General the personal invitation of President Sukarno and Prime Minister Mohammed Hatta of the new United States of Indonesia to attend the ceremony, to be held at Batavia (Djakarta) on 27 December, whereby sovereignty would be transferred from The Netherlands to Indonesia. He gave McNaughton credit for the birth of the republic and there is no doubt that had the General attended he would have been acclaimed. But McNaughton could not go; he was busy conducting negotiations between India and Pakistan at the time. One Canadian editorial (in the Windsor *Daily Star* of 24 December) referred to McNaughton as "The Father of the Indonesian Nation," who would always have a place in Indonesian history. "In the light of history," the editorial continued, "General McNaughton's contribution in this international field may excel his undoubted services to Canada as former head of the National Research Council and as commander of the Canadian army in the earlier years of the war." This was the United Nations first success and it was a very real one. Without any costly international peacekeeping force an armistice had been brought about; more important, a *political* settlement had been achieved.

Until the fall of 1948 McNaughton lived in a hotel in New York. Though his wife was there, and his work absorbing, he was not really happy, caged in the enormous city. Of course, there was the lighter side, for example, when Mrs. McNaughton and Mrs. Jessup (the wife of Philip Jessup, a United States delegate), sitting alongside each other while a Security Council meeting was in progress, were ordered not to knit. "Quite right, too," McNaughton growled. "They made people think of those days outside the Bastille, in France." Then there was a letter from a Canadian citizen telling McNaughton that he held certain Russian bonds which he had subscribed to during the latter part of the First World War. Could the General take up the matter of their redemption with Mr. Gromyko?

He kept busy at the United Nations; and the Permanent Joint

Board on Defence, which met frequently at military centres scattered over North America and farther afield, gave him welcome breaks from his city routine. He remained patient.

Patience [George Ignatieff wrote], together with fierce application of effort and concentration, characterized the General's approach to his work. Since sloth is incompatible with being able to cope effectively with the complexities and crises of the modern world, the General's untiring and enthusiastic dedication to whatever matter he had in hand set both a pace and an example for the whole Canadian Mission. One had to go "at the double" to keep up with the General.

"Genius," Frederick the Great once said, "is a transcendent capacity of taking trouble first of all." McNaughton had that kind of genius—whether he was mastering a diplomatic brief, mediating a dispute, working out some engineering problem to do with Canada's boundary waters,* or merely tying flies or mending my son's toys, all of which absorbed his restless energy, cooped up in New York.[80]

He attended dinners at Ottawa, as, for example, when Mackenzie King entertained Hugh Dalton, the British Chancellor of the Exchequer, during September, 1946. He also filled speaking engagements both in Canada and the United States whenever he was able. But Paris, in the autumn of 1948, perhaps provided the biggest change. Mackenzie King came over; though King took little part in it, there was a good deal of social activity, as well as solid work at meetings of the United Nations. Georges Vanier was in Paris as Canadian ambassador and McNaughton told him he was soon to have a house—Green Court, on Long Island, not far from Lake Success; "the latch string," he said "hangs out for you both."[81] The McNaughtons attended a dinner party at the Vaniers in honour of Robert Schuman, the French Foreign Secretary; with their daughter Leslie† and Mackenzie King they sat in the presidential box at the Paris opera. But, more to McNaughton's taste, he was one of eleven guns invited to shoot with President Auriol of France at Rambouillet and he dined at the château afterwards. Typically—having come to Paris to work, not play—he had to borrow a gun but even so he managed to bag six brace of pheasants.[82]

---

* McNaughton was not yet with the International Joint Commission, which deals with boundary waters. He was probably making plans at this point how best the St. Lawrence Seaway impasse could be overcome. This is described later.

† A year later Mackenzie King proposed the toast to the bride at Leslie's wedding.

John W. Holmes, who "went to Paris from Moscow in September, 1948, and worked with the General in the Security Council in Paris through to December of that year," has described this period of McNaughton's life:

> Mrs. McNaughton and Leslie were on hand and they kept the General going to all the receptions. As you know, he was no cocktail party diplomat but he was always ready to go where the action was. I recall once asking him where he was dashing off to for a reception and he said that Mrs. Mac and Leslie had persuaded him he should go to this one. It was some important Frenchman. He didn't know what he did but his name was Jacques Fath. I told him what Jacques Fath did but I think he went anyway.[83]

McNaughton flew with his wife to Berlin, then under blockade, and saw the air lift in action. From Berlin they flew to Oldenburg to visit the grave of their twenty-two-year-old son, Ian.* At Oldenburg they saw the military "McNaughton Club," which had been named in the General's honour.

The United Nations meetings were held at the Palais de Chaillot. One lovely day, late in September, while McNaughton was chairing a committee, Gerry Riddell, who was then head of the United Nations Division of External Affairs, slipped a note to him: "Mr. King wishes to come to Committee this morning and realizes that you will be sitting in the chair and making the statement." Later McNaughton received another scribbled scrap of paper from Riddell: "Congratulations—couldn't possibly have been better if a whole regiment of advisers had worked on it for a month. You've got the debate off to exactly the right start, and I hope it manages to keep at the same informative, objective, tightly-worded level." It was this same Riddell who, after McNaughton's first year on the Security Council, took the trouble to offer his appreciation: "You have been representing Canada on the Security Council with great distinction and we all have every reason to be proud of you. I have more reason than most other people to know how difficult the job is, and also to know

* Paraphrasing Bunyan, McNaughton requested that these words be added to the stone:

Here let the slow bell toll;
On *the other side*
The trumpets sound for him.

how well our interests have been looked after. . . . During the year my admiration for you has continually increased, and to it has been added a very sincere affection."[84]

Unfortunately there is very little room in this narrative of McNaughton's crowded life for social detail, and we must return to the Kashmir dispute.

In May, 1948, in retaliation for an advance by Indian troops, regular forces from Pakistan had crossed the border and entered Kashmir, thus stepping up the danger of full-scale war. Two weeks before that, on 21 April, McNaughton's ideas had been incorporated in a Security Council resolution, of which he was one of four sponsors, that would provide a five-member commission to work towards the restoration of peace and the holding of a plebiscite. The resolution was adopted. The resultant five-member United Nations Commission for India and Pakistan (UNCIP) reduced the danger of full-scale war by achieving a cease-fire. This became effective on New Year's Day, 1949. The following July a cease-fire line was established–which still exists–and here again McNaughton's experience of terrain was drawn upon to decide where it should run. There was more progress. Both India and Pakistan accepted UNCIP resolutions dated 13 August, 1948, and 5 January, 1949, which, taken together, covered all the stages from cease-fire to the holding of a plebiscite. Both parties agreed to the nomination of a plebiscite administrator and the appointment of U.S. Fleet Admiral Chester W. Nimitz met with mutual approval. It was on the vexatious question of the demilitarization of the State of Jammu and Kashmir that negotiations bogged down.

It was again McNaughton's turn to be president of the Security Council for the month of December, 1949, after which Canada's membership on the council would cease. It so happened that the first meeting of the Atomic Energy Commission in December was also under McNaughton's chairmanship. He had already been asked to remain in New York as Canada's permanent delegate to the United Nations or, if he wished, he could take up a commissioner's vacancy which existed in the Canadian section of the International Joint Commission, with the understanding that he would be chairman later. The former, without the heavy responsibilities which he had shouldered on the Security Council, would, he said, "have put me on the cocktail circuit" and that did not appeal to him. The latter, as had been suggested by Louis St. Laurent,*[85] would give him a chance to do something about the St. Lawrence Seaway which had still not been built; it would mean a return to engineering and McNaughton welcomed that. He had, therefore, decided to discontinue his work at the United Nations, apart from representing Canada on the Atomic Energy Commission as he had done since 1946. McNaughton's tenure on the Security Council would end with Canada's participation at the turn of the year.

On 17 December, 1949, UNCIP reported that mediation between India and Pakistan could perhaps be more effectively carried out by a single individual rather than a commission. Arne Sunde of Norway—hoping, as he said, for "a short cut"—proposed McNaughton of Canada for the task, and by a vote of nine in favour, with the U.S.S.R. and the Ukraine abstaining, the proposal was adopted. McNaughton interviewed both parties—repeatedly—set down in detail the position they adopted, and then hoped to reconcile their stands. The principles already accepted by both sides would not, of course, be abandoned. He then drafted a suggested basis of agreement that should be observed for the demilitarization of the state on either side of the cease-fire line. This included the minimum forces required for maintaining law and order, and their disposition; also the dates by which forces should be progressively reduced to various levels. All this would be undertaken under UN supervision and a UN representative would be available to both parties to exert his good offices in case of dispute. Once this programme of demilitarization had been accomplished to the satisfaction of the UN representative, the plebiscite administrator would set to work on lines already accepted by India and Pakistan. What McNaughton was after was the reduction in the state of armed forces, whether Indian or Pakistani, so that the balance of security on either side of the line would always be the same. Once accomplished, this should lead to conditions in which a free and impartial plebiscite could be held.

The task of drafting the proposal was not easy. Every word of the

* St. Laurent, after Mackenzie King's resignation, became Prime Minister in November, 1948. On his retirement, King sent a message to McNaughton: "I shall ever gratefully remember what your friendship and helpful co-operation have meant to me personally and officially over the years of our association and what you have meant to our country over that period of time."

prospective agreement was examined with great care and discussed during the drafting stage with the parties concerned. Its provisions were repeatedly redrafted in an effort to obtain the maximum degree of equity and acceptability to both sides. Between 9 and 30 December McNaughton arranged, and attended, no fewer than forty-five meetings, some starting as late as eleven at night, and there were others attended by members of McNaughton's staff. All this was necessary, although time-consuming, and at least McNaughton, as Sir Benegal Rau of the Indian delegation told him, had the respect of both sides. Not only did he meet the accredited representatives of India and Pakistan; he also had discussions with Sardar Mohammed Ibrahim Khan, president of the Azad (Free) Kashmir Government, and Sheikh Abdullah. As Arne Sunde said of McNaughton in the Security Council meeting of 29 December, "At the time when most people were relaxing and enjoying a holiday you went about your important task with complete disregard for your personal convenience and in a spirit of wholehearted devotion to the public interest of the community of nations."[86]

At 11 P.M. on 22 December the draft agreement was delivered by hand to the respective parties at the Barclay and Ritz-Carlton hotels. And then a little-known side of McNaughton emerged. He was with his team at the Canadian delegation. The carbons of the proposal were on the desk. It was shortly before Christmas and the General opened a bottle of Seagram's rye for his tired staff. "All we need now," one member remarked with satisfaction, "is the agreement of the Governments of India and Pakistan." McNaughton, without a word, reached for the bottle and snipped off two lengths of yellow and black ribbon from around the neck. He placed the ribbons at the top left corner of one of the carbons, struck a match, and fixed them there with a blob of wax. He applied his own personal seal, pressing it heavily down. Then he took a pen and marked a firm I on the yellow of one ribbon and a P on the other. "Now," he said with a chuckle, "we have it."

A meeting of the Security Council was scheduled for the 27th and McNaughton wanted the reaction of the governments of India and Pakistan by then. There was delay, so the meeting was put off until the 29th when the verdict had been received. The Indian government rejected the proposals; Pakistan, with minor amendments, accepted. India's rejection of the proposal was not made public at this stage;

later, at Canada's request, it was changed to one of submitting amendments. Thus the negotiations did not break down.

At the meeting of 29 December Sunde of Norway suggested that McNaughton, "of whose insight into and knowledge of the matter full cognizance should be taken," remain as mediator even after the expiration of his term as president. Sir Alexander Cadogan of the United Kingdom agreed. Ambassador Ernest Gross of the United States also wanted McNaughton, "a statesman of the highest calibre," to remain. Yakov A. Malik of Russia, however, pointed to the progressive steps by which McNaughton's authority had already widened: first the council had agreed "to vest the President of the Council with the function of studying the situation in greater detail and of submitting possible agreed solutions"; then he had been "vested with the function of preparing proposals for the consideration of the parties. Today," he went on, "we have heard a proposal . . . which amounts to assigning certain functions to the President . . . even after his mandate as President . . . has expired" and after Canada had ceased to be a member of the council. "This indeed is unprecedented. It would put the representative of Canada . . . in a position which would be embarrassing or at least delicate. It is an innovation which is not provided for in the Charter, in the rules of procedure, or in the practice or the records of the United Nations." In fact it would add a third category to membership in the Security Council: permanent members, non-permanent members, and now "prolonged members."[87]

McNaughton himself made it clear that he could continue only if certain conditions were satisfied: a resolution by the Security Council that he should do so; willingness of both parties to have him continue; and evidence from them that some prospect of agreement existed. Meanwhile, as Walter O'Hearn reported in the Montreal *Daily Star* on 19 January, 1950, McNaughton remained "waiting in the wings . . . to play his role in the United Nations. . . . In this connection there is speculation about Hon. Lester Pearson's visit to New Delhi following the Commonwealth talks at Colombo." What happened was this.

It seemed that India was moving towards partition in Kashmir but she could not do so openly because of her repeated assurances that she was committed to a plebiscite. The plebiscite idea had been accepted by both parties and it was the repeatedly announced goal of the Security Council. To resolve the impasse it was important to find out under what conditions Nehru would accept a plebiscite. The Colombo Conference was about to be held and informal discussions there with the representatives of India and Pakistan might be very useful. India was sensitive to world opinion and, for considerations of international prestige, had proved amenable in the past.* She might not wish to be considered responsible, as the result of this deadlock, for the breakdown of negotiations. Both the representatives of the United States and the United Kingdom at Lake Success thought that the best course now would be to postpone the next meeting of the Security Council until after the Colombo Conference, and at Colombo to urge India to modify her position so that the negotiations might succeed. Lester B. Pearson of Canada would be at Colombo and the question of McNaughton's future role in the negotiations would give Pearson the occasion for discussions with the Indian and Pakistani delegations; Ernest Bevin, the British Foreign Secretary, was asked to approach Nehru as well.

The Security Council did not meet in January. Pearson was going on from Colombo to Karachi and New Delhi (where he would be present at the festivities arranged to celebrate the inauguration of India as a republic) and the results of his talks on the sub-continent were hopefully awaited. In New Delhi he would have an opportunity to use his in fluence with Nehru to convince him of the necessity for continued negotiations.

Bevin's report of his discussions with Nehru was encouraging and the British and Americans felt that Pearson might tilt the scale; but word from him was a long time in coming. News was finally received towards the end of January. He had obviously done little. He gave no account of any discussion with Nehru in New Delhi. He had spoken in the main with Sir Girja Bajpai (who had been with the Indian delegation at the United Nations and who had returned with Rau to New Delhi for the celebrations) and had told him that if the supposition that he, Pearson, might bring the parties nearer agreement was being stated as the reason for the lull in activities at Lake Success, he did not wish to be regarded as the cause of the inactivity. If India were to be induced to modify her stand, the United Kingdom, the

---

* As an example of that, she had rejected a resolution of 21 April, 1948, but on 5 January, 1949, she had accepted a further resolution that had contained all the vital recommendations of the earlier resolution and had even given the plebiscite administrator extra powers.

United States and Canada would have to demonstrate solidarity of purpose; Pearson's attitude at Delhi did not give the clear impression that such solidarity existed. Nor did Pearson think that any useful purpose could be served by having McNaughton stay on as mediator. In all, McNaughton said, Pearson had not been helpful. To remove McNaughton, without pressing India to modify her position, could only strengthen India in her obdurate attitude. After that the Indians might well believe that they had only to reject a proposal for it to be abandoned and another substituted more favourable to themselves.[88]

Both the Americans (as Acheson, the United States Secretary of State made clear) and the British were concerned when they learned that McNaughton was being withdrawn; they deplored the hiatus that would be caused by his leaving at this time. Nevertheless, on 6 February, after submitting to the president of the Security Council a complete report of his activities, the General left New York to take up his duties with the IJC in Ottawa. It was his last act on behalf of the Security Council. He continued, however, to keep himself in formed regarding the progress of the Kashmir negotiations.[89]

After McNaughton had left New York, John W. Holmes states, the Department of External Affairs "was uncertain what kind of Mission we would keep in New York after going off the Security Council and I was sent down to take charge of the Mission temporarily . . . "

I recall that Sir Owen Dixon [who was going to Kashmir to act as mediator between India and Pakistan] came through New York in the spring of 1950 . . . and the British and American representatives at the U.N. asked me if the General would come down to New York to join them in a briefing of Sir Owen, as even some months after his departure the General was still looked upon as a real expert in the Kashmir negotiations. Needless to say the General accepted with enthusiasm and he came down briefly. I had lunch with him, Sir Owen Dixon, Sir Alexander Cadogan, and Ernest Gross, the U.S. Representative (or Deputy Representative). At the beginning of the lunch Ernie Gross said, "Well, General, it must be very peaceful dealing with disputes between Americans and Canadians after the Indians and Pakistanis."

"Oh, no," said the General, and then he launched into a long explanation of the bitter disputes between, I think it was, the Northern and Southern Mormons over the particular river dispute he was working on. It was obvious that the enormous enthusiasm he put into Kashmir had now been transferred to the Milk River, or whatever it was, and I had a little difficulty steering him back to Asia before we reached the coffee.

As soon as he switched gears, of course, he was precise and lucid and helpful and, as always, encouraging and optimistic. Cadogan and Gross in their own ways were entirely different people from the General but the affection and respect they had for him were quite obvious.[90]

On 14 March, 1950, the Security Council, by a vote of 8-0,* adopted a new resolution. Submitted by Britain, the United States, Norway and Cuba, the resolution called for a programme of demilitarization, based, as the New York *Times* put it, "on principles first outlined by General Andrew G. L. McNaughton of Canada." India, it reported, "has placed important stress on the idea that the principles are what count and that she is not bound to details of the McNaughton plan."[91] "The Security Council," the London *Times* stated, "must now either persuade India or Pakistan to accept a compromise over procedures or confess its inability to deal with the dispute." The seeds of compromise, if compromise was possible, might well be contained in the McNaughton proposal. "General McNaughton has proposed that Indian and Pakistani forces should progressively withdraw; that the local levies supporting 'Free Kashmir' in the western part of the State and the administration of Sheikh Abdullah in the Vale of Kashmir should be disbanded; that the United Nations administrator, Admiral Nimitz, should work with the 'Free Kashmir' authorities on one side of the present dividing-line and with Sheikh Abdullah's Government on the other."[92] But the new resolution, according to McNaughton, contained a fatal flaw.

The council had decided to replace the India-Pakistan Commission with a mediator, as Nehru suggested, who would work between the governments at New Delhi and Karachi. McNaughton agreed with the idea of a single mediator but not with the idea of his being removed from Lake Success. Without the details of the method of demilitarization being settled by the Security Council in New York beforehand, the mediator, whoever he might be, would find himself in India and Pakistan with the dispute on his hands. Even with a prior settlement of how demilitarization would work, the task of the

---

* India and Jugoslavia abstained; Russia was boycotting the council at the time because of the presence of the Chinese Nationalist member.

mediator would not be easy; without a clear-cut plan behind him, in McNaughton's view, it would prove impossible.

Both India and Pakistan were consulted before a mediator was appointed. Pakistan wanted a man of strong character, international reputation and having, if possible, an intimate knowledge of the problem. As his chief task would be to arrange for demilitarization he required a military background. A preference was stated for Nimitz, McNaughton or Eisenhower. The latter's reputation and ability, the Pakistanis said, would overcome the disadvantage of his coming fresh to the dispute. In the end, however, the Australian, Sir Owen Dixon, was appointed; but, as McNaughton had feared, it was found impossible to achieve an agreed solution and the stalemate still exists. It might, McNaughton always felt, have been otherwise had appropriate action been taken at the time of the Colombo Conference in January, 1949.

McNaughton's work brought him many tributes. Louis St. Laurent drew attention to his "eminent and distinguished services for Canada and the world";[93] Mackenzie King sent "heartiest congratulations on [his] splendid service";[94] Ernest Bevin referred to his "constant help and wise advice";[95] Sir Alexander Cadogan put on record "my admiration of all that you have done, my gratitude for the constant loyal co-operation which you have given . . . and my regret and sense of personal loss at your departure from the Council."[96] Sir Zafrullah Khan, Pakistan's Foreign Minister, spoke of his "influence for good"; "General McNaughton," he went on, "is a God-fearing man who does his duty whatever it may be and who, in the course of two years, has acquired immense influence and prestige in New York."[97] John Holmes, who is now the director of the Canadian Institute of International Affairs, in 1969 summed up McNaughton's contribution: "I have always felt it was a important formative period in post-war Canadian foreign policy, as the General really got us started in what we later began to define as the middle power role."[98] But there is no better way of winding up this short account than by quoting George Ignatieff who worked with the General and saw him constantly, day by day:

Camus, in concluding his remarkable book *The Plague*, says: "It could only be the record of what had to be done . . . by all who, while unable to be saints, but refusing to bow down to pestilences, strive their utmost to be healers." A. G. L. McNaughton was among those who, undaunted by the pestilences, perils and challenge of a world convulsed and transformed by war, revolution and unprecedented change, strove his utmost in his day and generation to be a "healer."[99]

Chapter 5

North American Defence

The Canada-United States Permanent Joint Board on Defence, to which McNaughton was appointed as Canadian chairman in August, 1946, had come into being during 1940. On August 17 of that year Mackenzie King, on President Roosevelt's invitation, met Roosevelt in the President's private railway car on a siding near the station of Ogdensburg, New York, for an informal discussion. The PJBD had resulted and McNaughton gave the reasons for its establishment:

The fall of France in the spring of 1940 and the withdrawal of British forces to the United Kingdom, which was then the only bridgehead against Nazi-dominated Europe, created some anxiety as to the eventual security of North America and it became necessary for Canada and the United States to study the problems of the defence of this continent even if the time had not arrived when the United States should declare war on Germany. In August 1940, the President of the United States and the Prime Minister of Canada issued a joint statement establishing a Permanent Joint Board on Defence, with terms of reference to "commence immediate studies relating to sea, land and air problems including personnel and material. It will consider in the broad sense the defence of the north half of the Western hemisphere." This Board was designed as a permanent and sensible arrangement to work out and prepare the measures which should ensure the continued existence of our two countries in safety on this continent until such time as the world had moved forward to a new order based on friendship and good will between all nations. I cannot overemphasize that this Board does not represent any alliance formed as a threat to other peoples.[1]

No matter how in formally the arrangement had been entered into, it was, nevertheless, an alliance and the use of the word permanent in the title of the board left no doubt that the partnership would continue to some extent in the postwar years.

In McNaughton's time the board consisted of five members from either country–a chairman, senior officers of the army, navy and air force, and a member of the foreign service as secretary. It is still organized in two national sections. These sections have offices situated in the respective capitals, each with its own administrative machinery. "When the Board meets formally, the Chairmen preside jointly, but with a single purpose, which is to ensure that a proper answer is given to every problem in the interests of North American security." At meetings in Canada the Canadian chairman presided, as did the U.S. chairman at meetings in the United States.

The procedure is by way of discussion and agreement, never by vote, and in all its history every conclusion has been unanimous.* Very often, discussions and interchange of information serve to bring to attention some point on which there may have been a lack of understanding in the relations between the Armed Forces of the two countries and when this happens the responsible representatives undertake to have the matter studied by their respective services and to report the results. This is usually sufficient to dispel the problem, which otherwise might, in the process of more formal diplomatic communication, have grown in magnitude.[2]

The board, in fact, was powerful. The official United States historian† attests to that:

Operating problems were, in theory, handled through the military attachés and, after its establishment in 1942, the Canadian Joint Staff in Washington. In practice, the Board did not limit itself to making policy recommendations, and in both Ottawa and Washington the sections of the Board, through their members drawn from the military departments and the Departments of State and of External Affairs,

* There was one exception. At a meeting of the board in November, 1941, the Canadian Air Force member did not agree with the opinion of the rest.

† Colonel Stanley W. Dziuban, who has written what is still the best account of the work of the PJBD during the war years. For further details see his *United States Army in World War II, Special Studies, Military Relations Between the United States and Canada 1939-1945.* Articles by Hugh L. Keenleyside and C. P. Stacey, both published in the *International Journal,* also deal with the board and its later work.

functioned informally and unofficially as executive agencies. Additionally, the substantial volume of correspondence between the two sections of the Board, and between the individual members and their counterparts, formed a major alternate channel between the military and political departments of the two countries.[3]

Moreover, the chairmen of the national sections had the ear of the Prime Minister and the President respectively:

The responsibility of the sections of the Permanent Joint Board was to the highest level of authority in each country. In the United States, formal recommendations were presented directly to the President, usually by the chairman or the secretary acting for him. Approval of a recommendation constituted the basis for the necessary implementing action by the appropriate executive departments. The Canadian Section of the Board reported directly to the Cabinet War Committee, over which the Prime Minister presided. Its approval of Board recommendations constituted a directive for their execution.[4]

What the board did was to gather around one table the persons appropriate to the examination of a problem, to make recommendations, which were then presented at the highest level, or to seek further information on which a recommendation might be based. All this was done informally and far more expeditiously than through other channels, and progress could be followed closely:

Supplementing the joint meetings of the two sections of the Permanent Joint Board was a continuous and substantial correspondence and telephonic liaison between the pairs of corresponding members of the Board. Through these means the Board followed up implementation of the conclusions and recommendations decided on at its meetings, paved the way for new proposals, and in a variety of ways facilitated the joint defense measures of the two countries.[5]

The original chairmen of the board were Mayor Fiorella LaGuardia of New York and, on the Canadian side, Colonel O. M. Biggar. McNaughton, it will be remembered, had been closely associated with Biggar—a most competent man with a long and distinguished record of public service—during the St. Lawrence studies between the wars. In 1946 Biggar was unwell and McNaughton replaced him as Canadian chairman. LaGuardia remained to head the United States side of the board until his death in September, 1947. McNaughton said that under the leadership of LaGuardia and Biggar "the basic plans for the defence of Canada and Alaska were drawn up . . . arrangements were made for such important defence undertakings as the Alaska Highway, the Northwest Staging Route for ferrying aircraft to Russia and China [and] the Crimson Route across Hudson's Bay, Baffin Land, Greenland" to move planes, men and supplies to Europe. "It was at the instance of the Board also," before the end of the war and in the immediate postwar period, "that steps were taken to transform the international character of the various installations of these joint undertakings to ensure that full ownership and clear title to all establishments in Canadian territories should be vested in Canada."[6]

The shape that the defence of Canada would assume when hostilities ceased had been a preoccupation with Mackenzie King throughout the war. Canada must not recoil from Europe as she had done after the First World War. Abhorrence of war, isolationism and neglect of armaments had not paid off between the wars. The weakness that these had caused had merely encouraged aggression on the part of others who had built up their strength. The best hope of preventing war lay in collective security; a combination of nations who felt the same would lead to strength: "The strong man armed" McNaughton was fond of saying, "keepeth the peace." Canada should be one of a large company of free nations but her sovereignty must be respected; she must follow her own policies, co-operating alike with Britain and the United States. Canada, the Prime Minister said, must be "a nation wholly on her own *vis-à-vis* both Britain and the United States . . . we can never expect to have any recognition . . . in any other way."[7] This, however, would be a difficult course to chart, for there were dangers to Canadian sovereignty from both these nations.

As early as April, 1941, King saw a "political danger" in the United States:

> Their forces are so much more powerful than our own and so completely needed to protect their own country, as well as ours, because of the gateway which Canada opens to an enemy, that the defence of this continent is bound to be increasingly that of the United States itself. Just what may result from this line it is difficult to say. I personally, would be strongly opposed to anything like a political union.[8]

He had good reason for his fears, for can it be doubted, if Canada had not been willing or able to carry out the defence measures that the United States considered necessary on Canadian territory, that the Americans would have done so unilaterally in the interests of their own security? Canada co-operated with the U.S.A. during the war but, even so, difficulties arose through consultations and negotiations being conducted by the United States in an unnecessarily tactless manner. The U.S. official historian has not glossed this over; he has pointed them out with the fair comment that "the record of these disagreements is a necessary and useful one, for their existence and causes should be noted and should serve as guideposts for the future."[9] At the end of 1942 , in the war Committee, King referred to "the efforts that would be made by the Americans to control developments in our country after the war, and to bring Canada out of the orbit of the British Commonwealth of Nations into their own orbit." He opposed "anything of the kind," wanting to see Canada develop to be, "as our country certainly will, the greatest of nations of the British Commonwealth."[10]

There would be dangers in a Commonwealth relationship too if the British tried a return to centralization. Canada never had a voice in the policy decisions of the war. The British had spoken on behalf of all the Commonwealth members in their discussions with the United States, despite Canada's enormous contributions in men and money and her status as the fourth largest supplier of weapons, munitions and equipment to the Allied cause. (Only the U.S.A., Britain and Russia surpassed her in industrial output.) Britain's cavalier treatment of Canada throughout the war was not far-sighted and in 1943, when King's request that the Canadian part in the invasion of Sicily be mentioned in communiqués seemed to have been ignored by the British but acceded to in Washington, the Prime Minister reflected, "If the day ever comes that the people of Canada should incline in some political affiliation to the United States rather than to Britain, it will be the British Government itself and their officials [who] will be responsible for it."[11]*

One British representative, though unwittingly, seemed intent on

* King was in error when attributing the fault to the British military authorities. The responsibility lay with Eisenhower, as is made clear by the Canadian official historian (see Lieutenant-Colonel G. W. L. Nicholson, *The Canadians in Italy*, pp. 73-4). Nevertheless, King's words do reveal his attitude.

driving Canada along that road. Speaking unofficially before a Toronto audience towards the end of January, 1944, Lord Halifax, the British ambassador to the United States, issued an appeal for a closer union of the Commonwealth countries. He urged them to come together to form a fourth power that would match in strength the individual might of the United States, Russia and China. What he advocated, in fact, was a centralized Commonwealth, "closer unity of thought and action" in fields which he mentioned: foreign policy; defence; economic affairs; colonial questions and communications. McNaughton, when he read the press reports, was appalled[12] as, indeed, was King. It was as if the Statute of Westminster–which Halifax suggested was a declaration of interdependence rather than independence–had never been enacted. Both interpreted the speech as an appeal for the turning back of the clock, a return to "domination from Downing Street," and a bid for Canadian manpower and resources to help augment those of the British Isles so that Britain could compete with the titans of the postwar world.

Canadian autonomy had been too hard-won to be relinquished. Canada must remain free as an independent middle power–bound by the old Commonwealth link, it is true, but bound of her own volition. On 31 January, when Parliament reconvened, King set forth his views in statesmanlike fashion. It is quite clear from what he said that he was aiming at a form of collective security, but international in character–far wider than could be achieved by the Commonwealth alone.

Should we not, indeed, must we not, aim at attaining the necessary superiority of power by creating an effective international system inside which the co-operation of all peace-loving countries is freely sought and given?

It seems to me not to be a matter of matching manpower and resources, or, in other words, military and industrial potential between three or four dominant states. What we must strive for is close co-operation among those great states themselves, and all other like-minded countries. Behind the conception expressed by Lord Halifax . . . there lurks the idea of inevitable rivalry between the great powers. Could Canada, situated as she is geographically between the United States and the Soviet Union, and at the same time a member of the British Commonwealth, for one moment give support to such an idea?[13]

At the meeting of Commonwealth prime ministers in London four months later King again stated Canada's position. There would be no return to isolationism after the war:

I feel that Canada will greatly need [the] strength alike of the British Commonwealth as a whole and of the United States in protecting her position. Of all countries, we are really the most vulnerable because of our extensive territory, resources and the like. . . . The one thing we must seek to avoid is . . . centralization and [we must] strengthen all we can the goodwill between all parts and extend that policy to other nations of the world.[14]

King, it is evident from his diary, was much more incensed over Halifax's speech than his words in the House revealed.[15] In his reaction to British centralists may be found the reason for his support of three measures either proposed or concurred in by McNaughton as Minister of National Defence. One, as already noted, concerned the refusal to permit the corps in Italy to move to the Balkans and its subsequent return to Canadian control. As for the second—that Canadian troops should not participate for any extended period in the occupation of Germany—Churchill had warned in 1943 that the extent to which the dominions would be consulted with regard to the terms of an armistice would depend on the contribution they were prepared to make to an army of occupation; and King had replied favourably at the time. But McNaughton changed all that. Both these measures, which King backed firmly, brought a decided coldness into British-Canadian relationships. The third—Canada's declaration that she would decide what she would contribute to the Japanese war, and where—was described by King as "the strongest assertion made thus far of Canada's position as a nation, demanding an equal voice on matters which pertained to her own forces." The British, he said, had been "casually assuming" that they would make the dispositions.[16]

We have noted that the Canadian division for the Far Eastern theatre was to be organized on American, not on British lines and equipped with U.S. weapons to simplify supply. "This division was to operate as part of a United States Corps in the invasion of Japan."

One of the primary reasons for the decision to associate the Canadian Division with a United States Corps was to obtain experience with the United States system of Army organization and U.S. equipment in view of the obvious necessity for the future to co-ordinate the defence of North America as a firm base against possible contingencies. Actually, the organization of the 6th Canadian Division was well advanced when Japan surrendered and the need passed for its completion.[17]

But McNaughton would have gone even further. Not only would the army component have operated with a U.S. corps; he would, if he could, have integrated the Canadian naval and air contingents with the Canadian division so that Canada's total force in the Far East would have fought under American command.

As Minister of National Defence at the time I would have preferred a self-contained force of all Services, but with the existing composition of our Navy, this would not have been practicable to organize in the time available and so we had to plan to make our contribution by separate Services. Ships of the Royal Canadian Navy and the Units of the Royal Canadian Air Force were to operate with the Royal Navy and the Royal Air Force respectively.[18]

The pattern remained consistent—co-operation with the United States in North American defence and co-operation with the Commonwealth, but with domination by neither–until, towards the middle of 1945, an attractive alternative for Canada seemed to present itself. Between April and June of that year the representatives of fifty nations drew up and approved at San Francisco the charter of the United Nations under which the Security Council of the new organization was given wide powers concerning collective security. The Security Council would have the right to enforce world peace by drawing upon the armed forces of member states to an extent that would be agreed. As both the United States and Britain were members of the United Nations, Canada would gladly contribute, as a sovereign middle power, her share of the armed force considered necessary for the maintenance of the peace; for by doing so she would not only ensure her own security–she would do so within the framework of a world-wide organization without any danger to her sovereignty.

No agreement on the proposed United Nations force had been reached in August, when McNaughton joined the PJBD, though a good result then still seemed possible. Until that force became a fact Canada would co-operate with Britain as an independent power within the Commonwealth and she would maintain her defence ties, because of her geographical position and economic environment, with the United States. But the latter relationship, if Canada were to protect her nationhood, needed careful handling; in fact, the task would call for the utmost firmness, tact and diplomacy. The selection as chairman of McNaughton, who had proved to be a doughty champion of Canadian rights in his dealings with the British, seemed the right one to King.

McNaughton, in an important speech at Detroit shortly after his appointment, left no doubt about Canada's position:

> Canada is a member of the British Commonwealth and she holds to that association with all the firm conviction which has marked the course of our history since early times. She holds to that association, not as any dependent colony of a central authority in London, but as a nation in her own right happy to co-operate in all matters which make for peace and orderly progress in the Commonwealth, and in the world; but with intention made clear to all that she will make her voice heard in deciding the objects of that co-operation, and that then she herself will determine the form and the scale of the action she will take—that her decision in these matters, so far as human judgment runs, will be based on the realities of the situation and the best interests of all concerned, her own included.
>
> No one who has ever had close experience of the actual working of inter-Commonwealth relations will question that all matters will be dealt with objectively by Canada, or will doubt that worthwhile causes will be supported; and that the effectiveness of this support, when given, will far transcend anything which might have been considered possible in the older outmoded Imperial form of organization from which our present constitutional position has steadily evolved down the years.
>
> But Canada is also a nation of the American continent and as a consequence we share many interests with you—interests which under modern conditions continue to develop with ever-increasing acceleration particularly in matters of defence, where, with the remarkable progress in the application of science to war which we have witnessed during the last two decades, distance has lost its former attenuating effect on the problems of international relations. We must realize that continents have now become geographical units in which most questions, both of peace and of war, must be stated and that effective arrangements for the defence of the territory of one nation have become a matter of vital concern to all other nations of that continent.
>
> There are some who see a difficulty in reconciling our position as a

nation on the American continent with our membership in the British Commonwealth, but I do not share this anxiety.

In the first place I believe that fundamentally the real interests of the member nations of the British Commonwealth are very close to the United States. In the second place, I think we are a practical people ready to deal with each question as it arises on its merits and on the basis of the facts. We see no fundamental reason for conflict of interest and you can be quite sure, with our deep concern for the welfare of each of these great associations of peoples with whom we are so intimately linked, we will devote ourselves on every occasion to promoting unanimity of view.[19]

The bogy of a centralized Commonwealth system was finally laid at a meeting of Commonwealth prime ministers early in 1946, when Canada's position was accepted. The British Chiefs of Staff at last conceded that a centralized system of Commonwealth defence would not be acceptable; and in the British White Paper on defence, whose publication followed the conference, it was clearly stated that "Methods of collaboration between the various members of the Commonwealth are governed by the principles enunciated in the Statute of Westminster. . . . Admittedly the Dominions have a close interest in problems that affect the Commonwealth and Empire as a whole, but each of them has a special and distinct outlook on world affairs, dependent on its geographical position and its political and economic environment, and Dominion Governments must retain full liberty of action."[20]

McNaughton was extremely cautious in his early dealings with the Americans. During September of 1945 he presided for the first time at a meeting of the board in Montreal where he and LaGuardia warily assessed each other. The Americans had welcomed the news of his appointment and expected good results: "The designation of General McNaughton to chair the Canadian Section of the Board indicated that Canada did not expect the Board to lapse into a subsidiary role. With the addition of this eminent and experienced soldier-statesman-scientist, the Canadian Section was prepared to deal with the highest questions of politico-military policy."[21]

But personalities are important. On the face of it, there would seem to be little in common between the pudgy, short, third-consecutive-term Mayor of New York, with his flamboyant wide-brimmed hats, and the tall, lean, intense McNaughton; but a mutual regard and warm personal relationship soon grew up. The answer may lie in the drive of this squat little man. Though five years older than McNaughton, LaGuardia, the son of an Italian immigrant–lawyer, airman in the First World War, and later Congressman–was one of the busiest public officials in the United States; "he blithely ignored time, distance, fatigue, opposition and the supposed limitations of human endurance."[22] He was, moreover, decisive, and totally lacked pomposity. When LaGuardia received the Medal of Merit, the highest United States civilian award for war services, for his work on the PJBD, McNaughton congratulated him. "Thanks so much for your thoughtful and courteous note," the Mayor replied, "very happy to hear from you. I guess my name just happened to be around when they had one medal left over."[23]

McNaughton had reason for caution in September, 1945. The U.S.S.R. had emerged from the war as the only nation in the world able to challenge the military potency of the United States and though as yet the Soviet attitude was uncertain, there was an increasing awareness that Russia posed a threat to American security. The question of atomic bombs and their delivery was the chief preoccupation in the military circles of the United States. The Germans had developed the V2, a rocket with a range of more than two hundred miles, which could deliver approximately a ton of explosive to the target. Rockets of longer range could undoubtedly be developed to carry an atomic charge and, no matter how costly, they would provide a relatively cheap method of erasing an enemy's cities. Meanwhile there was the bomber threat, for although the U.S.S.R. had needed no heavy bombers in its fight with Germany, these could be built while long-range rockets were being designed on the Russian drafting tables. They would, have range enough to reach the United States on a one-way trip (an American plane, the B-29 had proved that) and to deliver an atomic bomb to the target would be worth the sacrifice of a plane. The United States had a monopoly on the bomb and also a delivery system in being; Russia had neither the bomb nor heavy bombers, but there was no room for complacency. There must be air defence in depth, required for radar tracking and the interception of fast bombers by fighter planes, and for that the United States needed Canadian territory just as Russia needed the territories of Poland, East Germany and Czechoslovakia.* The United States wanted close Canadian participation for the operation of northern airfields, radar, navigational aids and in the field of conventional warfare as well.

Canadian territory stood between the U.S.S.R. and the United States across the North Pole. And, as the U.S. official historian pointed out, a "significant body of Canadian public opinion was expressing the view that it was unwise for Canada to act jointly with the United States in measures that might antagonize the USSR." It was even suggested at a conference of the Canadian Institute of International Affairs that "Canada should abandon the Permanent Joint Board in the postwar period as constituting an irritant in relations with the USSR."[24] But that, in McNaughton's view, would have been the height of folly "if we were to protect our sovereignty."

> The Ogdensburg Declaration had to come into play. We had to hold our territory. We mustn't draw back from anything to be done but as it was for U.S. defence they would have to pay a large part of the costs. We had to discharge our obligations to make sure that nobody attacked the U.S. over our territory. If we had not done so there was the danger that the U.S. might have taken over the Canadian North in the interest of their own security.[25]

And there would be a by-product from the co-operation of Canada in this—the increasing development of the North which McNaughton had done so much to foster in the years between the wars. It had made sense to learn about the North at that time; how much more so now when the danger was more pronounced?

Canada had to be very careful, however, not to be swept completely into the United States orbit and in September McNaughton answered an American proposal, which had been made in June by Major-General Guy V. Henry, the army member of the U.S. section, that there should be complete uniformity of equipment, organization and training between the forces of the two countries. Henry had cited the two instances where Canada had adopted the U.S. model: a small force for Kiska and, at a later date, the ground force intended for use against Japan. These were isolated episodes, McNaughton said, in no

* Russia already occupied East Germany and Poland; Czechoslovakian territory was acquired in 1948, as we have seen.

way comparable to the long and intimate association that Canada had had with the forces of the Commonwealth. That association was likely to continue. But we have seen that during the Second World War McNaughton had looked principally to North America for the vehicles and weapons he needed. The views he put forward were essentially the same as those he expressed in a public speech a month later.

In Britain available labour was scarce and urgently needed for maintenance and other purposes and, most important, we had learnt from hard experience that designs worked out in the British industrial environment were often unsuitable for mass production according to the North American technique. Screw-threads differed–rivets spaced for insertion by hand could not be placed with our automatic machines–and we literally could not do in quantity the elaborate hand fitting which was the customary practice in the British Isles. Even technical terms had different meanings on either side of the Atlantic and some there were who were opposed to new devices for they threatened the supremacy of the established interests.

He then turned to the atomic bomb, "a decisive weapon in this year and day of grace, but as well it represents nothing more than a transient advantage." He was looking towards a wider association for Canada than close integration with the United States, in which Canadian plants would be crushed by the industrial might of the Americans. "If we are to retain our place as the wardens of peace," he said,

the most important thing we have to do is to maintain our research and development and to continue to produce the newer and better weapons, at least in prototype, and to ceaselessly experiment to master the technique of their employment. . . . It is a vital requirement that the Allied nations who think with us remain collectively the strongest military force on earth, and each and every one of us must make our contribution to that end.[26]

McNaughton saw no possibility of the invasion of Canada's North for at least a decade. There was, therefore, no immediate need for complete uniformity between the Canadian and American forces, and the way lay open for Canada to model hers on any pattern she might choose. But there were some fields, because of the common industrial environment in North America, where he could see uniformity developing on an *ad hoc* basis. Ideally, for the security of all the countries making up the English-speaking world (as well as for the furtherance of international peace) there should be co-ordination of military supply between those countries. An important move towards this would be the standardization of the military equipment of the British Commonwealth and the United States. So while Canada would not as yet adopt American standards, she thought it desirable that there should be uniformity in the military organization, training and equipment of the United Kingdom and the United States and would assert her influence towards that goal.

Throughout the early part of 1946 McNaughton continued to tread warily in his dealings within the board. As the United States official historian shrewdly points out, "Military co-operation with the United States had not yet become a clear-cut facet of contemporary Canadian foreign policy. This policy was fundamentally one of full support for the search for security through the United Nations, and one of minimizing bilateral or multi-lateral regional arrangements as detracting from the maximum potential of the United Nations for peace."[27] Canada was not going to commit herself to United States defence schemes as yet. "There were," McNaughton said, "too many unknown factors in the international situation to enable us to study many of our defence questions with any degree of finality. The number, the nature and the scope of the special agreements to be negotiated on the initiative of the future Security Council and of the regional arrangements for international security for which provision has been made in the Charter of the United Nations, had yet to be defined. Nor again could the question of defence co-operation within the British Commonwealth of Nations be yet said to be entirely clear. In those circumstances no clearly defined picture could be drawn."[28]

The process was going on of taking over the joint installations set up during the war to ensure that "full ownership and clear title" should be vested in Canada. Agreement had been reached concerning the disposal of the northern airfields, but other United States property, both movable and immovable, remained. Early in March McNaughton suggested that the transaction be completed, in one parcel, by 31 March, the end of the Canadian fiscal year. It was a broad challenge which LaGuardia welcomed. Though there were minor difficulties,* these were overcome, and the agreement was signed on the 30th. Canada paid twelve million dollars for property whose cost had been almost sixty million,[29] and this was met from the United States dollar balance which had mounted during the war. It was necessary for Canada to keep costs down; in the interests of her sovereignty, however, she reimbursed the Americans and thus obtained clear title. But Canada was a nation of only thirteen and a half millions compared to one of a hundred and fifty millions; in developed resources and in annual national income the difference was even greater. It was therefore important to reduce the costs of providing security, in the interests of both countries, to a figure that was within Canada's means.

"I drew the attention of the Board," McNaughton recalled, "to a number of defence installations that were constructed in Canada at United States request during the war but eventually became a charge against Canadian public funds. Prompt action in war, of course, must be a paramount consideration but the cost of some of these facilities was out of proportion to their value as factors towards the achievement of victory."[30] Canada had to be made fully aware of what the United States considered necessary in the future. With regard to the use of the facilities taken over, Canada met American wishes wherever possible. The Alaska Highway, the Northwest Staging Route and other installations in that region were received by McNaughton on behalf of Canada at a ceremony held at Whitehorse on 3 April, 1946, the day after the General's appointment to the United Nations Atomic Energy Commission had been announced. The Americans considered the highway, which passed to Canada free of charge, of great importance for the defence of Alaska so that Canada, although under no legal obligation to do so, agreed to maintain and keep it open. Renamed the Northwest Highway System, its maintenance became a task of the Canadian Army with a headquarters established at Whitehorse for the purpose. Similarly the North Atlantic Ferry Route (Project CRIMSON) was kept open largely for speeding the return of American servicemen for demobilization. And Canada co-operated in other ways. She invited the United States to send observers to Exercise

---

* The stated value of a bulldozer at a northern site, as an example, was not considered by McNaughton to be realistic. It would cost a considerable amount of money to bring the equipment out to a place where it could be used. The real value, therefore, was the agreed value less the cost of transportation.

MUSKOX, held between February and May, when a mechanized expedition using snowmobiles, with the troops supplied by air, covered more than three thousand miles of northern territory which had previously been thought an impenetrable barrier against attack. Much was learnt of a military and scientific nature, which was made available to the Americans. And in September it was agreed that a "winter warfare" experimental centre at Churchill, for the cold weather testing of arms, aircraft, fuels, lubricants and other supplies and equipment, should be established for the use of both nations.

As 1946 wore on, however, it became increasingly clear that Canada's hopes for the effectiveness of the United Nations in providing peace and security would not be realized at any early date; and though this remained a primary objective of foreign policy it was realized that, in the interim, national defence was not enough. Canada accordingly entered into a closer relationship with the United States for the defence of North America. The co-operative defensive arrangements, formally announced in a joint statement of the two governments on 12 February, 1947, were not inconsistent with collective defence sanctioned by the charter of the United Nations. General McNaughton has described them:

> This statement defined the measure of agreement which had been reached for co-operation in our defence policies and set forth the following principles, which in addition to providing for the interchange of individual members of the respective armed forces set out the requirements for co-operation in exercises; the development and tests of materials of common interest; the encouragement of common designs and standards in arms, equipment, training and new developments; the mutual availability of military, naval and air facilities; the expeditious transit through the other's territory of military aircraft and . . . vessels. It is prescribed that all of these arrangements shall be without impairment of the control of either country over all activities in its territory.[31]

This was little different from the relationship between Canada and Great Britain, except in that instance there was no joint defence board. In the Commonwealth, while no agreement existed, there was an historical relationship that provided, as Louis St. Laurent said, for "the organization of the armed forces on common lines, a large range of common arms and equipment and the exchange of service personnel and military information on an extensive scale." Each country "regards the other as a potential ally in the event of a general war" and Canada, independent within the Commonwealth, recognized that fact.[32] There was now a "new relationship of complete equality" between the two nations and with that had come "a closer bond of friendship and a greater fund of good-will than ever before existed between Canada and Great Britain."[33]

Nevertheless, Canada recognized that the Commonwealth, by itself, did not constitute a system of collective security. Nor could the United States and Canada alone ensure the maintenance of international peace.[34] Consequently Canada's defence relations with Britain and the United States were still based upon the assumption that an effective United Nations could ultimately be established. A long step forward in this direction seemed to have been taken with the signing of the Treaty of Brussels by Britain, France, Belgium, The Netherlands and Luxembourg on 17 March, 1948. Mackenzie King hailed it. "This pact," he said, "is far more than an alliance of the old kind. It is a partial realization of the idea of collective security by an arrangement made under the Charter of the United Nations," and Canada, he continued, would play her full part in such movements "to give substance to the conception of an effective system of collective security."[35] McNaughton, in Paris at the United Nations in November, 1948, was invited to address *L'École Supérieure de la Guerre*. He spoke very strongly in favour of broadening the Brussels Pact to include Canada and the United States in a North Atlantic alliance for which negotiations had begun. A little more than a year later, on 4 April, 1949, Canada, together with Belgium, Denmark, France, Iceland, Italy, Luxembourg, The Netherlands, Norway, Portugal, Britain and the United States, signed the North Atlantic Treaty, one of whose objects was mutual defence against aggression in Europe and North America. With that Canada attained much of what she had been striving for: membership, as a sovereign power, in an association for collective security which included Britain and the United States, but with no fear now of being dominated by either.

The detailed story of the PJBD throughout this period and afterwards is an interesting one, but it is beyond the scope of this book. All that can be done is to give a general account and then focus more closely on McNaughton as chairman of the Canadian section. He remained with the board from 1945 until 1959 and thus saw all the developments which will be described. It might be thought that the many roles that McNaughton filled until 1950 were contradictory, but this was not so. On the one hand, as a humanitarian who feared the consequences of atomic warfare, he strove indefatigably as a member of the Atomic Energy Commission of the United Nations for the elimination of atomic weapons while, as a scientist and head of the Canadian Atomic Energy Board, he planned for the peaceful use of atomic energy in the service of his nation. In the Security Council he did his utmost to lessen the risk of war. Yet, as a soldier by training and a co-chairman of the Permanent Joint Board on Defence, he still had to think in terms of atomic warfare and to help plan measures that would reduce its destructive effects. There is nothing unusual, however, about struggling for the best and, at the same time, making plans in case the worst should happen. Those plans, the heart of North American defence policy, were based on defensive measures and retaliation. Defensive measures would reduce the general vulnerability of the region (and hence temptation) and preserve the power of striking back. The ability to retaliate in kind would deter the aggressor.

We have seen that there were three main phases in the steadily deteriorating relations between the nations of the West and Soviet Russia after the war. The first phase ended in 1949 when the Soviets tested their first atomic bomb, thus ending the American monopoly. There was a second phase from 1949 to 1952 during which the Americans tested a hydrogen bomb in November, 1952, but in which it was recognized that, in the event of war, the United States could no longer obtain a cheap or rapid victory. The third phase, which is with us still, began in August, 1953, when Russia also exploded a hydrogen bomb; from then on each side had the capability of destroying the other. The delivery system, as has been mentioned, was also of great importance. Though Russia first tested intermediate ballistic missiles, with a range of fifteen miles in 1956, planes were not displaced as the means of delivery until 1957. From then on inter-continental ballistic missiles have had to be reckoned with.

During the first—reasonably complacent—phase there were ground forces whose principal role was the elimination of enemy lodgements on Canadian or Alaskan territory. These were air-borne and air

At Eniwetok Atoll for H-bomb tests, 1956

Aboard a St. Lawrence icebreaker, a part of North American defence

Attending a Permanent Joint Board on Defence meeting aboard USS *Wisconsin*

transportable. In addition coast defence units were provided and a special Canadian force maintained the Northwest Highway System and McNaughton's long-established Northwest Territories and Yukon Radio System. The RCAF and the USAF worked in close co-operation, the main task being the protection of the larger population centres by means of interceptor squadrons directed from the ground by radar units. Bombers were assigned to support the ground troops in their mobile striking role, and there were aircraft-ferrying operations, search and rescue, transportation, air survey and "submarine hunting" tasks. Naval vessels joined in the latter, and with the North Atlantic Treaty of 1949 the RCN gradually converted to the anti-submarine role. Ships, for the first time, took a serious interest in the North and 1948 marked the first venture by the RCN into Canadian Arctic waters.

It was important to know the North and its conditions. Intimate knowledge of the terrain, including approaches by land and sea, would bring in estimable advantages to the defenders. Thus there had to be air survey and extensive mapping; hydrographic surveys to chart the waters; a network of observation stations for recording weather and other basic information; and a series of expeditions throughout Arctic Canada and Alaska. (One such expedition, of course, was MUSKOX.) There must be improved facilities for flying–aids to navigation, landing strips and better means of communication–so that defence in depth could be pushed to the outer perimeter as far as practicable, still retaining at the centre a mobile striking force. As Mackenzie King said, echoing McNaughton, there would be no "Maginot Line," no "large-scale defence projects":

> As the Government views it, our primary objective should be to expand our knowledge of the North and the conditions necessary for life and work there with the object of developing its resources.[36]

Thus the Northwest Territories and Yukon Radio System, as well as being used for defence, continued to serve governmental and commercial interests in the North. The Northwest Staging Route from Edmonton was used for civil aviation as well as serving air force planes. And Fort Churchill, a military experimental station convenient to a wide variety of both Arctic and subarctic features (so that troops could acquire experience of working together in cold weather over terrain similar to that prevailing throughout a large proportion of the northern approaches to the continent) tested transport, clothing and supplies under winter conditions, which was of general benefit to all who lived in the North. A railway ran to Churchill for easy transport of materials for testing.

Towards the end of 1945 McNaughton linked to the PJBD the activities of the Arctic Institute of North America, which had been established by private scientific interests and from which much could be learned. He was also, throughout the period of his co-chairmanship of the board, a member of the Canadian Government's Advisory Committee on Northern Development where, according to its chairman, his "long and very active interest in all things concerning the north" were of extremely great value and where he contributed much "in furthering and guiding the great developments that have been, and are taking place."[37] This "by-product" of joint defence, as he termed it, was always a source of great satisfaction to McNaughton.

All three services, it will be seen, had been concerned to a large degree in these defensive measures. In 1946 a first step towards the integration of the three Canadian services was taken when a single Department of National Defence replaced the wartime expedient of three departments. McNaughton favoured the move, as well as the integration of the services wherever practicable–services that were common to all, such as medical and dental services, chaplains, pay–but he opposed the unification of the armed forces. He had made his views known in Washington in 1942, and he was to repeat them again in a letter to a senior American naval officer in 1958:

> I cannot conceive how anyone with experience either in military or political organization can maintain that the individuality of the fighting services should be destroyed and that one person, however able, should be vested with the supreme authority over all. I do not think any individual has yet been thrown to the front who could adequately grasp the complexities and deal with matters effectively and with the dispatch required.[38]

For good or ill, Paul Hellyer, when Minister of National Defence some years later, embarked on that experiment. McNaughton's comments take on an added significance now that the unification of the services is an accomplished fact.

A year later, in March, 1947, a Defence Research Board was

created with two main tasks: to conduct research; and to co-ordinate development for all three services. Not unnaturally, McNaughton took a friendly interest in that. Dr. Omond Solandt, its first chairman, on his retirement wrote to thank the General:

Your support has smoothed over many difficulties that would otherwise have been insuperable and I know that all our staff are grateful to you for all you have done to help. . . . Ever since I started work on tank problems in England in 1941 I have felt in a vague way that I was part of your team and working under your general guidance. . . . Not only D.R.B., but all Canadians owe you a great debt of gratitude for your unfailing efforts in the national interest.[39]

After 1949 the Canada-United States Military Co-operation Committee, which had been set up in 1946, considered what further measures were required for the security of North America. In February, 1951, Brooke Claxton, the Canadian Minister of National Defence, announced a programme "to make this continent safe from attack" which, after the Russians had tested a hydrogen bomb, went into high gear. "You of Canada and we of the United States," President Eisenhower announced in Ottawa towards the end of 1953, "can and will devise ways to protect our North America from any surprise attack by air. . . . These days demand ceaseless vigilance. We must be ready and prepared. The threat is present. The measures of defence have been thoroughly studied by official bodies of both countries. The permanent joint board on defence has worked assiduously and effectively on mutual problems. Now is the time for action on all agreed measures."[40]

Canadians are aware of the Pinetree Line, an interlocking chain of radar stations—the first comprehensive detection system in North America—and a joint undertaking which became fully operational in 1954. They also know of the Mid-Canada Line, an exclusively Canadian undertaking ready by 1958, and the DEW (Distant Early Warning) Line, financed by the United States, which became operational in July, 1957. Inside this radar shield—which no hostile bomber force could penetrate without detection—were various lines of air defence, "a complex of huge radar scanners and screaming jets."[41] From these air defences the idea emerged of one command, controlling all the air defence installations in Canada and the United States, and the result was the North American Air Defence Command

(NORAD) established by agreement dated May, 1958. Later that year came the introduction of "semi-automatic ground environment" (SAGE), which has been described as a combination of high-speed computers and high-performance radars. "A giant SAGE computer . . . receives data from radar sites and in a split second furnishes the commander with a visual display of all air traffic over the Canadian land mass and provides possible tactical solutions to meet hostile air threats."[42] All this, however, was not proof against a missile attack but, as the Canadian Armed Forces report, "present indications are that an anti-missile system is feasible but extremely costly."[43]

In the defence of North America the problems faced were strategical, not tactical. The waging of local wars had a place in the overall concept and McNaughton's views are interesting in the light of Vietnam. Commenting in 1958 on statements by Dr. Solandt that "we must win local wars if freedom is to survive" and that it was important to "make it clear that we are prepared to resist the spread of Communist domination by every means in our power" he wrote:

> I agree we must win the local wars we undertake, but we must be very discriminating in those we *do* undertake as otherwise our strength will be dissipated. I do not believe that on the basis of conventional forces on any practicable scale we can in fact maintain a cordon round the Communist states. Somehow, I think, if we are less aggressive in maintaining the cordon the need for doing so will lessen and perhaps there may be a gradual approach to stability in relations between West and East which would be a real step towards association in peaceful endeavours in many fields.[44]

Some of the work of the PJBD was not well known: the standardization of weapons and equipment, the sharing of defence production, and the co-operation and dispersal of industry. All these matters, and many more, were tackled by the board, though any attempt to discuss them would transcend the limits of this book. Only examples can be given here.

McNaughton's interest in the standardization of weapons sprang from his experience in two world wars. In the first, troubles with the Ross rifle and its ammunition had pointed to the need for a common weapon. During the Second World War, when weapons, vehicles and equipment had come from the United States, Canada and Britain, the problem had been much more complex. Yet the key to the whole question, as McNaughton told the board, was simple: the standardization of screw-threads. "It may sound nonsense," he told the writer, "but you remember the story of how a war was lost, and all through the lack of a horseshoe-nail? It was the same with screw-threads, although we didn't lose the war."[45]

The British system of screw-threads was based on a thread angle of fifty-five degrees; the American on an angle of sixty degrees. An accurate fit between components having such a widely differing thread angle was completely impossible, so that there could be no interchangeability between North American and British manufactures without subsequent machining or hand-fitting. And the Canadian Army, with weapons and equipment originating on both sides of the Atlantic, had been caught in the middle, forced to carry excessive scales of spares simply because parts could not be interchanged. As for Canadian industry, it had been necessary to work to British and American thread forms—with the attendant confusion of gauging systems and drafting practice—through two world wars.

A common system, important in war, would also confer great benefits in peace. In the past, international trade in any kind of mechanism had been inhibited by this lack of interchangeability in screw-thread parts. Suppose, for example, that the British wished to market a particular type of plane in the United States. That plane, quite conceivably, might contain more than a hundred thousand bolts, nuts or screws. True, spares for these could be provided with the plane, but there would be a psychological deterrent—the Americans would think twice about a product that might be difficult to maintain. And the same applied to automobiles, tractors and other engineered products. With uniform standards such limitations on the international commerce of Britain, Canada and the United States would disappear.

That there would be advantages had been appreciated for many years. In 1919 Britain, the United States and France had conferred about it and had recognized the need for continuing studies. Apart from that, there had been no result. Not until the fall of 1945 had there been real progress. At that time, in Ottawa, a standard type of thread had been accepted and it seemed likely that agreement on a thread form would quickly follow. McNaughton, who was present at the parley between representatives of the three nations, pointed out to the British that "had there been a standardized thread in 1939, the United Kingdom would have saved billions on war orders placed in the United States."[46]

McNaughton waited patiently for results, but nothing happened. Industry in all three countries, apparently, was not particularly interested in pushing the project to a conclusion inasmuch as it would inevitably mean some change in practice in peacetime manufacture. But from the point of view of the armed forces it was vital to force the matter through and McNaughton determined to utilize the board to overcome this inertia. Through LaGuardia, and with the instigation of President Truman, McNaughton managed to enlist the support of the United States Secretary of War and the Secretary of the Navy, who urged action on the National Bureau of Standards, and the American Society of Mechanical Engineers. The result of all this was a conference between representatives of the United States, British and Canadian governments on 24 October, 1947, at the National Bureau of Standards.[47] J. G. Morrow, chairman of the Canadian Standards Association, gathered new hope from the meeting. "Your efforts," he told McNaughton, "have resulted in bringing this matter to the fore much sooner than could otherwise have been expected, and I am writing to . . . express appreciation."[48]

An arduous struggle, however, still lay ahead and McNaughton kept the pressure up. He corresponded directly with engineering associations in Canada, Britain and the United States, urging and co-ordinating their activities. They, in turn, passed to him copies of the correspondence they were conducting on both sides of the Atlantic.*[49]

Finally, on 18 November, 1948, a "Declaration of Accord" on the uniformity of screw-threads was signed by representatives of the three countries at the National Bureau of Standards of the United States in Washington, D.C. McNaughton, who was in Paris, was invited to attend the ceremony by J. G. Morrow, who wrote:

I sincerely hope that you will be on this Continent at the time of these meetings and that you will be our official leader. I have looked

---

* He also received many letters from private individuals in England who sought to interest him in weapons and developments. One example might be given. Lieutenant-General Sir Frederick Morgan wrote to General Turner on behalf of an Englishman with an invention for the improvement of jet engines: "If you think fit, it will of course be just grand to have General McNaughton run his eye over this. If by any chance one could get his blessing it will be worth more than all the money in the Treasury."

forward to this situation when you would have the senior part and I trust you can arrange it.[50]

But McNaughton could not be present. General Assembly meetings and meetings of the Security Council were scheduled until the first week in December. Morrow went to Washington with Neil P. Peterson who headed the Canadian Committee on Screw Threads and who had worked devotedly for years. There C. D. Howe, the Minister of Trade and Commerce, signed the accord on behalf of the government of Canada. Nevertheless, said McNaughton, "they paid me the compliment of asking me to go and I should have dearly loved to have signed that document. The standardization of screw threads may sound trivial, but it was one of the greatest things we ever did."[51]

A Tripartite Standardization Agreement, largely on weapons, followed. Under that, it is interesting to note, the Sabot projectile, designed on McNaughton's principle of 1942, "was the first Canadian-developed armament item to be adopted by the three countries. . . . In 1952 this Sabot project was completed."[52] "General A. G. L. McNaughton," the Windsor *Star* commented, "is given credit for the original design. . . . This calls attention to the remarkable versatility of this great Canadian."[53]

Little publicity attended McNaughton's activities on the board. The work was done quietly and efficiently, almost in the background, and though the newspapers kept the public generally informed of projects and progress there were few first-class rows or drama on which the press could bite. Security also played its part. This was a constructive period during which Americans and Canadians marched together in the interests of both, their dealings marked by consideration and co-operation. There were differences, as there were bound to be, but they were faced in a forthright manner and smoothed away. On these occasions, where Canada's interests were concerned, McNaughton spoke plainly, so bluntly, in fact, that W. H. Barton of External Affairs—and for several years the board's Canadian secretary—was shocked when he first attended its meetings. "The General," he said, "was fond of the word 'servitude.' 'That would impose a *servitude* on Canada,' he would say, 'one that we can in no way accept.' But the

Americans didn't seem to mind. They were used to him. He was a hard negotiator and they respected him."[54]

McNaughton got on well with all the American chairmen from LaGuardia onwards: Dean Acheson, who replaced LaGuardia in November, 1947, before becoming Secretary of State under President Truman; Major-General Guy Henry, who succeeded Acheson at the end of 1948; and Dr. John A. Hannah, president of Michigan State University, who took over in April, 1954. Acheson, who had first met McNaughton in Washington in 1934, valued the appointment and his collaboration with the General on North American defence.[55] Hannah's regard is perhaps best conveyed in the words he used in 1955 when conferring an honorary degree on McNaughton at Michigan State:

Soldier, patriot, international statesman, you served our neighbor nation of Canada with great honor and distinction in two world wars when our two nations were allied in the cause of justice and freedom. More recently, your extraordinary talents have providentially been turned even more directly to the advancement of our mutual interests through your leadership of such agencies as the International Joint Commission and the Permanent Joint Board on Defense. . . . We have come to share your countrymen's admiration of your generous spirit, wisdom, and precision of judgment, and are proud to salute you as a truly great citizen of the New World.[56]

Louis Johnson, when United States Secretary of Defense in 1949, congratulated McNaughton on the "grand job" he was doing; but he enjoyed one tribute from a relatively junior American officer, Lieutenant-Colonel William M. Fondren, who had attended some board meetings. Writing from his home in Texas during 1951, Fondren enclosed a few artificial flies:

My young son Billy (age 11) has recently taken up fish fly tying. He has heard me speak of you so often, and, among other things, your prowess as a fisherman, that he asked me to send you a few of his samples. . . .

General, it has been a real privilege for me to have known and been associated with you. I wish you continued success in all the fields of endeavor in which you play such a vital part.[57]

Good as his relations were with the Americans, however, there were inevitably times of strain; for McNaughton would concede nothing that would prejudice the interests of Canada. In general, this worked out well, for the Americans, anxious for Canada's full co-operation, were flexible. Canada, McNaughton made clear, would indeed co-operate to the full provided her sovereignty continued to be respected. "Canada means to continue," he said, "as we have always done, to carry our full and proper responsibilities for the defence of our own territory" but, he emphasized, "all arrangements within our territory will remain strictly under our own control."[58] Within the board, McNaughton expressed Canada's views in the form of general principles. He might say, for example, "It is the intention of the Canadian Government to maintain control over all flying and other military activities in or over Canadian territory." Then there would be discussion and consultation. The development of policy followed, so that a mutually acceptable *modus operandi* emerged which, when set on paper, became an agreement between the two governments at a future date. Thus the board's work was immensely practical in clearing problems: Who would build the radar defences in the North? Who would pay for their maintenance? What were the conditions of tenure? All these, and similar questions, were answered to the satisfaction of both parties. The ground required for American projects, such as the DEW Line, was provided by Canada but remained under Canadian title. Canadian contractors and labour were to be used for construction as far as possible, and the Canadian government reserved its right to take over the manning of any or all of the stations at any time. Furthermore, as McNaughton said, the emergency that had come about with Russia's nuclear parity would not last forever. No one could say when it would be over, but when it was "the measures taken would end . . . then United States forces would as promptly as practicable leave Canadian territory, taking with them their movables; but . . . buildings attaching to the soil would pass to the ownership of Canada." Land would be provided free of charge but its use by the United States would end with the emergency, and "there would be no payments by Canada to the United States" for property left at that time.[59]

Diplomacy was needed; the United States was often much more imperious than Britain had ever been in proposing measures for defence. As John Hannah, Assistant U.S. Defense Secretary and McNaughton's opposite number on the PJBD, said in an address to the United States Chamber of Commerce in Washington on 20 April, 1954:

I am afraid a great many Americans are not adequately aware how fortunate we are to have as our nearest neighbor on the side in which the danger lies an ally on whom we can depend absolutely. . . .To have the benefit of an early warning of attack we must place our picket lines far to the north, within Canadian territory. . . . Happily, the Canadians understand these things; they see that the defense of our two countries must be operated as a close partnership. They appreciate fully that, for both of us, the only defense is a continental defense.

Canada, he went on, had insisted on paying a fair share of the costs, but Americans sometimes forgot that and made things "difficult . . . our American impetuosity sometimes tries them severely. I am afraid that too often we forget that we are only visitors in their domain, welcome, to be sure, but with obligations to behave ourselves as welcome visitors should behave.[60]

Newfoundland, however, undoubtedly occasioned the greatest strain between the two sections of the board. There the United States occupied certain bases, which had been leased by the British in March, 1941, at a most critical stage in the war. The bases had been leased for a period of ninety-nine years for military co-operation and defence and under the agreement, in very broad terms, the United States enjoyed certain privileges in the base areas that were not essential for the military operation of the installations: rights to civil and criminal jurisdiction; the right to establish United States postal facilities; and certain customs-free privileges and exemptions from taxation. On 31 March, 1949, Newfoundland became a province of Canada, and the Canadian government viewed these broad extraterritorial and non-military rights with alarm.

Canada thereupon sought modification of the Bases Agreement under the recommendation, made on 20 November, 1946, by the PJBD, that "defence co-operation projects in either country should be agreed to by both Governments, should confer no permanent rights or status upon either country and should be without prejudice to the sovereignty of either country." She also drew attention to the joint statement issued by Prime Minister King and President Truman in February, 1947, which called for co-operative arrangements that would be "without impairment of the control of either country over all activities in its territory." The United States, on the other hand, pointed to the existence of a legal agreement. Firm rights had been granted under the Bases Agreement of 1941, and why should any of these be surrendered unless, by so doing, definite advantages would accrue?

Here, then, were grounds for conflict, and a five-day verbal battle did, in fact, ensue at a meeting of the board in Newfoundland early in January, 1950. Some of the rights accorded under the 1941 agreement, McNaughton said, were inconsistent with Canada's status as a nation; they encroached on Canada's sovereignty; they meant that Canada's control over activities on her own territory was impaired. To accept Article IV of the lease agreement, which concerned jurisdictional rights, could not possibly be justified to the Canadian Parliament or the Canadian public. The problem was fundamental—that of Canadian sovereignty.

The Canadian proposals, the Americans said, were difficult to accept in time of peace and were believed to be "unworkable and unacceptable" in time of war. Nevertheless, as McNaughton put it, "a formula under which military requirements could be met without disturbance to Canadian political thought and in harmony with the new and altered status of Newfoundland as a part of Canada" was found before the meeting's end.[61] The United States section agreed to the suspension of legal rights under Article IV of the Bases Agreement for a trial period of five years and a recommendation—later implemented by both governments by an exchange of notes—was made; the old concessions have not been revived in the years that followed.

Not only did McNaughton concern himself with ensuring that an "honourable partnership" existed between the forces of the two nations; he was equally concerned that there should be an honourable partnership in the field of industry. The reciprocal availability of military, naval and air facilities in each country had been assured since early in 1947 but the same could not be said of weapons and equipment. "In World War II," he said, "it was shown conclusively that we need not doubt our ability to invent, develop and produce any article of war short of the larger battlecraft which were, in size, beyond the capacity of our plants. We did not undertake to make everything because this would have been uneconomic in the case of supplies of satisfactory types otherwise available in sufficient quantities."[62] After the war it would have been even more uneconomic for Canada to try to manufacture everything that was required for her own small forces; she wished to buy certain items from the United States and in return to sell others that were being manufactured in Canada, some in excess of her own military requirements. There were, however, obstacles in the way of the procurement by the Canadian forces of arms, munitions and other matériel from the United States; Canada, legally, could have only those items that were declared surplus to the needs of the U.S. armed forces, which made little sense. Under joint defence arrangements each should have been able to procure from the other whatever was needed and McNaughton worked steadily for this during his first years with the board. He finally achieved his objective in 1950.

Throughout 1947 to 1950 we find McNaughton harping on a constant theme: it was essential to have a close association between Canadian and United States war industry, and especially there must be plans for industrial mobilization. On no account, especially in Canada, should industrial efficiency be allowed to decline. He called for full co-operation and adequate co-ordination between the industry of the two countries. "The best protection for the countries of North America and Northwest Europe, as well as for the rest of the world," he still felt, "would be an effective universal organization of security under the United Nations, but until this can be brought about, the continued production and further improvement of all our weapons and the maintenance of industrial efficiency and preparedness are vital to the prevention of aggression by making impossible any prospect for its success."[63]

There was the danger, of course, that the Canadian defence industry would not survive, that it would be squeezed out of existence by the industrial strength of the United States. Once the restrictions on Canadian purchases of American military supplies were lifted, America could quite easily fill the needs of both countries. McNaughton used two arguments to counteract the danger he foresaw. The most reliable defence against weapons of mass destruction, he pointed out, was dispersal—of people, of defences and, especially, of industry. The ideal was to offer no profitable targets or as few as possible, to prevent a crippling surprise blow. Thus industry should be decentralized. Canada's industry must remain in being as an alternative to that in the U.S. Secondly, he drew the board's attention to the realities of Canada's exchange position. Canada had always paid for her military supplies; if Canada were to standardize her equipment with the United States to the desired extent, and to build up her armaments, she would need to sell defence equipment and supplies to the United States so that she could make purchases in return. Therefore, there should be reciprocal purchasing arrangements. Not only would this further the implementation of the standardization programme; it would also assist Canada in the development and maintenance of defence industries important to the security of the continent and—harking back to his first theme—help decentralize the industrial pattern of the continent. All this would be encouraged by the placing of U.S. military orders in Canada. American industry would not suffer, and Canada would use the dollars she obtained to buy military equipment in the United States.

McNaughton's concept found agreement within the board. Similar overtures to Britain in 1949,* however, brought no firm response and, not unnaturally, Canada moved closer towards the United States in reciprocal arrangements. In April the U.S. Munitions Board expressed the view that "increases in reciprocal purchases of military equipment" were desirable, a view that was accepted by the U.S. Secretary of Defense one month later, and for the year 1951 an amount of between fifteen and twenty-five million dollars was agreed upon. C. M. Drury, the Canadian Deputy Defence Minister, paid tribute to the PJBD, whose recommendation had brought about this result. The principle of reciprocal purchases, he said, far transcended in importance the amount involved.[64] From that principle grew a billion-dollar business in the years ahead. A defence production-sharing agreement was made in 1959. Up to September, 1966, the United States had placed contracts worth $1,430,000,000 in Canada; during the same period arms contracts valued at $1,230,000,000 had been placed by Canada in the United States.

By 1951 McNaughton could be satisfied with what had been achieved. A joint United States-Canada Industrial Mobilization

* When it was pointed out that Canada's productive capacity greatly exceeded her own military requirements. A sound and continuing estimate of production demands that Britain and the United States might be likely to place in Canada would be helpful for intelligent planning.

Committee had been set up in 1948 whose duty it was "to exchange information and co-ordinate the views of the two governments in connection with planning for industrial mobilization in the event of an emergency." Principles for economic co-operation having been drawn up by this committee, a close association between Canadian and U.S. war industry had developed. A co-ordinated programme of requirements, production and procurement, had been established, as well as a system of controls over scarce raw materials and supplies. There would be consultation between the two countries before one imposed any control that would affect the other. Technical knowledge—even productive skills—would be freely interchanged. Barriers impeding the flow of goods (essential for common defence) across the border had been removed as far as possible. Most important, arrangements had been made for consultation concerning financial and foreign exchange problems. "A fundamental conception of these measures of co-operation," McNaughton concluded, "is that, apart from the gifts which we would each make to our allies, there should be balance in trade in arms between us."[65]

Canada did make gifts to her NATO allies and by doing so she moved firmly away from the British types of weapons which traditionally she had always used; from then on the equipment of the Canadian Army was standardized with that of the U.S.A. As McNaughton put it:

The opportunity for standardization came with the redeveloping of the armed forces of our allies in Europe where UK types of equipment are in wide use and which these nations desired to continue [to use] so as to avoid the problems and delays of retraining, at least during this critical period. We have been able to present the equipment for a full division each to the Netherlands and Belgium and a further similar gift has now been made to Italy. To Luxembourg we have given the equipment for a Field Artillery Regiment complete, which is what they indicated to General Eisenhower they wished to have. These are all free gifts to help our allies and our Parliament has provided the Army with the funds required to replace with U.S. types. Some of this we are purchasing in the United States . . . and some we will manufacture ourselves. When these arrangements are complete all the forces of Canada and the U.S. in North America will have the same basic military equipment and our factories will be able to supply each other with greater economy and facility and thus, as was foreseen by the PJBD, there will be important consequent advantages to each.[66]

One other matter in the field of defence production deserves to be mentioned; this concerned the provision by Canada of electronic equipment for the radar installation in the North. It took place at the instance of McNaughton who, as early as November, 1947, had suggested to the Americans that "the possibility of using existing facilities in Canada for the production of material required in any joint defence programme should be kept in mind."[67] Not that facilities for the production of such sophisticated equipment existed in Canada then, but they could be provided. In February, 1951, when the construction of the Pinetree Line was under consideration, McNaughton firmly announced that Canadian industry could provide most of the electronics equipment by the target date; but the Americans were sceptical. Nevertheless, if Canadian industry could match American production in quality and price, and have the equipment ready on time, as McNaughton said it could, there was no reason why the contracts should not go to Canada. In November, however, the Canadian air member announced that Canadian industry could not meet the dates stipulated for completion owing to the lack of firm contracts from the U.S. Air Force and, consequently, a lack of funds. He wondered, therefore, if the contracts should not be cancelled. McNaughton would not hear of it. To cancel the contracts, he said; would mean that a developing industry would be forced to close, and that would "jeopardize the continued availability of Canadian productive capacity," and certainly such a decision would run counter to the desire of both countries to disperse vital production facilities of electronic equipment. The Americans expressed concern at the news, but the contracts continued. By September, 1952, the Canadian electronics industry was already well established and had proved that it could produce equipment of high quality at competitive prices. Then, McNaughton insisted, if the industry was to continue to be able to make an effective contribution to defence requirements, it must be awarded a share of new contracts. Though the United States chairman, General Henry, doubted whether the question of the future allocation of such contracts was a proper matter for the board, McNaughton had his way and by 1954 the principle "that electronic equipment at installations on Canadian territory should, as far as practicable, be manufactured in Canada" had been firmly established.[68]

Canada did very well as a result of these contracts. Of thirty-three stations on the Pinetree Line, for example, eleven were to be financed and manned by the RCAF and seventeen by the USAF, while five were to be paid for by the Americans and manned by Canadians. The provision of electronic equipment at the seventeen American-financed stations paid the costs at the Canadian eleven. However, as McNaughton pointed out, Canada was not out to make large profits in these defence industries, and she therefore agreed to pay the costs of manning the five remaining stations; this amounted "probably to something of the order of seven million dollars per annum." The country was glad to make this alteration when, in McNaughton's words, "we found we had made a better bargain than we had thought equitable and desirable." Inwardly, he was very pleased, as he confessed later: "We certainly did not wish to have our military personnel paid from any outside source, even if indirectly."[69]

Simonds, whom we last met at the time of McNaughton's abortive visit to Sicily and whose relationship with McNaughton had been chilly ever since, was a post war Chief of the General Staff. He dismissed the whole business as "gadgetry." McNaughton, he claimed, had "a more ready access to senior cabinet ministers" than had the Chiefs of Staff, which was undoubtedly true. "Whatever the terms of reference may be," he went on, "many of the expressed views of ministers bear the unmistakeable imprint of the influence of an intellect more remarkable for its bias toward pseudoscience and gadgetry than for its record of sound military judgment."[70] McNaughton saw it differently. "Canada," he said, "had now established an electronics industry comprising over four hundred firms" and there was the assurance of enough work to keep it effective.[71] In quality, the Canadian products could stand comparison with any in the world, as Fraser F. Fulton of the Canadian-based Northern Electric Company made clear in a letter to McNaughton. Writing of radar equipments, he said: "At the U.S. working level they are quite forthright in their complimentary remarks and state that the relative lack of troubles, the amazing interchangeability of parts and better operating characteristics . . . lead them to consider the Canadian produced equipments are vastly superior to those they themselves have received from American production. I think you should take a great deal of satisfaction from comments of this nature since they completely vindicate the stand you took. . . ."[72] The industry continues to thrive today.

As a footnote to the decentralization of industry we might mention aluminum. The aluminum industry had been centred at Shawinigan in Quebec before the development of Kitimat in British Columbia. McNaughton now looked to further dispersion in Labrador and promoted the Hamilton Falls power development to supply the necessary electricity. "He had always," wrote a correspondent, "in the back of his mind the magnitude of the power potential of the river."

In 1953 he was visited by members of the Brinco [British Newfoundland Corporation] which had recently been granted extensive concessions in Newfoundland-Labrador. Like many others who came to McNaughton's office they were looking for his advice. In this case it was for his view on what priorities should be given to the exploitation of the resources that had been granted to Brinco.

McNaughton's advice was that the development of the power of the Hamilton Falls should be given a high priority. Until that time I do not believe the Brinco Group were aware of the full value of this great asset.[73]

On 9 September, 1953, A. W. Southam, manager of Brinco, called on McNaughton in Ottawa where, in the words of Commander R. H. Stokes-Rees (who, as a representative of the aluminum industry, was present), the General "explained your own potential to you." Another meeting with representatives of Brinco followed on 14 October at which McNaughton acted as spokesman for the Canadian government. He expressed his personal interest and the interest of the PJBD in promoting an increase in aluminum production, and described his role as that of a catalyst to stimulate the development of Canadian resources. The Grand Falls site, he said, was the only possible one on the Hamilton River for the production of cheap power. He had flown over the area and studied the geological data later. Not only did he suggest how the site should be developed—"a low dam above the falls could divert water through a tunnel or canal to a powerhouse located in a canyon running into the Hamilton Canyon a few miles below the falls"—but he also voiced the belief, contrary to others which had been expressed, that it would be possible to run a railway due west of the falls to the Iron Ore Railway, rather than going southward for fifty miles before turning west.

Labrador was a new country and the muskeg there was different from that north of Lake Superior. The geologic history was not the same.

The last ice cap over Labrador had melted "in situ" and had produced eskers and as a result the lakes were shallow and had hard gravel bottoms. The difficulties encountered in railway construction north of Lake Superior, where the muskeg was deep and full of mud, would not be encountered. But this could be tested on the ground. He felt sure it would be found that there was shallow muskeg over firm ground.[74]

This is no more than an indication of McNaughton's wide-ranging interests, his thoroughness and his grasp of essentials. As is well known, the Hamilton Falls site, now renamed Churchill Falls, is under development. When completed it will deliver 7,000,000 horsepower, about 5,250,000 kilowatts and 750,000 kilowatts more than the huge Soviet installation at Bratsk, Siberia. Certainly Churchill Falls will be one of the largest hydro electric power plants in the world.

The board never met in fixed places such, for example, as Ottawa or Washington. Instead it travelled. "We sought always," McNaughton said, "to go to where the problems were because in that environment solutions were most often to be found."[75] He often flew to board meetings and this was useful when, from 1950 on, he was also chairman of the Canadian section of the International Joint Commission. A diversion from the flight path here and there, a little more flying time, and he could study river basins, potential power sites and stretches of international water from the air. "These journeys," he said, "were not directed especially, of course, to a study of international rivers . . . but looking back in the light of the water problems which later have come my way I think that the greatest remaining value to me has been the by-product that we managed at one time or another to have a look at most of the great power rivers of northern North America in the Pacific, Arctic, Hudson Bay and Atlantic watersheds."[76] Not much escaped McNaughton's attention.

Many of the journeys he made were of absorbing interest, and he loved them. We have mentioned that the board settled the dispute over the Newfoundland bases, on the spot, in Newfoundland. Similarly, defence problems concerning the northwest of the continent were discussed in the Alaskan communities of Juneau, Anchorage and Fairbanks, as well as at Whitehorse in the Yukon Territory; the board met in HMCS *Ontario, en route* from Vancouver to the North and back. There were other "seaborne" meetings, in the United States ships *Franklin D. Roosevelt* and *Wisconsin.* The latter conveyed the board to Guantanamo, Cuba, and then to the Panama Canal Zone, for a tour. There were meetings at a guided missiles centre in Texas, at an air force research centre in Massachusetts, at a proving ground in Maryland, and at air force bases in Alabama, Colorado, Florida, Louisiana, North Carolina and Missouri. In 1950 McNaughton, combining Permanent Joint Board of Defence and International Joint Commission business, conducted R. B. McWhorter and E. B. Weber, who were both United States commissioners with the IJC, to the Quebec-Labrador iron deposits in the vicinity of Knob Lake. His purpose was to arouse their interest in the construction of the St. Lawrence Seaway, which had still not been built, both for defence and for the other benefits that would accrue. He succeeded. In a letter to the General, McWhorter expressed "his conviction that Quebec-Labrador ore was needed by the industries of both countries and that, therefore, the construction of the St. Lawrence Seaway should be commenced with the least possible delay."[77]* In 1951 McNaughton descended three hundred feet aboard the submarine USS *Cutlass* off Key West, Florida, and for doing so became an honorary member of the Ancient Order of Deep Dunkers. Five years later he set off for a Pacific rendezvous to witness a hydrogen bomb test on Eniwetok Atoll; there we find him, as well as being stimulated by the scientific atmosphere and the strange, sunny environment, experimenting with flies on tropical fish.[78] The thermonuclear test, as it might well have done, impressed him. Those who had doubted the claim of the atomic bomb to be the absolute weapon could not shrug this one off. With these bombs, for the first time the energy of fusion–itself the source of the sun's energy–had been unleashed.

Perhaps the most interesting experience of all was in March, 1954, when McNaughton flew over the North Pole on an historic journey from Fort Churchill-Resolute-Greenland; for the first time Canadian voices were recorded over the Pole. Of this trip, the General said afterwards, "We had been able to view the terrain where some of our most important problems lay."[79] He travelled with his opposite number, Dr. Hannah; the Minister of Defence, Brooke Claxton (with

---

* The St. Lawrence Seaway question is dealt with in more detail in the next chapter.

whom he enjoyed the most cordial relations); Douglas Stuart (the United States ambassador); General Charles Foulkes (chairman of the Canadian Chiefs of Staff Committee); Gordon Robertson (Deputy Minister, Northern Affairs and National Resources) and Leonard W. Brockington, who described the flight over the pole for CBC. Here is a fragment from Brockington's superb account:

We saw a little green open sea and once or twice a radiant Arctic rainbow where the sun and the mist joined with the sheen of the ice crystals. We heard no sounds except the noise of the engines of our plane although we could be sure that in the frozen seas beneath us the ice cracked and growled and roared. . . . I think we all felt in our hearts a sort of religious hush as through the crystalline atmosphere we passed over the frozen brittleness and the icy stillness beneath us. The mountains which we saw stood immaculate in white adoration. Where once there had been chaos and a multitudinous mighty movement of land and sea, time had merged it all into the greatest stillness and the greatest loneliness in all the world. Once or twice we seemed to be looking on great cathedrals of ice, where the mighty organs were frozen and all the choirs silent. . . . Everywhere were majesty and awe, solemnity and sublimity.[80]

McNaughton, on the other hand, dwelt on the practical aspects of the journey. "Great progress," he recorded, "is being made in learning to live and to move and to work in the Arctic"; this would be useful "both for peace and for defence."[81] Keith R. Greenaway* navigated the North Star aircraft on the lonely flight, and an excellent job he did. Brockington, in a letter to McNaughton, spoke of the General's "youthful vision of Arctic development" and felt there was "a fine apostolic succession from you to the . . . young pilots and navigators of our plane."[82] The trip had its lighter side; on a visit by snowmobile from Resolute to an Eskimo settlement, Brooke Claxton's picture was found adorning the wall of an igloo. The "pin-up boy," as might be expected, came in for a good deal of teasing.

All this might be thought to be pretty strenuous for a man approaching seventy; but McNaughton, buoyed up by his boundless enthusiasm, was spry enough. In 1954, in addition to his PJBD and IJC responsibilities, he accepted the chairmanship of one of the

* The author of *Arctic Air Navigation* and co-author, with Moira Dunbar, of *Arctic Canada from the Air*.

National Research Council committees—the Associate Committee on a National Aviation Museum—whose purpose was to foster the establishment of such a museum, which in turn would be the first step towards the foundation of a National Science Museum. The National Aviation Museum opened on Dominion Day, 1961, with McNaughton continuing to serve as honorary president of the National Aviation Museum Association, "a fitting recognition," wrote E. A. Côté, of Northern Affairs and National Resources (under which department the museum came until it passed to the Secretary of State in 1964), "of the contribution that you have made to the creation and growth of the National Aviation Museum."[83]

McNaughton, as we have seen, had been appointed to the PJBD by Mackenzie King. He had continued with the board under the succeeding Liberal administration of Louis St. Laurent. The Conservatives then came to office and the General, feeling that John Diefenbaker, the new Prime Minister, would wish "to appoint to this office someone who has been in close association with yourself," offered his resignation in June, 1957.[84] Diefenbaker did not accept it; McNaughton's offer, as he interpreted it, was of "continued services to suit the Government's convenience" and so the matter rested for more than two years.[85]

The easy arrangements of the past, when McNaughton had reported directly to King and later, under St. Laurent, to the Defence Committee of the Cabinet, were under Diefenbaker, it seemed, in question. The first sign of this appeared in the form of a letter from the Chairman, Chiefs of Staff, who passed on some comment from Pearkes—McNaughton's old subordinate and now Minister of National Defence—who had noted that the last meeting of the board had been held at Fort Churchill, the next would be held at Esquimalt, British Columbia, and the one following in Cuba; the Minister suggested that, for the sake of economy, future meetings might be held at only four locations, Ottawa, Washington, Montreal and New York.[86]

McNaughton, in a letter to the State Department, pointed out that meeting places had been arranged at the suggestion of each of the six services, Canadian and United States, in rotation; each, in turn, had been able to show without inconvenience matters it felt important for the board to see. The establishments shown had been related to problems that were, or would be, under discussion in the board. It was true that the board had travelled widely; it needed to do so to go to the places where the problems were, "and it is in the atmosphere of practical relevant knowledge thereby obtained that solutions to the many difficult questions which have been presented have been evolved."

Considering the magnitude and far-reaching significance of the matters which have been dealt with, and which, I believe, without exception, have been resolved to the satisfaction of Canada, I must say that the suggestion of narrowing-in the range of travel by the Board would not result in economy; in fact, I believe, in the overall results, the loss of intimacy in knowledge and of effectiveness in its use would prove very costly indeed.[87]

The manner in which the question had been raised could only be interpreted by McNaughton as unwarranted interference by the new Minister of National Defence, who had not even had the courtesy to communicate with him directly. He did not let this pass without comment. "To the best of my knowledge," McNaughton wrote, "this is the first occasion, in my considerable experience with the Board, when either such observations have been made or such a channel for communication (not even directly from the Minister of National Defence) with the Chairman of the Canadian Section PJBD has been used. . . . You will recall," he went on,

that the PJBD was established by the President of the United States and the Prime Minister of Canada . . . and ever since the chairmen of the Sections have enjoyed freedom of access direct to the heads of their respective governments and by their authority to ministers in charge of departments who are concerned with business before the Board. As I believe that the essential conditions for the Board's continued usefulness rests on this foundation, you will I am sure appreciate my conviction that such a change should not come about by acquiescence in a procedure but only, if it is to be done, by direction of the Prime Minister himself after he has had an opportunity to consider the kind of service which he desires the Board should continue to render.[88]

The General then pointed out the dangers of making such a change, which could only be detrimental. He was not concerned with prestige—he had offered his resignation—but the continued efficiency of the board would largely depend on the stand he took.

> . . . it certainly is open to the two countries to have a Joint Board, the Sections of which would operate under direction from a Chief of Staff or the equivalent, but in such a case I do not believe it would be useful, or indeed that it would be either possible or desirable, to draw on the kind of experience for the chairmen which has heretofore been made use of—on the United States side, for example, Mr. La Guardia [*sic*], then Mayor of New York; Mr. Dean Acheson, who was afterwards Secretary of State; and the present chairman, Dr. John Hannah, President of Michigan State University , who continues also on occasion to serve the President of the United States and the Secretary of Defense personally and directly in numerous missions of special character.[89]

Of course he made his point; no change was made. Despite the word "Defence" in the board's title, McNaughton was convinced that the PJB should never be subordinated to the Defence Department. Two years later, in conversation with Sidney Smith, Secretary of State for External Affairs, the General reiterated his views:

> I . . . told him of my concern that PJBD should continue to be regarded as an arm, and an important one, of the External policy of Canada, and that it should not come to be regarded as something essentially for the Defence Department. . . .
>
> I said that when I took over the chairmanship of the Canadian Section in 1946 the secretary only had been provided by External. I had begun by inviting External to send a senior officer and later had arranged that this change in composition of the Canadian Section would be made formally by Order-in-Council. Then I had developed the practice that the External representative would come to be regarded as the senior next to myself and in my absence, regardless of his individual precedence, I had arranged that he would preside.
>
> I said I hoped that these arrangements would continue; and I said they were very important as a means to ensure that defence arrangements between Canada and the United States were not allowed to be made either formally or, worse, informally direct between the military staffs. If this were done we would wake up some day to the existence of commitments of serious import, the significance of which probably had not been recognized by the military officers at the time they were entered into.[90]

McNaughton was well able to handle differences with the Defence Department. These did not lead him to seek release from the chairman's duties, but rather his increasing responsibilities within the International Joint Commission. It was becoming very difficult for McNaughton to co-ordinate his activities in the two fields. One would think that it might be pleasant for him to be done with public service, to sit gossiping about the past in his comfortable Rockcliffe home, but the General was not built that way; he was never preoccupied with the past. Always he looked to the future and he felt he had a duty there. The future of the Columbia River concerned him most in 1959, when the thorny negotiations were entering on a crucial phase. He decided he would concentrate on the essentials, as he saw them, and give up the PJBD, where his work was largely done.

On his seventy-second birthday McNaughton broached the matter with Sidney Smith but Smith died shortly afterwards. The General then sought release directly from Diefenbaker. He spoke of the great importance of the Columbia and how necessary it was "to evolve the terms of a settlement with the United States which would protect the proper interests of Canada."[91] On 21 April, 1959, the press announced that "General A. G. L. McNaughton, 72, one of Canada's most experienced negotiators in Canada-U.S. relations, will retire this summer as Chairman of the Canadian Section of the Permanent Joint Board on Defence. He will be succeeded by L. Dana Wilgress, veteran Canadian diplomat. . . ."[92] Douglas LePan of External Affairs, who passed the official notification to McNaughton, added a personal footnote: "I cannot sign this official letter without adding a word of personal appreciation of the unique value of your continuing services. You command the admiration and affection of all of us here."[93] The changeover took place in July.

McNaughton looked back on his joint chairmanship of the board with satisfaction. His relations with the Americans had been largely smooth, a great deal had been accomplished and it had been a happy and interesting period of his life. At the time of his resignation he had already had nine years experience of negotiations with United States representatives within the International Joint Commission, and the contrast there was marked. He himself commented on this in 1961:

I have found it very interesting to contrast the free understanding co-operative association of the PJBD with the very different experience in another joint Canada-U.S. body, the IJC. However, I think this is to be expected, for the fundamental reason that in the Board we were moving with common objectives for the defence of North America. In the IJC the situation is far different for here we are rivals for the beneficial use of the waters along the boundary and for the immensely valuable resources they represent. It is not to be expected therefore that agreement in this area will be as easy to attain as in Defence.[94]

In this rivalry for the use of international waters McNaughton was the strong champion of Canada and he must now be studied in that role.

Chapter 6

The International Joint Commission

On New Year's Day, 1950, McNaughton joined the Canadian section of the International Joint Commission as commissioner. Seven months later he was appointed Canadian chairman and he remained as such until April, 1962. During his twelve years as chairman a score of questions came before the IJC, not all of which, for reasons of space, can be examined. These included gigantic projects such as the development of the St. Lawrence River for power and navigation; the best use of the Columbia River for power in the West; and, on the other coast, a study of the international tidal power potential in Passamaquoddy Bay. Some of the tasks that the commission dealt with did not match these great undertakings, as when the IJC turned from the Columbia, with its potential storages of millions of acre-feet,* to little Sage Creek, the spring run-off of which amounted to no more than a few thousand acre-feet. Perhaps it is a measure of the greatness of this international body that no problem is too large or too small to be placed before it.[1]

"The boundary between Canada and the United States, including Alaska," McNaughton said, "is an immense affair which runs for some 5,655 miles. . . . Everywhere, in practically every mile along this vast expanse of borderland, there is a differing problem of water, the fundamental character of which varies from place to place. . . ."[2] Some of these problems reached the IJC in the form of "references" (asking for recommendations) or "applications" (for decision and subsequent

* One acre-foot is the amount of water that would cover one acre to the depth of one foot.

order embodying the commissioners' conclusions). McNaughton, in a light-hearted review, mentioned some of those that were before the commission in 1958.

Out on the east coast was the tidal power project in Passamaquoddy Bay and with that there were many difficulties. Not only was there a plea for sardines—the local sardine industry, which would be affected by the development, was up in arms—but there was no large local system into which to feed the power. To feed it direct to local industries would present problems because the tides get fifty minutes or so later every day. "I don't know of any industry," commented McNaughton, "that is prepared to put itself onto a lunar schedule." It seemed that there would have to be auxiliary plants and these were under study. Another project in the East that was under reference was that at St. Croix. "This is a small watershed. The St. Croix is a small river. However, a 10,000 kilowatt plant in this region is relatively more important than a half million plant on the Columbia in the opinion of the [local] people. [So] next month we will go down to that region and present our engineering reports to the people, both in Maine and New Brunswick, at a public hearing where we try to draw out the opinion of all concerned and everyone has an opportunity to freely express his views. . . . Everything which is said is taken into account by the Commission in reporting to the Governments." He mentioned the St. Lawrence—which will be dealt with in detail later—and one of his remarks is worth noting. Flooding, and the levels of the lakes, had always been a problem. "We have many people hollering about the inflow to the lower lakes," he said. "People downstream ask why we don't stop the outflow from the upper lakes. It has even been asserted that people can drown at Duluth in order to protect Chicago."[3]

Farther west both the Red and Souris Rivers were being considered by the IJC. "The 'Red River of the North' is a very interesting river. It rises close to the Divide with the Mississippi waters and flows north. Of course in these northern latitudes a river flowing north over hundreds of miles is difficult to handle because in the spring it gets warm first in the headwaters and when the ice upstream does melt—suddenly—then in the northern regions, where the ice still holds, heavy flooding results. . . . The Red River on the rampage becomes a very powerful river indeed!" The Souris, rising in Saskatchewan, flows south through North Dakota in a big loop and then turns back across the border into Manitoba. A lot of the flow, south of the border, had been taken up by the American Wildlife Service to develop bird sanctuaries and game reserves. "Now it's nice to provide water for ducks," was McNaughton's comment, "but it is not so nice for people in Manitoba when you realize that every duck that is raised . . . requires three acre-feet of water in evaporation, and people in the north need this water for which there is no other source. We have to protect ourselves against the use of water in that way. . . ." And on the west coast was the biggest reference of them all—the Columbia River.[4]

The references or applications which we shall look at from the many before the IJC are the St. Lawrence, Libby Dam, the Waterton and Belly Rivers, and the Columbia. It will be appreciated from the foregoing that these studies amount to but a fraction of the work the IJC was doing under McNaughton's chairmanship, during which time, as we have seen, until 1959 he also remained head of the Canadian half of the Permanent Joint Board on Defence, with frequent meetings anywhere on the North American continent and sometimes farther afield. The IJC references that will be studied are complicated enough; remember that there were others and that for nine years, while chairman of the Canadian section of the IJC, McNaughton headed the Canadian section of an equally important international group.

McNaughton was primarily brought into the IJC by Louis St. Laurent on King's advice, to do something about the St. Lawrence Seaway impasse. He was the ideal choice. He was a hydroelectrical engineer who was fully conversant with the St. Lawrence Basin from his work with Biggar between the wars. The opening of the Welland Ship Canal in 1932 had marked the limit of progress then. Moreover, since the end of the war, he had proved a skilled negotiator both at the United Nations and with the United States members of the PJBD. And within the PJBD he had lost no opportunity of stressing the importance of a seaway to North American defence.

In 1950 the position with regard to a seaway was briefly this.*[5] We have seen that the United States Senate had, in 1934, failed to ratify

---

* The background of this account of the St. Lawrence Seaway negotiations was supplied by Miss Margaret S. Mattson, who placed her excellent major paper at the writer's disposal. All quotations, unless otherwise documented, have been taken from this.

the St. Lawrence Seaway Treaty signed two years previously by the Hoover and Bennett governments. Organized self-interest, in which the Association of American Railroads featured prominently, had led to that result; and the opposition set up by this and other groups was to block progress for another two decades. Nevertheless, in 1938 the United States government did take the initiative by submitting to the Canadian government a revised treaty covering navigation and power development on the St. Lawrence River. Faced with the opposition of Premier Hepburn, who did not feel Canada needed the power and improved navigation at the time, and with that of Premier Duplessis of Quebec, who supported Ontario's stand, the federal government could not proceed. More hopefully, a St. Lawrence Basin Agreement, designed to improve navigation and produce power in the "bottleneck" (the international section) of the river, was signed by both governments in 1941. This agreement also failed because it could not win enough support in the United States. From this time on, and especially after the Second World War, due to rapid industrialization and the need for power as well as the benefits that improved navigation would bring, Canada badly wanted the seaway.

A move within the United States early in 1948, urging the adoption of the 1941 agreement, seemed to promise progress; that year the United States Senate voted to commit a resolution of the previous year for further study. The resolution in question, a seaway bill, differed from all previous ones by suggesting that the cost of navigational works and the cost of operations could be liquidated by charging tolls. The introduction of tolls was an attempt to mollify railroad and eastern port transportation interests, as well as an economy-minded Congress. Though Canada had in the past rejected the idea of tolls, on this occasion the suggestion was not opposed; it was hoped that the United States government might be sufficiently attracted by tolls to approve a seaway. But once again vested interests in the United States gathered their strength; the Senate rejected the resolution.

Later in 1948 New York State and Ontario opened discussions in order to develop power alone (improved navigation could wait), and this could be done within the terms of the Boundary Waters Treaty of 1909. Subsequently New York and Ontario submitted applications to their federal governments; if approved, reference would then be made to the International Joint Commission. President Truman, however, wanted no separation of navigation and power; he wanted a seaway to combine both benefits. But Prime Minister St. Laurent, while agreeing that Truman's policy would be best, if capable of achievement, knew that there would be an acute power shortage in Ontario should the International Rapids section of the St. Lawrence not be developed soon. If both the navigational and power projects could not be accomplished, Canada would have to develop power. That was the Canadian view at the end of 1949.

McNaughton, long a proponent of the St. Lawrence Seaway, knew what it would mean for Canada. It was no longer primarily a matter of permitting ocean-going ships to pick up cargoes of prairie grain deep in the interior, though that, of course, still applied. Since 1947 high-grade iron ore was being mined on an increasing scale in Quebec-Labrador. In 1949 five American steel companies joined Hollinger of Canada and Hanna of Cleveland to form the Iron Ore Company of Canada. The building of a three-hundred-and-sixty-mile railroad followed to bring the ore to Seven Islands, Quebec. The target of ten million tons of ore a year could be doubled if there were an improved seaway to permit transportation of the ore to the Great Lakes steel plants. This would be of immense benefit to Canada, adding at least seventy million dollars a year to the national income. Not only that. Freighters could be used economically for two-way cargoes; after carrying ore to Lake Erie they would go on in ballast to the lakehead and there load with grain for Montreal.

Furthermore, Canada's appetite for power was increasing with industrial expansion. Few knew better than McNaughton what strides Canada had taken since the outbreak of the war towards becoming one of the great industrial nations of the world. Most of the industries that had grown up were to be found in southern Ontario, largely because of the low-cost electric power that was available for their use. Production in Ontario, valued at $1,746,000,000 in 1939, increased to a value of $8,600,000,000 in 1951. This growth, it was already obvious in 1949, could not continue unless the generating capacity kept ahead of the demand for power and that, it was equally obvious, would not be accomplished without deliveries of power from the St. Lawrence River. The Niagara Treaty, signed with the United States at this time, would stave off a power crisis until 1955. By 1956 St. Lawrence power from the International Rapids section would be required and for that, with a five-year construction period envisaged, the work would have to be authorized and must start before the end of 1951. This simple fact,

The Canadian chairman, International Joint Commission, 1950-1962

McNaughton and Senator A. O. Stanley sign the St. Lawrence Order, 29 October, 1952. Left to right, Eugene Weber, George Spence, Roger McWhorter, Lucien Dansereau
(LUCIEN DESJARDINS, PHOTOGRAPH "LA PRESSE," MONTREAL)

The development for power of the international section of the St. Lawrence Seaway. Left to right, Lucien Dansereau, McNaughton, Thomas E. Dewey, Robert H. Saunders, Leslie Frost and Robert Moses it is clear, governed the tempo of the Canadian negotiations.

McNaughton, before joining the IJC in 1950, had already advocated the construction of the seaway at meetings of the PJBD. His argument was that it was vital for defence: steel and added power were essential in war, when demands for both increased. The Korean conflict pointed to the need for larger steel and power reserves. And not only would a seaway produce power and facilitate the movement of ore; it would enable ships to be built in the sheltered harbours of the Great Lakes region where, in any case, sixty per cent of the component parts of vessels were manufactured. The speedy construction of ships had enabled the United States and Canada to survive in the past; a shipbuilding industry in the Great Lakes area, to supplement those along the coasts, would be very valuable in time of war. The Member of Parliament for Simcoe North, who opposed the seaway, commented shrewdly in the House on 6 December, 1951, that no doubt ship-building in the Great Lakes region was a defence argument used by the seaway proponents to sway the defence-minded Americans into action. The Joint Board, however, appreciated these arguments. In 1947 and again in 1948 it recommended the construction of the seaway for the navigational and power benefits it would bring. In 1951 the board pressed for immediate action and the Canadian government, not unnaturally, endorsed its stand. "It is the view of the Canadian Government," Brooke Claxton announced, ". . . that the Board is right and that the St. Lawrence development should be proceeded with. . . . The development . . . is required . . . to meet the urgent needs for power, transportation and iron."[6] Dean Acheson, Secretary of State, read the resolution of the PJBD into the record of Congress early in 1951:

In the preamble to this Resolution the Board made reference to the deterioration of the international situation indicating a period of protracted crisis, requiring a steady increase in our military strength. It pointed out that the St. Lawrence Seaway and Power Project would yield additional supplies of hydro-electric power already needed in the N.E. United States and Canada and which would be vital to an expansion of military strength. . . .

The board recommended "That the two Governments take immediate action to implement the 1941 St. Lawrence Agreement as a vital measure for their common defence."[7]

St. Laurent knew that sooner or later a seaway reference would reach the commission and the Prime Minister wanted McNaughton there when it did. Already in 1951 the General could report that the IJC also stood on record as having recommended both a seaway*[8] and power and thus, he said, "There is now on the public record the advice of the two bodies which have been created by Canada and the United States jointly to consider matters of mutual interest and concern . . . both for peace and for defence. The International Joint Commission and the . . . Permanent Joint Board on Defence . . . have both recommended–repeatedly–that the St. Lawrence project . . . should be built, and promptly."[9] In his dual role, McNaughton was strategically placed to tug the essential strings, and an early and successful outcome was vital in the interests of Canada. When he discussed his proposed appointment to the IJC with St. Laurent in 1949, the power applications made by Ontario and New York came up. The idea of splitting the project into two parts, power and navigation, had been born early in 1941 when Mackenzie King was under pressure from Roosevelt, who feared a shortage of power before the war was done, to join in the development of the seaway. King anticipated a public outcry in Canada if men, machines and money were to be diverted from the making of munitions to "dig a ditch," and there was some discussion about whether it would not be better simply to develop the international section for power and leave the seaway until the war was over. The mechanics of this were not explored, however, as both governments signed the agreement of 19 March for the complete development, Roosevelt having no objections to an announcement that the agreement had been entered into by Canada at the instigation of the United States.[10]

With the failure of the 1941 agreement, McNaughton considered it essential to split the power from the navigational aspect and he based his reasoning on this: for the development of power in boundary waters, because of a precedent set by the United States in 1914, *presidential* instead of *congressional* approval was enough. Thus the support of the President alone, after IJC approval, would ensure the success of the New York State Power Authority's application to develop power in the International Rapids section. That would

---

* Though "the Commission's jurisdiction does not extend to the works for navigation in the St. Lawrence Seaway," said McNaughton, "the Commission was invited to facilitate this undertaking."

effectively circumvent the opposition groups in the United States which had thwarted the wishes of a succession of presidents in the past. Then, side by side with the development of power, Canada might construct a seaway on its own side of the river for which the sanction of the United States government would not be required. "International considerations are involved," the General said in Toronto in 1951, "and so in respect to navigation and in respect also to power there are projects of immense importance which need to be freed from the existing international inhibitions so that they can be undertaken at the earliest practicable date and carried out for the benefit and welfare of the peoples both of Canada and of the United States."[11] On a later occasion McNaughton said: "Since it seemed that the nations could not solve the problem as a whole we had better separate it into two parts–power and navigation."[12] St. Laurent requested an Interdepartmental Committee (which had been established in 1947) to advise on the procedures to be observed if New York State and Ontario were to develop power in the International Rapids section; the navigation project would be considered separately. This separation of the projects was the first step towards an all-Canadian seaway plan, which in tum would lead to the solution of the whole problem.

An all-Canadian seaway would cost an estimated $217,000,000; though a costly proposition, with tolls it would be self-liquidating. Without the power development, however (entailing an expenditure of some $400,000,000 by Ontario Hydro and its American counterpart), the cost of the seaway to the federal government would be many millions more and it was concluded that it would then be unattainable. Power and navigation together were feasible. As McNaughton said, "Since . . . 1941 the Canadian economy has strengthened in a most remarkable manner and we have come to appreciate that we do not need to be dependent on others for financial assistance in carrying out the St. Lawrence projects for power and for navigation. On the contrary, we can well proceed with them on the basis of ordinary useful commercial ventures the full costs of which, including amortization, can confidently be expected to be returned . . . together with considerable profit, both direct and indirect."[13] The Canadian government considered it possible to achieve the power development if there was willingness on Canada's part to carry out the navigational works, and in the summer of 1950 St. Laurent suggested to Lionel Chevrier (Minister of Transport) that Canada should build the seaway alone. Chevrier launched a publicity campaign for the all-Canadian plan. On 21 November, in a CBC broadcast, Chevrier stated that once the IJC had approved the power development, the government would be free to begin the navigational works on the Canadian side of the international section of the river and in this he was supported by St. Laurent and C. D. Howe (the Minister of Trade and Commerce). The campaign aroused a good deal of enthusiasm in Canada but the Americans seemed to doubt Canada's ability to proceed alone.

The purpose of the campaign was twofold: to obtain the support of the Canadian public; and, more important perhaps, to convince the United States that Canadian patience was not in exhaustible. Joint development of the seaway under the terms of the 1941 agreement was still the course that Canada preferred but an independent move might well provoke the Americans into definite and timely action. The President's ability to act alone in approving the production of power, thus skirting the powerful opposition in the United States, was a major factor in the Canadian plan. Yet the Secretary of State for External Affairs, speaking in the House on 14 June, 1950, advised against an application for power, cautioning that this would not remove the obstacle of having to obtain federal authorization and approval in the United States to permit New York State to participate with Ontario in the power development.*

We know that President Truman favoured the construction of a seaway. We have also seen that he opposed the separation of the power benefits from those of improved navigation. He thus supported joint development under the 1941 agreement and, encouraged by the activity on the other side of the border, he pressed Congress to participate with Canada no fewer than three times between January and July, 1951. "Each President in turn," said McNaughton, "from President Harding through Presidents Coolidge, Hoover and Roosevelt to President Truman, had endorsed the project and sought the approval of Congress in one form or another for its execution."[14] But the traditional opposition in the United States had not diminished; the cost of the Korean conflict, moreover, was produced as evidence that seaway expenditure at the time was undesirable. Truman's appeals brought no response. The best hope for Canada

* Only the approval of the Federal Commission would be required.

now, it seemed, lay in persuading the President to support the power aspect only, with an undertaking by Canada to construct the seaway, which Truman wanted; after his recent experience with Congress he might be ready for this. As George Bain wrote in the *Globe and Mail* on 28 July, 1951: "It is the belief now that, if Canada was to announce its intention of going ahead alone on the waterway, approval of an international power development would follow without too much difficulty."

Late in September, with excellent timing, St. Laurent visited Washington to enlist Truman's support. He obtained it. While the President still favoured a joint enterprise he would support the Canadian plan if early prospects of joint action seemed impossible. The Prime Minister, significantly enough, agreed with Truman that the best plan would be for both countries to construct the seaway within the terms of the 1941 agreement; failing that, Canada, with Truman's support, would proceed alone.

No sign of any changed attitude could be discerned in the United States; for Canada, governed by the need for additional power in 1956, time was running out. The Speech from the Throne in October, 1951, therefore, announced Canada's intention of going ahead with the all-Canadian plan. This drew bitter comment from the United States. It was a "gambler's chance," the New York *Times* stated on 7 October, that Canada's seaway would meet the "economic and military requirements of the North American continent." In December the Canadian government introduced two bills, one for providing a deep waterway between Lake Erie and Montreal and the second for the approval of an agreement between the federal government and the Province of Ontario with regard to power development in the International Rapids section of the St. Lawrence River. Studies indicated that it would be possible to liquidate the seaway costs, plus interest charges, within fifty years. On 12 December the bills passed unanimously and on 11 January, 1952, the United States government received notice of Canada's decision.

Reaction in the United States amounted to consternation and disbelief. The New York *Herald Tribune* termed Canada's determination to construct the seaway "pure poppy-cock"; it was "physically impossible" for Canada to build the seaway alone.[15] *Saturday Evening Post* commented equally contemptuously: "Any time a country wants to spend its own money, with no contribution from the United States—that we must see, if strictly from amazement."[16] But Truman pointed to the realities of the situation. In a message to Congress at the end of January, 1952, he said:

> The question before the Congress . . . no longer is whether the St. Lawrence Seaway should be built. The question before the Congress now is whether the United States shall participate in its construction, and thus maintain joint operation and control over the development which is so important to our security and our economic progress.[17]

In the interests of Ontario power, April of 1952 was regarded in Canada as the latest date to which the government could postpone the decision to apply to the International Joint Commission for approval of the power project. Consequently the government, while not spurning American participation, would back whichever alternative—the joint or all-Canadian—would produce the earlier results. Nothing tangible had transpired by 1 April, and in the middle of the month Lester B. Pearson visited Washington where he achieved an agreement with Truman for the first indispensable step towards power development. Each country would submit to the IJC an application to change the water level of the St. Lawrence River. This, it was stated—although it would be the first international move towards an all-Canadian seaway—would not prejudice any action the United States might take towards joint development. It was now clear that Truman would sanction the power project after the IJC had approved the plans.

American press commentary then changed about. "The St. Lawrence," a *Herald Tribune* editorial of 17 April announced, "is a common border and by every concept of statesmanship, neighbourliness and international relationship it ought to be developed as a common responsibility." And following an interview with McNaughton by an editorial writer for The New York *Times* in November, the newspaper, which had consistently opposed the seaway, recanted. In an editorial on 8 December this influential paper flung its full support behind the seaway: "The opposition from railways, shipping lines and ports no longer makes sense. . . . The plan will be as good for us as it will be for the Canadians and the time has come to go ahead full steam."[18] But, as yet, Congress remained unmoved. Lobbying against the seaway was extremely strong, the most vicious opposition coming, as McNaughton put it, "from owners of jerkwater railways who have no conception of public good."[19] An exchange of notes and the submission by the governments of Canada and the United States to the International Joint Commission on 30 June, 1952, of an application for approval and operation of certain works for development of power in the international section of the St. Lawrence River pointed once again, however, to the realities of the situation. In addition, as Bloomfield and Fitzgerald point out, the Canadian government undertook "to construct and operate all the works necessary to ensure uninterrupted 27-foot navigation between Lake Erie and Montreal and agreed that such works would be completed as concurrently as possible with the completion of power works."[20]

At the same time as it received the Applications for the approval of works for power and to facilitate navigation in the International Section of the St. Lawrence [said McNaughton] the Commission received a Reference from the Governments in regard to the levels of Lake Ontario requesting recommendations as to what might be done to bring the levels on that lake under effective control. . . . In this connection, it is of interest that the structures to be erected in the St. Lawrence for power and to facilitate navigation include the great regulating works which are to be built at Iroquois to meet the needs of power. . . . These works, with their associated channel enlargements will . . . for the first time in the history of Lake Ontario and the St. Lawrence, permit the Commission to exercise physical control over the levels both upstream into the Niagara River and downstream in the great [Canadian] power plants existing at Beauharnois and to be constructed to make beneficial use of Lachine; also to protect the vitally important interests in Montreal Harbour and the lower river.[21]

This will give some idea of the significance of the works at Iroquois for power, navigation and flood control; because of their importance McNaughton in the following years was to fight to retain them under Canadian control.

All that was required now for the realization of the all-Canadian seaway was the approval of the IJC and the nomination by Truman of a United States power company to work with the Ontario Hydro-Electric Power Commission in the International Rapids section. Canada's dreams and hopes for twenty years seemed about to be fulfilled. Nor, with an all-Canadian seaway, was there "the least wish to deny to the trade and commerce of the U.S. any facility which we may open to them without hurt to ourselves."[22]

In October, 1952, the IJC approved the applications. This was remarkably fast as, to obtain this approval, it had been necessary to hold public hearings at various interested places. The St. Lawrence Order was not signed by all the commissioners; one United States commissioner presented a minority opinion. Still, remarked McNaughton, "agreement by five out of six . . . is not an unsatisfactory basis for proceeding."[23] Meanwhile the Canadian Parliament completed the legislation required to authorize the Ontario Hydro-Electric Power Commission to construct the power works. It remained now for Truman to designate some United States power authority to work with Ontario, and one of his last acts as President was to urge the Federal Power Commission to expedite the application made by the New York State Power Authority. Thus the IJC approval, the New York-Ontario power project and the all-Canadian seaway promised the speediest means to achieve Canada's aims, and on 4 November, 1952, the Canadian government informed the government of the United States that Canada did not intend to take further action to ratify the agreement of 1941. The announcement coincided with Dwight D. Eisenhower's presidential victory. Shortly afterwards rumours of an eighteen-month delay in designating a power company in the United States, through litigation over granting a power licence to New York State,* reached the Canadian government.

Accordingly, in January, 1953, Canada addressed a strongly worded communication to the United States government through the American ambassador. This read in part:

> While the Canadian Government is, of course, prepared to discuss, in appropriate circumstances, joint participation in the Seaway, the demand for power in the area to be served by the international rapids power development is so urgent that the Canadian Government is most reluctant to engage in any discussion which might delay the progress of the plan now under way for the development of power in the international rapids section of the St. Lawrence River at the

* The groups opposing the seaway in the United States were determined to appeal the decision of the Federal Power Commission if the New York State Power Authority was granted a licence to develop the American share of St. Lawrence power.

earliest possible moment.

Once an entity is designated and authorized to proceed with construction of the United States share of the power works, if the United States wishes to put forward a specific proposal differing from that put forward by the Canadian Government for the construction of the seaway in the international section which proposal would not delay the development of power under arrangements agreed upon in the exchange of notes of June 30, 1952, and approved on October 29, 1952, by the International Joint Commission, the Canadian Government will be prepared to discuss such a proposal.

The Canadian Government would naturally expect the discussion to be such as not to cause any serious delay in the completion of the whole seaway.[24]

Early in 1954 McNaughton said publicly, "We do not question the propriety of this course [the challenge to the power 'entity' in the United States] which is an entirely domestic matter but we have requested, and received, assurance that these proceedings will be pressed to a conclusion with all dispatch. Already the Federal Power Commission order has been cleared by the . . . Court of Appeals and the matter now goes to the U.S. Supreme Court . . . . In consequence, with every confidence in the ultimate outcome, the Canadian entity is in fact proceeding with preparations to launch the power works. . . ."[25] As usual, McNaughton was optimistic and on this occasion his optimism was fully justified. "The forces in the United States that were opposed to the seaway," he said later, "could only delay the power project temporarily.* They knew that we meant business and that the seaway was going through whether they liked it or not. And our note of January 1953 called the shots. Our friends across the border could come in with us if they wanted to, but it would be on our terms. We had by-passed the obstructionists who had stood in our way down the years and *we* had taken over the diplomatic initiative. It wasn't long before the Eastern railroads, the Atlantic interests and the other people who were afraid of losing business down there knew that they were beaten; they crossed over to the winning side."[26] Meanwhile the urgency of obtaining power from the St. Lawrence had become

* In November, 1953, Eisenhower designated the Power Authority of the State of New York. On the same day the United States Federal Power Commission issued a fifty-year licence. The issue of the licence was appealed but it was re-affirmed by the United States Court of Appeals for the District of Columbia. A final appeal to the United States Court was refused in June, 1954.

slightly less pressing. The division of waters at Niagara, including arrangements for the preservation and enhancement of the scenic beauty of the falls,* had been settled by the IJC in 1950, and by the middle of 1954 water was already flowing "in the new canals, penstocks, and through the [Canadian] turbines."[27]

A year later the United States Senate passed the Wiley Bill to permit United States participation with Canada in the navigational works required for a deep seaway. In May, 1954, the House of Representatives passed the Dondero Bill to the same purpose. The Wiley-Dondero Act, signed by Eisenhower on 13 May, set up the St. Lawrence Development Corporation to construct the navigational facilities on the United States side of the international section of the river. Two boards, the St. Lawrence River Joint Board of Engineers and the International St. Lawrence River Board of Control (whose establishment had been provided for by the IJC), would consist of an equal number of representatives of Canada and the United States to give effect to the instructions of the International Joint Commission.

This late decision by the United States government to participate in the seaway received a cool reception in Canada. The Americans, moreover, having at last made up their minds to support a joint seaway, left no doubt under the terms of the Wiley-Dondero Act that they intended to call the tune by having all the navigational works in the international section sited within United States territory. A joint plan, of course, held out monetary advantages which Canada could not ignore and Lester B. Pearson, having seen the American mood change from opposition to the seaway to joint participation, advised McNaughton at the IJC to accept the situation. He obviously considered there had been victory enough and it would be well not to goad the Americans too far.

McNaughton, however, was unwilling to accept a rigid interpretation of the Wiley-Dondero Act, especially as the engineering studies had

---

* The rate of erosion was slowed down by the construction of a power dam, 1,550 feet long, extending like a breakwater across the current from the Canadian shore above the falls. By means of gates to control the flow, the erosive action above the Canadian falls was reduced and the flow over the American falls was increased. In addition, the edges of the Horseshoe Falls were trimmed so that the water would flow more smoothly. McNaughton, who spoke at the dedication of the remedial works in 1957, was glad of the power; but he was equally pleased with the beautification of the place. He was interested in the Indian name *Niagara*, which means "Tobacco Smoke"—a quiet name for this wild torrent but not inappropriate when one views the cloud of mist that always rises over the now smooth curtain of falling water.

convinced him that the Iroquois works could be constructed more cheaply and efficiently on the Canadian side. And he had another reason for his stand, for he was concerned about an all-Canadian seaway at some future date.

> The Wiley Act makes no provision for this and the United States authorities have made it evident they wish these facilities disposed with and not to be reinstituted so as to leave no doubt that all shipping must pass the United States locks under United States control and pay tolls to assure the amortization of the project as required by the Wiley Act. I would think it very unlikely that Canada would agree.[28]

The issue lay deadlocked, despite American pressure and the nervous attitude of Pearson, until 6 July, 1954. On that day Robert B. Anderson, the United States Deputy Secretary of Defense, who knew where the opposition lay, called on McNaughton at his office in the Justice Building.

A United States group, led by Anderson, had come to Ottawa for discussions on 5 and 6 July with a Canadian group, led by Lester B. Pearson and including C. D. Howe, G. C. Marler (Minister of Transport) and Chevrier (who was now president of the St. Lawrence Seaway Authority).*[29] Because, as McNaughton put it, "of my position in the International Joint Commission, while I was closely familiar with the Canadian position, it had not been thought expedient that I should take a part in the discussions."[30] On the evening of 5 July the United States ambassador, R. Douglas Stuart, gave a dinner at the Country Club for the officials of both countries who were in Ottawa to discuss the problems of the St. Lawrence arising out of the Wiley Act. He also saw fit to invite McNaughton who, at the dinner, "gathered that the conversations during the day had been difficult" and that the Americans had "taken a rigid view as to what was required to conform with the Wiley Act." Later that night the ambassador drove McNaughton home and took the opportunity of their being alone to ask for his personal, frank opinion. McNaughton, he himself recorded,

* The appointment had been offered to McNaughton but he saw it as a red herring, contrived by someone in the Department of External Affairs who wished to remove him from direct negotiations with the Americans within the IJC and PJBD, in favour of diplomatic negotiations. He refused it.

> . . . emphasized the difficulties which would result from a narrow construction of the Wiley Act and . . . drew attention to the very wide public opinion in Canada favouring construction by Canada and which opinion had been offended by certain of the statements made in Congress and by certain of the reasons advanced for United States construction of the works in the International Section. . . . I expressed a pessimistic view as to the effect of the discussions on United States-Canadian relations.[31]

Next morning Anderson made an appointment through Stuart to see McNaughton; he arrived at the Justice Building in mid-afternoon. They talked cordially and generally for some time of the work of the International Joint Commission and Permanent Joint Board on Defence, especially the PJBD where, McNaughton said, "In 1947 and 1948 Mayor LaGuardia, then chairman of the U.S. section, and I had been able to come to agreement on general principles which had eventually been adopted by the two Governments." There had been concessions on both sides, he added, but "both countries were so closely involved that there was no question but that they *must* march together in the closest association." Having set the mood, McNaughton conducted his guest to the window "from which there is a view westward up the Ottawa River including the Chaudière Falls, the pool above, the mills and the power plants" and, after a friendly and discursive chat on power the talk naturally turned to the St. Lawrence. Anderson referred "in very generous terms" to the contributions McNaughton had made in the past to "the understanding and co-operation of our countries"; he was disturbed now over the current talks on the St. Lawrence and he again asked McNaughton to give him his views with "complete frankness as to what was wrong and what should be done to correct matters." This McNaughton agreed to do, warning him "that what I might have to say must be forthright if it was to be useful, and might be painful." Anderson encouraged him to proceed.[32] McNaughton mentioned the long history of Canadian endeavours to improve navigation in the St. Lawrence and into the Great Lakes. He referred to the frustrations that Canada had experienced regarding the improvement of the international section from the 1930s onward; the draft treaty of 1932; the draft agreement of 1941; the new approach through the IJC in 1952; and now the Wiley Act. Canada, he said, had studied the Wiley Act with close attention; it presented serious difficulties and appeared very rigid. But what had disturbed Canadian public opinion, "perhaps even to the point of detonation," was much more than the terms of the Wiley Act; it was what was being advanced in Congress. Canada had taken pride in giving in full measure the navigational privileges accorded to the United States by the Treaty of Washington of 1871 and the Boundary Waters Treaty of 1909. These had been administered with complete impartiality and Canada had been glad to have played the part of a good neighbour.

> Now we found that many people in Congress were demanding that the United States should build the International Section for the avowed reason that these links would be under their complete control. . . . These statements . . . showed an intention to have United States views imposed on Canada. . . . I emphasized that in my view there was not the least likelihood that Canada would accept the rigid prescriptions of the Wiley Act nor give the kind of assurances which it seemed the Congress had implied would be demanded. . . .
>
> A flexible approach without demands for commitments was needed and this might well bring a co-operative response where, as I repeated, the alternative might harden Canadian opinion and compromise the relations of the two countries most unfortunately.[33]

In fact, if joint participation meant that Canada was to be restricted in future seaway construction, the Canadian government might be well-advised to proceed alone. Agreement could still be reached between the two countries but this needed to be "entirely flexible and free from any suggestion of constituting a strait jacket"; it certainly should not be aimed at preventing Canada from building an all-Canadian waterway "at any time and at discretion"—in other words, at a future date when the need for duplication became apparent.[34]

"I have the conviction," McNaughton's notes (which he passed along to the Canadian negotiating team) conclude, "borne out by some observations made to me later in the afternoon by Ambassador Stuart (to whom it appears Secretary Anderson telephoned after leaving my office and before departing by air), that the views which I had expressed had made a deep impression and that they well might find some expression in the conduct of the further negotiations."[35]

Indeed they did. When the talks resumed in Ottawa on 12 and 13 August, the American attitude had completely changed. Anderson, who understood the Canadian position and sympathized with it, had evidently persuaded his government to do the same. Pearson, with the way cleared by the Anderson-McNaughton conversation, soon reached agreement and exchanged notes with the United States government on 17 August, 1954. These notes modified the former arrangements set out in the notes of 30 June, 1952, which referred to the all-Canadian seaway. Canada was relieved of "the obligation towards the United States Government to provide . . . the navigation works in the general vicinity of Barnhart Island on Canadian territory" and in the Thousand Islands section (which amounted to no more than dredging). But Canada insisted on building a canal and lock at Iroquois,* on the Canadian side, despite the Wiley legislation: "The Canadian Government wishes to state, however, that it will construct forthwith a canal and lock at Iroquois. . . ." And more than that, Canada would not be prevented from building an all-Canadian seaway at a future date when the need for parallel works became evident. The Canadian government, the note continued, ". . . intends, if and when it considers that parallel facilities are required to accommodate existing or potential traffic, to complete 27-foot navigation works on the Canadian side of the International Rapids Section."[36]

A duplication of the Barnhart works on Canadian territory was all that this would amount to now; with that, there would be through navigation on the Canadian side from Montreal to Lake Erie. There would be mutual consultation, the note concluded, before adding navigational facilities at either side of the international section of the river. Both countries were protected from discriminatory regulations affecting shipping interests or trade in the St. Lawrence River. With some of the works on one side, and some on the other, there was little doubt that this would be observed.

The United States, in its note of 17 August, accepted these stipulations. Thus Canada reaped the immediate monetary advantages of joint development while still retaining the right to build a future all-Canadian seaway. Though there were many in Canada who had strongly advocated an all-Canadian seaway from the outset, and deplored American participation, McNaughton did not share their views. In a letter to Colonel L. F. Grant, who had pressed strongly

* The importance of Iroquois to a future all-Canadian seaway will be appreciated by the reader. The work, undertaken when the ground was dry, would cost Canada a fraction of what would be entailed at some future date when the seaway was in operation with the water high.

through the Engineering Institute of Canada for unilateral action, he wrote:

> . . . we have come out of the discussions . . . without having conceded a single point which is vital and not possible of correction at our own instance at appropriate times of our own devising.
>
> On tolls, freedom of passage of Canadian ships and crews, maintenance of 14-foot system, completing a 27-foot system in Canadian territory, we remain complete masters of the situation.
>
> The Ministers . . . refused any commitment not to build in Canada at any time we may wish. They took note also of the Wiley Act direction to the United States to build a lock at Iroquois, informed the U.S. very abruptly that Canada had decided to build a lock there, and suggested that accordingly they might refrain. . . .
>
> Far from feeling disappointment at the results of the efforts which you and others have made . . . I feel you should be well pleased because without these efforts it would not have been possible for us to emphasize to our authorities, as we did, that public opinion in Canada was about to detonate.[37]

McNaughton had been instrumental in protecting Canada's position. When the expanding economy merits the building of an all-Canadian seaway, his foresight will be remembered.

The Vancouver *Province*, on 23 May, 1963, published an article by Paddy Sherman, which read as follows:

> . . . Anderson was made head of a team of U.S. negotiators. A man named Lester B. Pearson, then Canada's external affairs minister, headed Canada's team.
>
> They met on Aug. 12, 1953, and talked. Five days later the two governments exchanged notes that settled the whole affair after a half-century of wasted years.

In fairness to Mr. Sherman, that is how it may have looked.

The actual construction of the seaway and its official opening by Her Majesty The Queen and President Eisenhower followed in 1959.*[38] As an indication of what traffic the St. Lawrence Seaway might expect there was, even before its construction, a vast tonnage of commodities passing through Fort William, Port Arthur, Duluth and

* "Closures of the dam structure were made on 1 July [1958] and with the raising of the pool generations of power commenced."

other ports: iron ore, the wood products of adjacent regions and grain from the prairies measured in millions of tons. "The traffic which each year passes in and out of Lake Superior through the five locks of the St. Mary's River represents an amount which exceeds the total annual traffic of the Panama, the Suez, the Manchester and the Kiel Canals all put together."[39]

With the bottleneck caused by the International Rapids section removed, the commerce and trade of the basin will of course increase. In 1967, for example, it was reported that "this mighty artery of world trade" which "allows ocean-going ships to enter 2,300 miles into the heart of the world's most highly-industrialized region . . . has grown ever more vigorous." Since 1959, "tonnages carried along the length of the Seaway have soared well over double in volume. . . . For Canada, this surpassing of even the most optimistic expectations of Seaway growth and capacity is invaluable. Enormous shipments of grain downbound to fill unprecedented overseas orders, thriving inland commerce in iron ore and coal upbound, plus the many other facets of domestic and international trade . . . have strengthened the national economy to a tremendous degree."[40] And as the expenditures, whether for power or for navigation, are self-liquidating, the benefits afterwards "will continue for the peoples of the two countries in perpetuity and without cost other than to meet the small charges for maintenance and operation."[41]

To this great project, an American member of the International Joint Commission reported, "General McNaughton contributed more than any single individual." In the histories of the St. Lawrence Seaway, he went on, "it will be unnecessary to mention General McNaughton's name; it will automatically be linked with [it] because of the contributions that he has made."[42]

While the St. Lawrence negotiations were continuing the International Joint Commission had had before it, since March, 1944, references by the governments of Canada and the United States calling for investigations and recommendations concerning the further uses and development of the waters of the Columbia River system. The commission was to see if a greater use than was then being made of these waters would be "feasible and advantageous" and whether, in its judgment, "further development of the water resources of the river basin would be practicable and in the public interest from the points of view of the two Governments." Where projects were considered feasible and desirable the commission was required to indicate "how the interests on either side of the boundary would be benefited or adversely affected thereby and should estimate the costs of such works or projects . . . and indicate how the costs may be apportioned between the two Governments. . . ."[43]

To assist the investigation, the commission was empowered to use engineers and other qualified persons in Canada and the United States, and it was specially instructed to make use of information and technical data already available (or which might become available), "thus avoiding duplication of effort and unnecessary expense."[44] The IJC set up the International Columbia River Engineering Board which was occupied with its enormous task at the time McNaughton joined the commission.

The Columbia River—one of the greatest rivers in the world as a source of hydroelectric power—rises in Columbia Lake, British Columbia. It is 1,210 miles long from source to mouth, with the first 465 miles in Canada. From the Rocky Mountains, where it rises, the river falls almost three thousand feet to the sea; it crosses the border between British Columbia and the State of Washington at an elevation of 1,292 feet. Only fifteen per cent of the drainage area (260,000 square miles) is in Canada but that percentage, while small, accounts for one-third of the total water that flows in the river. Though there are two larger rivers than the Columbia in North America—the Mississippi and the St. Lawrence—the Columbia "is the greatest for power which depends on falling water. The St. Lawrence, for example, starts from Lake Superior at an elevation of 603 feet; Columbia Lake, the source of the Columbia, is at 2,652 feet."[45] The St. Lawrence, with a power capacity of seven million kilowatts,* was far outmatched by the Columbia with a potential almost five times greater (34,000,000).

The river is a "snowmelt" river and so is subject to great seasonal fluctuations in its natural state. In the spring, when the river is filled by snowmelt, the flow is sometimes forty times greater than in the winter months when frost grips the mountain sources. And when the sun shines on the glaciers in conjunction with early rains and soft spring

* One kilowatt is the power that will light ten 100-watt bulbs (the ordinary strength of lights around the house). Thus one kilowatt-hour of electrical energy will be consumed by ten 100-watt bulbs burning for one hour.

conditions, as happened in 1894 and 1948, the Columbia roars down in flood causing millions of dollars of damage, especially in its lower reaches in the United States. It was important, therefore, to harness the river. Dams were required to hold the spring flows back, not only to prevent the damage that would otherwise be caused but also to hold the run-off for the generation of hydroelectric power in the season of low-flows, instead of allowing them to pass unused to the sea. To keep the turbines humming in an economical manner there must be continuity of flow. And to produce electricity above the capacity of the minimum flow of the river the water would have to be stored for release during the dry season, to give increased flows at a steady rate. Thus existing power plants could produce more electricity at no extra cost.

At the time the Columbia studies were going on almost all the existing power plants were in the United States. In fact the hydroelectric potential of the Columbia system had been well exploited by the Americans and further development was planned; because of this there had been great industrial growth. This had taken place, McNaughton stated, "without simultaneously developing adequate storage," and now it was too late in the Pacific Northwest because, with the industrial and residential growth, "it has become difficult to persuade people to let their land be put under water for storage." As matters stood, McNaughton continued, "storage, and flow from storage . . . is the limiting factor on United States energy production on the Columbia, and that is a very serious position to be in."[46]

It was serious in view of the increasing industrialization of the Pacific Northwest which demanded much more power. The value of plants already producing power was estimated at $1,500,000,000; new construction, in excess of $3,000,000,000, had been authorized so that the amount of available power could be doubled within ten years. These plans would depend on storages which, according to McNaughton, the Americans could have built had they not been looking to Canada for them. It was too late for them to be provided in the United States "with perhaps the sole exception of Libby,* on the

* Libby was the major site, but others were possible on the Snake River.

Kootenai,"* which we shall examine. Thus the American interest in the development of the Columbia Basin lay in the provision of storages in Canada that would bring enormous benefits of flood control and increased power development to the United States. For that reason reference was made to the International Joint Commission in 1944. "By 1944," McNaughton said, "the United States had a very extensive knowledge of the water resources of their part of the basin. They had evolved at least the elements of a long-term plan for the systematic bringing into use of the resources which they thought would be available, including the flows which might come to them from Canada and the inundation of some of our valleys to make reservoirs to their advantage." But, he went on, "It is a Canadian responsibility—not theirs—that we should assess the value of these resources and require due recompense if, indeed, we are to consent to their use to someone else's benefit at all."[47] In this statement by McNaughton we can discern the grounds for the conflict that he was to wage concerning the development of the Columbia system for more than a dozen years.

"In Canada, in 1944," he continued, "our situation was far different. We did not even have accurate topographic maps of the region on which to plan; we had only rudimentary knowledge of run-off levels and flows. We knew little about potential dam sites for power and storage and less about the nature of the rock and its suitability for tunnels and foundations. We could not say with accuracy what area would be flooded by a dam to a given crest elevation or indeed how much water would be impounded behind it."[48] Canada, unlike the United States, had not constructed even one dam to develop power from the main stem of the Columbia. The reason for that lay in the economic growth of the Pacific Northwest which was already well-advanced. British Columbia—in its infancy, but a potential industrial giant—lagged far behind the United States. Industrial and economic development in British Columbia, however, would depend very largely on the future availability of cheap hydroelectric power. McNaughton was anxious to prevent what he described as "the dangerous condition", which had been steadily proceeding, "of the waters of rivers flowing from Canada being taken into use south of the border . . . without our having been aware of what has been going on until we wake up and find that a claim

* The Kootenay is known as the Kootenai after it crosses the border into the United States.

against our resources has been built up which may inhibit their use in our own country." The Americans had in the past, he said, claimed "vested rights" in Canadian waters because of their prior use of them.[49]

In the Columbia Basin the international boundary cuts across the flows of all the principal rivers and many of their tributaries. Not only does the Columbia itself rise in Canada and flow into the United States; two of its tributaries, which are important to our story, cross and re-cross the boundary. The Pend Oreille rises in the United States and flows into Canada, where it joins the Columbia just north of the border. The Kootenay rises in Canada, flows in a big loop through Montana and Idaho and then returns to Canada where it joins the Columbia a few miles north of the border. This "intricate pattern of boundary crossing and re-crossing," as McNaughton put it, gave rise to many problems under the Boundary Waters Treaty.[50]

An interesting feature of the Kootenay is this: at a place called Canal Flats in British Columbia the river, as it flows south from the Rocky Mountains, passes within a mile of the Columbia as it flows north at that point. Where the Kootenay comes close to the Columbia there is a depression in the ground, a kind of rough ditch, and at times, quite naturally, some of the waters of the Columbia may flow into the Kootenay and at others part of the Kootenay may flow into the Columbia. Both rivers are at exactly the same elevation, though flowing in opposite directions, and either could be dammed to cause it to flow into the other. Now this was known by McNaughton. When commanding Military District Number 4 in 1929, he had travelled into the Upper Columbia and Kootenay country where he had seen:

> the peculiar topographical condition which is presented when these two rivers . . . flowing in opposite directions in the same mountain trench, come close together at Canal Flats.
>
> I was very especially interested in this situation because of the possibility it presented of enhancing the hydro-electric power potentialities of the Upper Columbia to a most remarkable extent. . . . I confirmed that Canal Flats was a . . . key point in the planning of the hydro-electrical development of the Columbia basin. I think this was the most important information which came to me in the course of my sojourn in B.C. and certainly some twenty-five years later when the matter came before the I.J.C. this first-hand knowledge of the possibilities proved very valuable to the Canadian Section of the Commission.[51]

This knowledge was indeed useful when, on 15 January, 1951, the Americans submitted an application to the IJC to dam the water of the Kootenai at Libby, Montana, just south of the international boundary.

Libby, as has been noted, was the one place left for the Americans to store to any large extent the waters of the Columbia or its tributaries. The application was made on the basis of an interim report by the U.S. Army engineers, but it seemed odd to McNaughton that a request for permission to proceed with an isolated project should have been made at all while the whole of the Columbia system was under study. The dam, though to be built on United States territory, would raise the level of the water a hundred and fifty feet at the international boundary and cause it to flood back forty-two miles into Canadian territory in the form of a storage reservoir. This project could not be worse from the Canadian point of view. Storages, to be of mutual benefit, should be sited on the Columbia as far north as possible. Canada would then have the benefit of controlled flows for the generation of electric power over a great length of the river within her own territory as far as the boundary; the Americans would reap the benefit from there on, both for power and for flood control, as far as the river's mouth. The capacity of the existing American power plants would be tremendously increased without even the installation of an additional machine. The place for a dam, McNaughton thought, was at Mica Creek (on the "big bend" of the Columbia near its most northern point) and at an extremely high elevation; he was already referring to this as the "upper reservoir" in October, 1951.[52] Libby Dam, on the other hand, would hold back the waters of the Kootenay, not the Columbia, on the downstream side of the border. This could be of benefit only to the Americans, except where the river flowed back into Canada (the West Kootenay) after passing through Montana and Idaho. On that stretch of the river there were some Canadian power plants but the advantages to them of controlled flows from a dam at Libby would not be large. Not only that, the reservoir created by the dam, extending more than forty miles across Canadian soil, would force many persons to leave their homes, and highways and railroads between established communities would be disrupted. Though the Americans offered seven million dollars compensation, that was all.[53] It seemed that the United States hoped that American interests downstream would benefit from upstream Canadian storage without payment and, after all, why not? The Americans had already seen a similar application approved by Canada without benefit when, in 1941, permission had been given the United States to build Grand Coulee Dam on the Columbia River in Washington State; on that occasion the reservoir had extended up river across the international boundary and into Canada.

McNaughton remembered Canal Flats. There was, he thought, a better use for the Kootenay than allowing its waters to be stored behind Libby Dam. Suppose it were to be diverted into the Columbia? Then the Kootenay, added to the upper waters of the Columbia, would enable Mica to become a mammoth storage capable of controlling the Columbia floods and permitting the generation of enormous blocks of power in both Canada and the United States. And such a project would control the Kootenai flooding as well because it would permit the passage of the floodwaters of the Upper Kootenay into the Columbia, which would be of great benefit to the United States. There could be no damage to interests on the Kootenai, for none existed south of the border; and a live stream would be maintained. But, with diversion, American plans for expanding the production of electricity into the Kootenai district of course could not be carried out. McNaughton directed his engineers to study the feasibility of the scheme and meanwhile he opposed the construction of Libby Dam on the grounds that insufficient compensation had been offered.

As he put it, "A difficult question is that of recompense for natural resources used. . . . Storage for water is far more precious than the most precious of mines for the reason that the flow of water will go on forever whereas minerals are a wasting asset used up in being put to use." Almost one million kilowatts of power, he estimated, would be produced at site at Libby and in American plants downstream by building the dam but none of that would be possible if Canada refused to permit the water to be stored. Canada, then, was entitled to a fair return of the benefits that would accrue downstream from using Canadian natural resources and from storing them on Canadian territory for the United States. He thus claimed what came to be known as "downstream benefits," and there would be no progress on Libby, or anywhere else, he warned, "until there is agreement as to how the benefits are to be distributed between the countries on either side. I feel that I must forecast," he went on, "that this problem will not only trouble the International Joint Commission, but that it will trouble many other people also until the situation clears and the lines of equity appear and are accepted."[54]

Senator A. O. Stanley headed the U.S. section of the IJC. He was described by McNaughton as a "distinguished representative of the legal profession who formerly served for two terms as Governor of the great state of Kentucky and later as a member of the United States Senate."[55] He was also a tough negotiator. Who, he asked, had ever heard of the principle of downstream benefits? Up to that point the Americans "had every excuse for thinking that Canada was asleep; and since we weren't doing anything about the Columbia they could dispose of the river as suited them best."[56] It was even reported that a senator—unnamed—took McNaughton aside and proceeded to address the sixty-five-year-old head of the Canadian section as follows: "Sonny, you'll either agree to the building of Libby or we will break you."[57] McNaughton refused to budge from the position he had adopted. The American members would not even consider the highly unorthodox principle he had raised and so, with the commission in violent disagreement, the matter rested. "The free ride was over and American indignation was boundless."[58]

McNaughton was no less indignant. What he had proposed he considered fair. Canadian natural resources, he repeated, would be used in the Libby project. The flooding above the boundary would contribute to the head at the downstream site, and a regulated flow, made possible by the storage upstream, would add greatly to every power plant downstream without the need for an additional expenditure at these plants. In 1953, as the Americans remained adamant, he reported on the Libby project to the standing Committee on External Affairs in these terms:

> [It] is one in which we are invited to present, that is, to make a gift to, our friends to the south of the line of the right in perpetuity to a large flow of Canadian origin capable of being used in Canada. By this action, if we should take it, we would divert a resource of very great value from ourselves to the service of industry in another nation. . . .

At that time Parliament was behind McNaughton. As he told the External Affairs Committee, he had been "heartened and encouraged by the support which the Committee expressed for the point of view which we had put forward of Canadian rights to downstream benefits and which views were later endorsed by the House of Commons."[59] The government had a water policy in those days of Louis St. Laurent, which is more than can be said today. Men like Jean Lesage (Minister of Northern Affairs and National Resources) and C. D. Howe could be relied upon to support a Canadian stand. As Howe said, as late as 1955,

> At the present time the development and utilization in Canada of waters . . . is a matter of ever increasing importance in the development of the Canadian economy. This is true of the Columbia River system and its watersheds. The use of such waters has to be viewed not only in its present implications, and in the light of our relation with the United States, *but also in the light of future water and power requirements of Canada and Canadians.** The demand for hydro-electric power in Canada is insatiable; in fact our industrial economy is largely built on low-cost electricity.[60]

The Libby application was stalemated and so it remained for the next two years. During that time McNaughton moved from the idea of the simple Kootenay diversion at Canal Flats to one of diversion on a much more massive scale. For the *maximum* use of the Kootenay waters by both countries there should be a dam at Dorr, just north of the international border, to impound the flows not only of the Kootenay but also of its major tributaries—the Bull River and the Elk. These waters would fill the pool at Dorr and back up against another dam to be built north of Dorr at Bull River. The Bull River Dam, equipped with pumps to raise the Dorr water two hundred feet† into the Bull River pool, would drown out the watershed at Canal Flats and be retained there by another dam north of Canal Flats, on the down slope of the Columbia, at Luxor. Thus a large reservoir, Bull River-Luxor, would be formed at the highest elevation possible in the Columbia Basin releasing its water to flow down through Luxor and a site for power at Calamity Curve, generating electricity on the way, to

* Italics supplied.

† Water in the Bull River-Luxor reservoir would be at an elevation of 2,700 feet above sea level of which a head of some 2,400 feet would be utilized. Thus the ratio of head for generation (2,400) to head to be pumped (200 feet) would be 12 to 1, which represents an exceptionally profitable and economic return.

the key storage–mammoth Mica Creek. The Mica storage would be as far to the north as it was possible to get, near the big bend of the Columbia where the river turns south to flow towards the United States. At the big bend the river is joined by the Canoe and Wood Rivers, both powerful streams. Twelve miles downstream Mica Dam was planned, a rock-fill structure, more than seven hundred feet high, from which the water would be released in regulated flows down the main stem of the Columbia to the boundary and on through the western United States.

The capacity of Mica was to be about fifteen million acre-feet (Libby would have no more than six million with five million usable) of which about twelve million were to be released annually to regulate the Columbia flows and enable about 1.3 million kilowatts of electricity to be generated at Mica itself and a similar amount in Canada at power sites to be installed downstream, notably at Downie Creek and Revelstoke Canyon. Enormous blocks of energy (McNaughton thought in terms of twenty billion kilowatt-hours) would be added at plants in the United States. Furthermore the flooding in the United States, on the Kootenai and the Columbia, could be controlled.

The results of the "Dorr diversion," as it came to be known, would be these. The floodwaters of the Upper Kootenay and its tributaries were to be diverted to flow, in the *reverse* direction, to Mica Creek.* Though the United States would lose these waters along the Kootenai, the normal flow would be allowed to continue. Plans for Libby Dam, however, must be discarded. This would be more than compensated for by the benefits to the United States, both in power and flood control, down the main stem of the Columbia to the sea. Canada, standing to gain little from Libby, benefited under the scheme by power production on the way to Mica, at Mica itself, and at other Canadian sites along the Columbia below Mica Creek.

These works in Canada would be costly. Though of great benefit to the downstream state it had become quite obvious at the Libby hearings that the Americans had no intention of paying a share of them. After all, why should they? The regulated flows from Canada, in the course of nature, must reach them anyway, whether they paid or

* The waters could be allowed to flow, when occasion warranted it, south down the Kootenay to the United States, thus giving an "option play" in flood routing which is probably unique.

not. With this American attitude in mind, McNaughton turned to a two-diversion scheme.

The Dorr diversion was merely the first step in this. He then proposed to pass the stored floodwaters from Dorr, the Bull River-Luxor pool and Mica reservoir, in the amount of fifteen million acre-feet, by tunnel to the Fraser, a Canadian river from source to mouth. This seemed to be quite feasible by way of Eagle River, which flows into Shuswap Lake; the South Thompson which joins the Fraser, flows out of Shuswap Lake. And the headwaters of the Eagle River are only about seven miles from the Columbia at Revelstoke. As with the Kootenay, it was planned to maintain the normal flow of the Columbia so as not to affect the existing American plants; but there could be no expansion. The surplus waters of the Kootenay and the Upper Columbia would, in fact, generate power in Canada, in the Fraser system, close to the large markets around Vancouver. "This," McNaughton said, "is a very desirable condition representing the further addition of a large amount of power which otherwise it is unlikely could be developed at all."[61] He would base his case for diversion on the engineering possibilities and the legal aspects of what was provided by the Boundary Waters Treaty. While the engineering studies proceeded he turned his attention to the treaty.

The Libby application, then, was deadlocked; but there were others before the commission, including, of course, the St. Lawrence. Two, which had a strong bearing on the Columbia negotiations, must be looked at briefly. One of them, which was referred to the commission by the governments of Canada and the United States in 1948, concerned the Waterton and Belly Rivers, which flow from the State of Montana into the Province of Alberta.

Canada, in the downstream position, proposed works that would permit the entire use of these rivers for irrigation in Alberta; as the rivers originated in Montana, the United States asked for a common reference "to investigate and make recommendations regarding water requirements, existing and further uses, apportionment, conservation, control and utilization" of their waters.[62] The United States had never used the waters of these rivers and the Prairie Farm Rehabilitation Administration of Alberta, considering use by the Americans to be impossible, planned to do so. The United States then sought to follow suit on the upstream side.

It seemed this could be done by diverting the rivers and using the waters in another basin but this would entail tunnelling through a mountain–the construction of the "All-American Tunnel and Canal," as it was known. A geological survey, however, established the fact that the structure of the mountain was such that any attempt at tunnelling would fail because of falling earth and rock, and the United States engineers opposed the plan. Baulked in that, the United States then proposed an alternate deal to Canada: Alberta should have the waters of the Waterton and Belly Rivers; but, as compensation, Montana should have an equivalent amount of water from Canada's share of the St. Mary's River, to be delivered to a point along the boundary convenient to Montana, with the cost of delivery borne by Canada.[63] Thus America claimed that Canada should "surrender to it water in substitution for the other water which the United States could not make use of in the first place."[64]

An important article (Article 2) of the Boundary Waters Treaty–which, because of its consequence to the Columbia negotiations, will be examined in some detail later–gave the right to the upstream country to divert water in a stream, when such was possible, and give a remedy in case of damage. McNaughton, with his massive scheme for the diversion of the Kootenay and Columbia Rivers in mind, was careful to concede the American right to make a diversion in this small case of the Waterton and Belly Rivers. This, he saw, was the Kootenay (and the Columbia) in reverse, and, with the great charm of which he was capable, he provoked counsel for the United States into upholding Article 2 and making statements that could only help McNaughton's future case.

But although the Americans had the right to divert under Article 2, the right could not be exercised. How, then, McNaughton asked, could they claim compensation under that article for not diverting water, when they were obviously physically unable to do so? As he said: "Because of the mountainous terrain adjacent to the two rivers in Montana, their waters could not be economically utilized in the United States, but the conditions in Alberta permitted their utilization there. The plans of the United States . . . not being practicable, the question did not come under Article II of the Treaty."[65] And, in the opinion of counsel for Canada, when Article 2 was not applicable through nature preventing actual diversion, there could be no claim for compensation such as that of the United States.

Counsel for the United States rose to McNaughton's bait. Canada's position was against the spirit of the Boundary Waters Treaty and therefore selfish. "If, according to its requirements, Canada was to make use of all the waters of the rivers, it would be appropriation and not apportionment." In his opinion, the All-American Tunnel was feasible, and whether or not it was economically sound was not Canada's concern. *"Under Article II, the upstream country had exclusive jurisdiction and was not limited by any international servitude."*[66] That was the statement McNaughton wanted, and he was well content. "Counsel for Montana and the United States," he noted, "have argued forcibly that the rights of a downstream country extend only to payment in compensation for damage done and not to an injunction against diversion."[67]

The two sections of the International Joint Commission failed to reach agreement in the case of the Waterton and Belly Rivers. Each section reported separately to its own government. Canada, fully convinced that the United States could put the waters to no useful purpose whatsoever, advised Alberta to go ahead with the irrigation works it had in mind and this the province did.

Then, in May of 1951, came the other application that had a bearing on the Columbia question. It was submitted by the Consolidated Mining and Smelting Company of Canada and was for a dam on the Pend Oreille River, British Columbia, to be known as the Waneta Dam and Reservoir. The dam, if constructed, would create a reservoir flooding some three acres of United States territory in the Cedar Creek Valley of Washington State. The reservoir to be created by Libby Dam, it will be remembered, would flood forty-two miles into Canadian territory and drown *fifteen thousand acres* of land in British Columbia. American reaction to the Waneta proposal, therefore, was of great interest to McNaughton. It was as he had hoped. While the Americans approved the application, because of the encroachment on their sovereign territory they imposed stiff conditions. Any persons having claims for injury must have them satisfied, which of course was reasonable. But, the order of approval continued, the Consolidated Mining and Smelting Company would build the dam at its own risk because the "issuance should not be construed as waiving or otherwise impairing in any degree the right of the United States recognized in Article II of the Treaty, to construct, maintain and operate such works as it may consider necessary or desirable for the purpose of making the most advantageous use reasonably practicable on its own side of the international boundary by *diversion for power purposes or otherwise* * of the waters of the Pend Oreille River. . . ."[68] In other words, Waneta Dam might find itself deprived of water if the interests of the United States suggested that the waters of the river should be diverted elsewhere.

McNaughton learned much from the Waterton-Belly and Waneta cases. In the light of them the American application to construct Libby Dam had little chance of succeeding on the terms proposed. On 8 April, 1953, the United States, faced with the opposition of certain interests on its own side of the border (which had no wish to be submerged behind Libby Dam) withdrew its application.

In May, 1954, the United States placed a second application to build Libby Dam before the IJC. This time the objections of the local American interests, which had opposed the construction of the dam, were overcome by shifting the proposed site some miles closer to the Canadian border! Apart from that, the new application was very much the same as the one that had been withdrawn. McNaughton, with the experience of the Waterton-Belly and Waneta references behind him, brought into the open his diversion scheme: Canada, he said, would not approve the application "if current studies should indicate that more advantageous use of the waters concerned could be achieved by other methods, such as a diversion of part of the Kootenay River into the Columbia River in Canada." And, if diversion should not prove advantageous, there must be downstream benefits—"an equitable recompense to Canada for the use in the project of Canadian natural resources."[69] From the United States point of view, insistence on downstream benefits was bad enough; the projected diversion would be disastrous in that Libby could not be built, and it was obvious that a stern battle lay ahead. Governor Len Jordan of Idaho had replaced Stanley as chairman of the United States section at the Libby hearings and McNaughton, as he told Deputy Secretary of Defense Anderson in July, expected the difference of view of the two countries over Libby to be "sharply controversial."[70]

---

* Italics supplied.

In the course of the Libby hearings it became apparent that the Americans had to some extent modified their views on compensation and were disposed to accept the principle of recompensing Canada for upstream storage. McNaughton detected the trend early. Soon after Eisenhower had come to office there had been a change:

Under the previous United States administration . . . we were dealing . . . with the United States as such. That situation has changed very substantially because the policy of the present administration . . . is to encourage the States to undertake their share. . . . The consequence of that is, since there are at least five States in the United States that are concerned with the Columbia basin, the problem of downstream benefits as between States has become a very live issue in the United States. As long as they were dealing with us, dealing solely with Canada in the matter, they could afford to say, "We will not recognize downstream benefits; once the water comes into our country it is our water and anything we get out of it will be our power and we will not even consider your representations." But when you have four or five States of the Union, some of which are upper States and others lower States, the upper States have been telling the lower ones, "Well, if you will not give us the downstream benefits, we are sorry, but you won't have any storage." With the present United States administration and policies we are in a very favourable position to have our representations dealt with.[71]

The question was, to what extent would recompense be given? The United States signified its willingness "to consider equitable recompense for the use of Canadian resources, through the sale of power or otherwise, taking into account compensatory benefits in Canada," but explanations given by Jordan made it clear that the return the United States meant was to be based on the flows of the Kootenay in its natural channel. That, of course, would ignore the use that Canada could make of the water by a diversion or diversions such as McNaughton had in mind. The waters diverted from the Kootenay alone would add 400,000 kilowatts of power on the Canadian section of the Columbia, and the further diversion by tunnel to the Fraser would add 700,000 more. In flowing down the Fraser system, the waters of the Kootenay and Columbia would make possible the generation of an estimated 6,500,000 kilowatts of power or seventeen billion kilowatt-hours per annum. These figures are a measure of the

An IJC visit to the construction site at Barnhart Locks for the St. Lawrence Seaway, 1957

(ONTARIO HYDRO)

Old Associations; a visit to the Field Battery, Montreal, 1956. Left to right, Al Borretts, Joe Landry, McNaughton, and Percy Ferguson stakes involved; they give an idea of the sacrifice that Canada would make by permitting the Americans to use the waters of the Kootenay at Libby Dam. "They want us to give them a gold watch," McNaughton told the Standing Committee on External Affairs in 1954, "for the price of a bit of tinsel."[72]

The Americans, furthermore, suggested that Canada should be compensated in terms of money, not in power. McNaughton preferred a return of power and he had two main reasons for this: "Money has a tendency to lose its value over the years and any sums fixed now may soon bear no relation to the actual value of the benefits conferred"; and, more important, "The value of water for generating electricity will likely increase greatly as the years go on and it is not possible at the present time to estimate its true worth."[73] Though it was argued,* that the value of hydroelectric power will decline with the development of nuclear-powered generating stations, McNaughton did not agree. Certainly all available sources of hydroelectric power would be used up one day and thermal stations brought into use. In such a mixed power system water would still be used most economically for "peaking"–the periods of maximum demand–because water, stored behind a dam, can easily be released for the generation of power to meet those demands, and power can be produced just as cheaply for this peak load as for the base load. With thermal- or nuclear-power plants, it is otherwise. These cannot be started and turned off like a tap and it is very expensive to operate them for short periods. Thus, to obtain a true measure of the value of water in the future, hydro electric plants must be compared with thermal or nuclear plants used for short periods to meet peak loads, not with those always operating at full capacity.[74] In other words, since thermal generation is more inflexible, such plants in any mixed system will carry the base load; hydroelectric installations will be important for meeting peaks on demand. That being the case, water storage will be much more valuable then, probably three to four times what it is now, because the only alternative source of peaking power would be extra thermal-generating capacity which is much more costly.

For all these reasons, McNaughton considered the American offer of recompense for storage to be totally unsatisfactory and he took a

* In the Department of External Affairs apology for the final treaty, *The Columbia River Treaty and Protocol* (1964), for example.

closer look at his diversion scheme. Nevertheless, however inadequate the offer of compensation had been, it was clear that the United States had recognized the need for it, and this was of considerable interest to Premier W. A. C. Bennett of British Columbia, "a square-faced stocky man" in his fifties, "frankly flamboyant."[75] In 1954 the Kaiser Aluminum and Chemical Corporation, a powerful American concern, offered to build a low dam at Castlegar, in British Columbia, to store the water of the Arrow Lakes; twenty per cent of the additional power, generated downstream for the benefit of the United States, would be handed back to Bennett free of charge. Bennett, who saw a vast influx of electric power at no cost to British Columbia, found the offer attractive; McNaughton, who saw the natural resources of Canada being handed to the Americans at a fifth of their value, did not. Furthermore, a dam at the Arrow Lakes, apart from enabling Kaiser to generate power so cheaply for its Spokane plants that it could undersell the competing Canadian aluminum industry at Kitimat, would inhibit the full development of Mica, the essential site for Canada. McNaughton had no sympathy with the "quick-dollar" fever that afflicts so many Canadian politicians:

> The alarm bell on the Kaiser deal was rung by General McNaughton. . . . He had made an analysis of the arrangements, and it is my understanding on the basis of conversations with him, that he took his case to C. D. Howe and the decision was made to proceed with the International River Improvements Act.[76]

By this act a federal licence was required before work on international rivers could be undertaken; Bennett could not fly in the face of that. The act came in time to frustrate another proposal, an offer by the Puget Sound Utilities Council to build Mica Creek Dam for Bennett at an estimated cost of $250,000,000. British Columbia, however, would have to machine the dam for power, and no return of any part of the 1,100,000 kilowatts, which Mica would add to plants in the United States, was mentioned. Control of the flows was to be vested in the Americans for fifty years, with the option for another fifty after that.

The attitude of the federal government at the time is interesting. It reveals a recognition of the fact that the Columbia negotiations, after Libby, were about to enter upon a decisive phase; and that in this phase McNaughton's struggle to safeguard Canada's resources for the future needs of Canada was the policy of the Canadian government. Jean Lesage, writing about the International River Improvements Act in the July, 1955, issue of the *Electrical Digest,* said this:

> According to the Canadian constitution, works built on rivers in Canada and having an effect outside the country fall under the jurisdiction of Parliament even if they are entirely located in one province. Up to now, the Government of Canada has felt that it was unnecessary to exercise its jurisdiction and to legislate in this field. Conditions are rapidly changing, however, and . . . *a second period in the development of the Columbia River System is now starting during which, important international problems will arise.** Special legislation will be needed to cope with these questions and to provide guiding principles of policy designed for the protection of the public interest of the Canadian people.
>
> The Government of Canada has already made known the general principles which serve to interpret the national interest in this respect. They require that a *project must be compatible with present and future needs of the country and with optimum development of the site and the whole watershed.*† If no effective use of the water resources can be made in Canada, the improvement executed in Canada to permit downstream utilization must provide for benefits commensurate with the water resources thus made available.

That, it will be seen, was a most important statement of policy. Canada, it must be repeated, had a water policy as late as 1955, but none thereafter. Never could the construction of Libby Dam be "compatible with [the] present and future needs" of Canada—with "optimum development of . . . the whole watershed." Neither could such isolated projects as Arrow Lakes, which would merely store water for the benefit of the United States. Only if "no effective use of the water resources can be made in Canada," Lesage said, should waters be stored and released for American use, and only then against the payment of adequate compensation for the downstream benefits provided. The Columbia River Treaty, signed by the Diefenbaker government in 1961, reversed the policy of the St. Laurent government so obviously right for Canada, in almost every particular, as we shall see.

---

* Italics supplied.

† Italics provided.

Lesage had promised a stirring second round in the Columbia controversy and he was right. There had been no common ground in the hearings of the commission on the second Libby reference and, because the Kootenay is a part of the Columbia River system, it was decided that the outcome of the Libby project would await agreement on the development of the system as a whole. But one thing that came out of the Libby hearings, apart from acceptance by the United States of the validity of the principle of recompense, was the firm opposition on the part of the Americans to McNaughton's diversion scheme.

Already in 1954 engineering investigations had determined that it was quite feasible to divert the flows of the Kootenay to the Columbia by the Dorr diversion; reports on the second diversion of the Columbia to the Fraser were encouraging. McNaughton claimed that Canada had the right to make these diversions under Article 2 of the Boundary Waters Treaty of 1909; this was vehemently contested by Jordan of the United States. Whether Canada had the right or not became without doubt the biggest issue of the whole controversy and for the next three years, between meetings of the PJBD and consideration of other references before the IJC, McNaughton fought this battle. There can be no understanding of the moves and countermoves without an examination of Article 2 and to that we must turn our attention.

The essential parts of Article 2 of the Boundary Waters Treaty read as follows:

> Each of the High Contracting Parties reserves to itself . . . the exclusive jurisdiction and control over the use and diversion, whether temporary or permanent, of all waters on its own side of the line which in their natural channels would flow across the boundary or into boundary waters; but it is agreed that any interference with or diversion from their natural channel of such waters on either side of the boundary, resulting in any injury on the other side of the boundary, shall give rise to the same rights and entitle the injured parties to the same legal remedies as if such injury took place in the country where such diversion or interference occurs. . . .[77]

It will be seen that Article 2 is divided into two parts and the first part, which is expressed in clear language, seems to amount to this: "A state may do as it pleases with the waters in the territories over which it has sovereignty without regard to downstream interests" in another country.[78] The second part modifies the first by adding a proviso on compensation, but "there is a difference of opinion among lawyers about the right of 'injured parties' and the liability of those who exercise their right to divert under this article."[79] One interpretation is that this second part, "as a matter of grace or comity, confers upon those *citizens** who are injured downstream the right to come into the courts of those doing the injury to obtain the same rights and remedies available to citizens of that state."[80] In support of this interpretation are the words of Chandler Anderson, who put the words into the Boundary Waters Treaty in the first place and presumably knew their meaning. In a memorandum, explaining the 1909 treaty to the Foreign Relations Committee of the United States Senate, Anderson wrote:

> The purpose of this provision of Article II is to permit parties who are injured on the other side of the line to secure the same damages that they would be entitled to if the damage had been done within the same jurisdiction where the cause of the damage originated; but their claim is subject to the laws of the jurisdiction where the cause of damage arises, and they must come into the courts of that jurisdiction and prove their case on exactly the same footing as if the property injured was within that jurisdiction.[81]

William Pugsley (the responsible Canadian Minister) commented in the House of Commons in 1910:

> . . . as to all future cases the *citizens*† of either country are placed in exactly the same position as a riparian proprietor lower down the stream would be placed in regard to any diversion of water by a *private*‡ riparian owner further up the stream by which his rights would be interfered with. In other words, both nations . . . recognize that there would be the same obligation to make payment for that injury as if it was a question between citizens of the same country.[82]

It would appear, Professor Charles B. Bourne of the University of British Columbia concludes, that the "intention of the parties to the treaty seems to have been not to guarantee payment of compensation for all damage suffered, but to guarantee equality of treatment to all persons affected by a diversion, no matter on which side of the border

---

* Italics supplied.
† Italics supplied.
‡ Italics supplied.

they happen to be. For this purpose, the border is deemed to be no longer there and everyone had the same rights and remedies under the law of the place of the diversion. The words of the article seem to achieve this purpose very aptly."[83]

So much for the language of Article 2; the history of the article must now be considered and this will be done in the two parts into which it is divided.

The first part of Article 2 incorporates the so-called "Harmon Doctrine," an opinion expressed in 1895 by Attorney General Judson Harmon of the United States to the Secretary of State relating to Mexican protests over American diversions of the Upper Rio Grande. The Harmon Doctrine has been broadly defined in these terms: "the upstream state may divert at will and without restraint regardless of the effect produced upon the downstream state . . ."; that "there is no duty or obligation in international law on any state to restrain its use of the waters within its territory to accommodate the needs of another state"; and that "jurisdiction and control of a state over the waters of an international river wholly in its territory is exclusive."[84]

"The Harmon Doctrine," said Professor M. E. Marts of the University of Washington, "was a United States-inspired concept written into the Treaty to guarantee that the upstream state could have its own way. Apparently the Secretary of State's knowledge of geography was such that he must have thought that the United States was always the upstream state!"[85]

The United States argued that the Harmon Doctrine was a principle of international law, which it was not, and secured its embodiment in Article 2 of the Boundary Waters Treaty. Why did Canada agree to accept the doctrine—a principle that could be injurious to the nation's interests by permitting the United States to justify diversions and interfere with the flow of international rivers? Sir Wilfrid Laurier, Prime Minister at the time, answered that in the House of Commons on 6 December, 1910. International law, he conceded, was not embodied in the terms of the treaty, but, whether Canada liked it or not,

> the United States had taken the position that international law provides that . . . the upper [upstream] power has the right to use the water within its own territory as it thinks best. What were we to do? They might do so, and if they did so, they might do it to our injury

and we had no recourse whatsoever.[86]

Faced by this possibility, the second part of Article 2, which *modifies* the Harmon Doctrine by providing for compensation to injured parties, had then been written into the Boundary Waters Treaty of 1909 at Canada's request. Laurier continued:

Was it not wiser, then, under such circumstances, to say: Very well, if you insist upon that interpretation you will agree to the proposition that if you do use your powers in that way you shall be liable to damages to the party who suffers. At the same time we shall have the same power on our side, and if we choose to divert a stream that flows into your territory you shall have no right to complain. You shall not call upon us not to do what you do yourselves. . . . What wiser course could have been adopted?[87]

Throughout the years that followed the signing of the treaty of 1909 the Harmon Doctrine was relied upon to a considerable extent by the United States to protect its interests. One example is the "Chicago diversion." During the nineteenth century the city of Chicago dumped its sewage in Lake Michigan and, as this was the lake from which it drew domestic supplies of water, a remedy was found in 1889 by constructing a drainage canal, which drew the sewage out of the lake and flushed it into the Mississippi by way of the Chicago River. The canal, opened in 1900, originally diverted some four thousand cubic feet of water per second from the lake but in later years this amount was more than doubled. Not only was the level of Lake Michigan lowered; so was that of all the Great Lakes by six inches, and this endangered navigation in the St. Lawrence as far as the Atlantic Ocean. The beauty of Niagara Falls was also threatened. In 1912 Canadian counsel argued against any increased flow, only to hear the Sanitary District of Chicago argue in reply "that the United States had retained *full rights to divert under the treaty of 1909*,* and in any event sanitary purposes took precedence over navigation."[88] We have also noted, in McNaughton's time, American reliance on Article 2 to justify a future diversion of the Pend Oreille River and to carry out a diversion of the Waterton and Belly Rivers that proved impracticable.

Thus, on the face of it, Canada seemed to be on firm ground in

---

* Italics supplied.

backing McNaughton's scheme for the diversion of the Kootenay and Columbia Rivers under Article 2 of the treaty. "In the light of this provision," Jacob Austin of the British Columbia bar stated, "it is abundantly clear that no legal limit can be set to Canada's right to divert the waters of the Columbia as she sees fit and that no regard need be had to downstream uses or prior appropriations of any sort."[89]

McNaughton had come to that conclusion three years before those words were written. He said as much in a paper that he submitted to the Fifth World Power Conference held at Vienna in 1956. His paper, *Problems of Development of International Rivers on the Pacific Watershed of Canada and the United States,* is extremely important to any student of the Columbia controversy. It amounts to no less than the public unveiling of the McNaughton diversion plan. The broad strategy is clearly stated, as are the detailed tactics. The paper was supported by the most minute technical data of the rivers and their flows; the proposed location of dams; the siting of generating stations and their capacities for producing power.

In the Pacific watershed, McNaughton pointed out, "*Canada,* in almost all cases where major possibilities of diversion exist, is the upstream state." Thus "*Canada* for the first time in the history of her relations with the *United States* over the use of water becomes the beneficiary of the sovereign powers of diversion prescribed in Article II of the Treaty of 1909." In the past "the *United States* have quoted this Article, in cases where that country has been the upstream state, to the disadvantage of *Canada* and in the result an extensive body of precedent has become established through argument presented to the *International Joint Commission* principally by learned Counsel for the *United States* and for various States of the Union." The diversion by Canada of the Kootenay and the Columbia Rivers "is entirely consistent with the practice which has been urged by the *United States* in like circumstances."[90]

Though McNaughton strongly advocated the diversion of the Kootenay and Columbia Rivers "with consequent potentials for power development measured in terms of millions of kilowatts" for the greater good of Canada, he suggested ways in which storages within Canadian territory could be made available for American use. After all, his diversion plan would only conserve the surplus waters of the headwaters of these rivers. He suggested the construction of dams lower down to assist the United States. One might be where the

Columbia crosses the international boundary, to raise the water level there by forty-two feet which, added to the height of the Grand Coulee Dam, would add more than three million acre-feet to the storage capacity of the reservoir behind the dam. A second, north of the border on the Kootenay, would similarly raise the water level by thirty-seven feet. Finally, a measure of assistance might be given by permitting United States use of a four-million-acre-feet Canadian storage which could be built at Murphy Creek. But in all these cases, McNaughton warned, "*Canada* will no doubt require from the *United States* due recompense for natural resources used in the various projects and for services rendered."[91] This presentation by McNaughton on an international stage—a presentation in which he pointed to the wide choice of action open to Canada and the legal situation, which "has now been firmly gripped by the *Parliament* and *Government of Canada* for the Canadian people"—caused near-panic in the United States Senate.* A great deal was at stake, and though McNaughton argued that a diversion of about fifteen million acre-feet of water would take out of the Columbia only the water which at present flowed to the sea unused,[92] American plans for future plants would become so much wastepaper. Brigadier-General Emerson C. Itschner, Chief of the Corps of Engineers, United States Army and chairman of the U.S. section of the International Columbia River Engineering Board, said in the United States Senate hearings in 1958 that the loss in power "would be more than fifty million dollars a year. This amount would increase as the years go by." It was also suggested that it would "damage the fish industry of both countries by affecting the salmon in the Fraser and Columbia Rivers."[93] American concern over talk of diversion, therefore, is understandable.

Len Jordan, who has been described as "a tough guy" put into the IJC by the Americans "to stand up to McNaughton,"[94] attacked the Canadian diversion plan. The United States, he contended, had "an over-riding need of Columbia water for atomic energy production, fisheries, irrigation, navigation and power generation."[95] Some of the arguments which developed are interesting. For example, one of the United States spokesmen stated that America "has never, in practice, followed the Harmon Doctrine embodied in Article II of the treaty.

* See Senator Richard L. Neuberger, *Senate Hearings on Interior and Insular Affairs*.

Moreover that doctrine is not, today, a principle of general international law. Hence, Article II cannot be invoked to support such a diversion as the one contemplated by Canada." To that Canada replied: "Whether or not the United States continues to look with favour on the Harmon Doctrine is academic, *although that doctrine has been relied on in recent years by counsel for the United States Government appearing before the International Joint Commission.** It is not the Harmon Doctrine of absolute sovereignty, but a solemn treaty which has been adhered to for nearly fifty years, that determines the rules applicable in the Columbia case"; in other words, Article 2 of the treaty which, in its second part, modifies the Harmon Doctrine.[96]

The proposed diversion, according to the United States, was "unfair and unreasonable"; but was it, though? It seemed reasonable enough for Canada to claim fifteen million acre-feet out of fifty-four million acre-feet of Columbia River water that has its origin in Canada.[97] And even American legal authorities came out in support of the Canadian position: "On the stark language of the treaty, unilaterally interpreted," said Dr. Charles E. Martin before the Fifty-first Annual Meeting of the American Society of International Law at Washington, D.C. on 25 April, 1957, "Canada's right to divert is clear and unmistakeable."[98] Finally, Jordan admitted that. "Canada does have the legal right to divert," he said, "although it is not conceded that she can do so without regard to downstream interests,"[99] and on 4 October, 1957, the IJC agreed that the Canadian proposal, "in so far at least as the diversion from the Kootenay to the Columbia was concerned," would be submitted for study to the International Columbia River Engineering Board.[100]

With Jordan's concession, the question of diversion turned from Canada's rights in the matter to one of compensating "injured parties" on the United States side of the border. We have seen that the second part of Article 2 appears to give that right only to citizens. Would the United States, a sovereign state, have any claim against Canada? Professor Bourne finds it difficult to see.

> . . . how could the United States have any legal claim under the general principles of international law in view of the first part of Article 2? Does Article 2 not define the law for the parties to the

---

* Italics supplied.

treaty? Certainly, the history of the Harmon doctrine and of Article 2, which embodies that doctrine subject only to the qualification expressed in that Article, makes it difficult to believe that any tribunal could be persuaded that the United States has a legal right, found *outside* of Article 2, to compensation for downstream losses and injury occasioned by a diversion of Columbia River waters in Canada.[101]

Jordan, however, thought otherwise. At the semi-annual meeting of the IJC in Ottawa on 4 October, 1955, he threatened the Canadian commissioners with the might of Uncle Sam.

I think it proper to point out at this time [he said] that the injuries downstream occasioned by the annual diversion of 15,000,000 acre-feet of Columbia water to another basin will be suffered by a Sovereign . . . namely, the United States of America. Obviously therefore, the United States, as injured Sovereign, will not be limited to the redress provided for an injured party by article 2.[102]

He went further. At the same meeting he claimed that not only would vested interests–interests in the United States that were already using the Columbia waters–have a right to compensation if a Canadian diversion interfered with their flow, future needs for the water must be compensated for in the same way as for these vested interests.

The United States Government already has substantial investments in existing power plants in the Columbia Basin amounting to about one and one-half billion dollars; in power plants under construction, another billion dollars; plus another estimated two billion dollars for power plants expected to be built in the next ten years. Wide publicity has been given all of these projects. Canadian and provincial officials have been given all our engineering reports. Never at any time has secrecy shrouded our building or our planning. All of these projects were planned and all of the funds are committed in anticipation that the waters of these international rivers would not be utilized by Canada in such a way as to jeopardize downstream interests. . . . Diversion would result in very serious injury to downstream interests in the United States.[103]

"Even assuming," said Jacob Austin, "that prior appropriation [the fact that the water was already in use in the United States] will give a right to compensation, does this also include injury to a *prospective* * or *potential* †prior appropriation? To have a right to be compensated under article 2 there must be an *actual* ‡ appropriation of the water which is interfered with by the diversion in question." The United States, he concluded, "cannot protect itself against competition from Canada for the *unappropriated* § waters of the Columbia by contending that they thought about it first." In the end, however, the matter of compensation was never tested.[104]**[105]

The Americans were prepared to oppose diversion with every argument they could muster. They would fight for Libby, for with Libby built it would be a vested interest that, having put the waters of the Kootenay to use, would inhibit and perhaps prevent Canada from carrying out diversion. The United States, with Libby as the key, could virtually control the resources of the Columbia Basin to its own advantage. Not that the Dorr diversion in itself would be disadvantageous from the United States point of view; on the contrary, it would be beneficial. Bu it would enable Canada to have the kind of storage she wanted at Mica and that storage would permit Canada to carry out the diversion to the Fraser. That, for the Americans, would be a disaster. McNaughton did not hide his intentions. "The key to the whole development of the Columbia," he said, "is diversion. Without diversion a number of sites on the Columbia cannot be developed economically. The diversion will add almost a half-million kilowatts at Mica alone and it makes the whole group of plants in that vicinity thoroughly practicable and profitable. Also, as I have told my colleagues in the International Joint Commission, *the plan we have advanced results in the placing of the flows and the positioning of water in storage in Canada exactly the way we want it in the future.*" As for Libby, that to McNaughton was anathema. It was essentially a high-cost project, twice the cost per acre-foot of the proposed Canadian storages. "We don't understand why you keep pressing us about Libby," he said. "If there were no boundary, Libby would not be in

* Italics supplied.
† Italics supplied.
‡ Italics supplied.
§ Italics supplied.
** But the American negotiators managed to make their views prevail in the final treaty. If Canada ever does divert the Kootenay then the United States must be compensated in the amount of 275,000 kilowatt-years of energy per annum.

the plan for the best development."[106]* But he understood the American arguments very well. It was a question of national interests—Libby versus the Dorr diversion plus Mica Creek—and so the battle lines were drawn.

McNaughton had no doubts about the outcome. He prepared the House of Commons Committee on External Affairs by producing figures based on the final estimates of the United States engineers for Libby and of the Canadian consulting engineers for Mica. He brought maps with him, as well as profiles to show the heights of the surface of the water along the Kootenay, the Columbia, the South Thompson and the Thompson to its junction with the Fraser. These profiles showed sites where heads could be concentrated for the storage of water and the development of power. He produced hydrographs showing the mean annual flows of rivers and how these would be modified by diversions. His purpose was to show what was available and to try to ensure that it was used in such a way to obtain the best result for Canada. At those meetings McNaughton was both courteous and convincing. He outlined the legal aspects of the Boundary Waters Treaty and continued:

> I can, now, if you wish . . . go on with an explanation of the Treaty and of the provisions of the law on water which derive therefrom and which are most important, for they give the basis on which decisions will be taken as to what will or will not be done in fact.
>
> Or . . . I will be happy to submit myself to the questions of members to cover other aspects of the work of the International Joint Commission if they wish because in making my presentation I do not seek to narrow in any way the matters before you in which the I.J.C. is concerned in one way or another—the St. Lawrence project—water pollution—air pollution—the levels of Lake Ontario—progress at Niagara and on the St. John River—the irrigation problems in Alberta and Montana—and any others which are before the Commission on reference or otherwise. On all these matters I will be most happy to present information if you will indicate your wishes. . . .[107]

* Compare this with McNaughton's statement to the External Affairs Committee, 23 April, 1959: "If these matters were being considered from a strictly international point of view . . . there is very little doubt as to what the real development of the Upper Columbia would be. It would be to route the water by way of the Dorr diversion to take advantage of the additional heads which can be developed along the Columbia and make use of the flows to best advantage."

At meetings of the International Joint Commission in April, 1955, McNaughton again said that diversion was in the best interests of Canada. Having satisfied himself both as regards "topography and the law," he stated the Canadian position with firmness and clarity; but he did leave an opening for a fair bargain and "neighbourly consideration under appropriate conditions": the recognition of downstream benefits. He saw very clearly what the results of his stand would be:

There would be a rude shock to the American public. It would take some time for persons in authority in U.S. water matters to appreciate that a check had been given to their plans. Then there will be personal attacks on me for the impertinence of such an affront. There will be official representations of unneighbourliness and attempts to misinterpret the meaning of Article II. There would be a personal communication from General Eisenhower [talking of past friendship and future co-operation]. Then, gradually, it would penetrate that the Canadian position was justified and that we meant business and they had best seek an arrangement.[108]

All this, including the letter from Eisenhower, happened. But, having stated his case, McNaughton intended to remain firm with feet well planted and to stand upright under the storm of intransigence from the United States, which would pass. Time was on the side of Canada. The need for Canadian storages would lead the United States to make an approach, and a fair solution could then be offered.

"To regulate the flow of the Columbia River to the best advantage of power," McNaughton said, "the U.S. Army Engineers have suggested they need some 40 million acre-feet of storage. . . . The trouble is this; the only major storages available [in the United States] are Hungry Horse and Grand Coulee which together with some minor storages amount to only nine and a half million acre-feet; that is about a quarter of the stated minimum requirement. Well, you have 9½ and require over 40 and . . . it has now become difficult to persuade people to let their land be put under water for storage. Also special interests like salmon, which have the idea that rivers are created for fish alone, add scores of millions of dollars to the cost of dams.* The most the International Columbia River Engineering Board [indicates] as to what we might expect comes to a total of

* The building of Grand Coulee by the Americans had kept salmon out of the Canadian Columbia.

22.3 million acre-feet* in the whole of the United States portion of the basin, leaving out Libby as an international project."[109]

So where was the additional eighteen million acre-feet, which the Americans required, to come from?

With the Dorr diversion [McNaughton said] we will have an ability in Canada to store effectively and release 23 million acre-feet of the flows of the Columbia and Kootenay Rivers. I would like you to think of 23 million . . . in relation to the 22 million . . . that is all that is reasonably prospective in the United States and the minimum requirement of 40 million . . . which has been stated by the U.S. Army Engineers and other of your authorities. The situation is that you may add about 5 million in the next two or three decades to the 9½ million at Grand Coulee and Hungry Horse so you are going to face a period without enough storage even for flood protection on the scale demanded; also you will have enormous generating plants without enough regulated flow to operate them to the best advantage.[110]

Surely then, McNaughton suggested, the sensible thing for the Americans to do would be to accept Canadian water for a long enough period to permit them to solve their own storage problems: to come to terms with the salmon interests ("I am sure, in this scientific age, that a way can be found for salmon and power to co-exist") and the others that did not want to be flooded out. He would not abandon the diversion of the Columbia into the Fraser because of the future benefits that would bring; but Canada, he said, would defer it until the United States had found alternative storages or means of producing power. Until then Canadian water would be sent down to meet American needs.

We in Canada [he said] are in an early phase of development and we have felt that without compromise of Canadian interests we could be co-operative and help you for a period of say two to three decades until you can put your own situation in order. . . . In Canada the storages . . . are going to be reasonable in cost. This is because the plans have been made in an early stage of our development and the sites have been reserved by the British Columbia government and nobody can touch them. We will have an ability in Canada with the Dorr scheme to store effectively and release 23 million acre-feet of the

* For details see External Affairs Committee hearings 18 March, 1960, Appendices D and E.

flows of the Columbia and Kootenay Rivers. Power will be produced in Canada. At Murphy Creek the installed capacity may be 450,000 kilowatts; at Revelstoke, Downie and Mica together . . . about 2½ million with perhaps another quarter of a million or so at the smaller sites. We would build storages that would be to the advantage of the United States—on a cost and energy-sharing basis. And for about twenty-five years after we are actually making water available from Mica [which would be a few years in the building] we would allow those flows to go through the United States. We believe that when the U.S. public wakes up to the real need for storage you will get over the existing difficulties with fish—it is just too fantastic to believe you will not be allowed to solve these problems. We will be very happy to help you but very naturally just so long as it doesn't hurt our people in the doing.[111]

Canada's bargaining position, as McNaughton knew, was extremely strong. She had every legal right to make diversions and he was determined, at any rate, to divert the floodwaters of the Kootenay to the Mica storage. From there on, what happened to the fifteen million acre-feet of water was up to the Americans to decide. Was it to be diverted to the Fraser for Canada's own use or would it be spilled in controlled flow down the Columbia during low-flow periods? If the latter, how much were the Americans prepared to pay for flood control and the extra billions of kilowatt-hours they would gain annually? McNaughton, according to the pattern of previous agreements on boundary waters (though not in international rivers) looked for an equal sharing of both costs and benefits, fifty per cent to each; he also intended to safeguard Canada's right to carry out the Fraser diversion twenty-five or thirty-five years hence, when all the hydroelectric power that could be produced would be needed.

Despite every objection the Americans could think of, both as regards diversion and downstream benefits, McNaughton was "a difficult man to get around. He was a tough bargainer; he would say what he thought. But no American member of the Commission ever believed him to be anti-American. They understood him—he was one of their kind—and they admired his stand."[112] Democratic Senator Mike Mansfield of Montana, a powerful proponent of Libby Dam, put it this way: "General McNaughton is one of the most determined men to come up against in negotiations. He is an extreme nationalist—and I say this in the good sense." And in so far as the Columbia studies went, "he was the best informed man on both the Canadian and American sides, and the best informed by far."[113]

Gradually the American opposition began to change. In 1954 McNaughton's assertions of Canadian rights had produced only irritation in the United States. "If only it weren't for McNaughton," Michael Barkway of the Canadian *Financial Post* was told in Washington, "we could quickly get ahead." In 1955 McNaughton's claims were described by the Americans as "utterly unrealistic," though it was beginning to be recognized that he had the government behind him and most of the Canadian people. In 1956 Len Jordan found it necessary to defend himself from Canadian accusations that his "Sovereign" speech was a repudiation of the 1909 treaty. This, he said, "distressed" him. He had "never denied, and hope I never will deny, that Canada has a perfect right to divert the Columbia waters. Of course it has a legal right to do so." Asked about his statement that the United States, as a sovereign party, would demand redress, he replied: "I don't know what the redress might be for the United States. It might just be a stiff note from the State Department." As for diversion, Jordan asserted, "I want the Canadian engineers to press on with their studies of the possible diversions of the Columbia. I want to see those studies completed. And when that's done I believe we can make a bargain that will be fair to everybody." That year, it should be mentioned, Len Jordan was under attack from Senator Neuberger who believed that "Mr. Jordan's formerly uncompromising attitude on downstream benefits has driven Canada into 'an extreme assessment' of the value of upstream storage to the United States."[114]

It is significant that Jordan, after bitter resistance when it was first proposed, sanctioned the inclusion of the Dorr diversion in the studies to be carried out by the International Columbia River Engineering Board. Itschner,* chairman of the U.S. section of the board, also agreed that it would be unrealistic not to include the study of this diversion. In all, three sequences of projects using waters of Canadian origin were selected for detailed study and these might be briefly described as follows: one (including Libby) of most benefit to the United States; a second, showing the value to the United States of the best uses of the resources in the interests of Canada (the Dorr Diversion); and, third, a compromise between the other two.

A note of wry approval began to appear in the American press

* McNaughton kept a wary eye on Itschner who was a most competent engineer.

concerning McNaughton's stand:

> Canada's diversion scheme was touched off when the U.S. did not offer Canada any share in power that could be generated downstream from . . . Libby Dam. . . .
>
> As a result of the squabbling the Canadians are now threatening to divert a portion of the Columbia into the all-Canadian Fraser River. . . .
>
> Ironically, the U.S. by its position on a half-century-old treaty has backed itself into the spot where Canada would be entirely within its legal rights.
>
> . . . the U.S. figured without the bargaining instincts of the fiery gentleman who heads the Canadian Section of the International Joint Commission. . . . This man is General A. G. L. McNaughton, sometimes known as the General McArthur of Canada, and a hero of two world Wars. . . .
>
> To those who observed that the waters had to flow downstream to the U.S. anyway, General McNaughton eyed a made-to-order geographic picture that made the ancient law not necessarily so. . . .
>
> The logical move would appear to be for the U.S. to offer the Canadians more downstream power benefits in return for the upstream storage in Canada. One question, of course, is how much. "What's acceptable to the Canadians remains the $64,000 question," comments a U.S. engineer. . . .
>
> "I'd be stalling, too, if I were in Canada's strong position," says one U.S. observer. "Time is on our side," asserts a Canadian. For . . . British Columbia has developed less than 10% of its hydro-potential. Meanwhile the U.S. faces a vanishing potential during the next 15 years.[115]

All this had not been accomplished without some blunt words at commission meetings. "I would be quite wrong," McNaughton admitted, "to leave with you any impression . . . that all is peaceful progress . . . for in fact, in some aspect or another, every problem is marked by a healthy clash of wits and we would not have it otherwise; on occasion the temper of debate may be pretty hot. It would indeed be most unrealistic to expect a regimen of passive discussion where real tasks of long-range significance to the two countries have to be undertaken."[116] The General was not always tactful. "He calls a spade a spade," said Michael Barkway, "even when he's talking to the Americans; and this deeply shocks some of the more sensitive diplomats who flourish around here. To many hardier souls it is extremely refreshing."[117] The Americans did not resent McNaughton; it was Canadian diplomats who felt they did. "They would bellyache: 'Really, you know, it's impossible the way McNaughton treats the Americans. We *can't* talk to the Americans *like* that! ' "[118]

McNaughton was doing very well, but the idea of trying to reach agreement through diplomatic talks, instead of the International Joint Commission, took root somewhere in the Department of External Affairs. "It was accepted by Mr. Pearson . . . and by Mr. St. Laurent who formally proposed the talks to President Eisenhower in March, 1956."[119] The negotiations must be private, it was explained, "because only then can there be give and take. Canada—that is—wants secret talks so that the U.S. can yield without a public outcry."[120] McNaughton feared those closed doors (especially when he heard that the Canadian "give" would amount to a statement that there were gaps and deficiencies in the Canadian argument under Article 2 of the Boundary Waters Treaty); they might equally well muffle the cries of Canadian diplomats who were being skinned alive. The move was not welcomed by the press; it was pointed out that the General was also a diplomat and, moreover, one with a sense of humour. "In the middle of a burst of fact and fury he'll suddenly pause, his eyes will twinkle, and he'll confide, 'It's great fun, isn't it?' "The new Canadian representatives were also "able men," but "there's no question that they're less tough." It was possible "they might settle for something which . . . might not be the best possible deal."[121]

The official chronology of the next twelve months, issued by the Department of External Affairs, reads as follows:

March 25-28, 1956. Prime Minister St. Laurent and President Eisenhower met at White Sulphur Springs and agreed that the Columbia River problem should be discussed at the intergovernmental level.
May 23, 1956. Press release issued by the two governments announcing that diplomatic talks would take place with respect to the waters of the Columbia River.
March 8, 1957. Meeting at Washington. Canadian side led by Minister of Northern Affairs accompanied by B.C. representatives.[122]

What happened was this. On 29 February, 1956, McNaughton saw Lester B. Pearson. Subsequently he left "with a distinct feeling of uneasiness that we were about to, or had in fact already, embarked on a course of discussion with the United States which might well compromise our existing favourable position under Article II. . . ." On 24 March, however, he had an informal chat with Jean Lesage who "made it evident that he stood, and would continue to stand, for the principles of Article II; and he agreed . . . that within these principles we had ample facility under which we could make a 'fair bargain' with the United States." McNaughton told him that he "would not give away one iota" and he gathered that Lesage had no intention of doing so either. He drew renewed hope from that conversation.[123]

Next day the Prime Minister met President Eisenhower for a three-day discussion. It was decided then to take the Columbia problem out of the hands of the International Joint Commission, the reason being, St. Laurent announced in the House on 9 April, that "the Chairmen of the two Sections [of the IJC] had publicly expressed views so diametrically opposed to each other that there was little probability of their being able to make the kind of progress we would hope would be made in arriving at a solution of these problems."

"This evidence of lack of support," McNaughton said, ". . . caused me to consider sending in my resignation."[124] As he said, the position he had taken at meetings of the commission had been thoroughly checked with the government beforehand and he had stated it with the approval of all concerned. He could not modify his stand "except by direction" and that he "had neither sought nor so far been given."[125] He did not resign, however:

> If I remained I might have some opportunity to protect Canadian interests against abandonment by those who did not fully understand the position. . . . The Prime Minister looked at water law from the riparian point of view, which was his experience in the East, and quite evidently he failed to understand the reversed position under the 1909 Treaty on which Canada's advantage depended.
>
> . . . Mr. Pearson had more than once stated to me that he did not understand the water problems in the Columbia and I mentioned that he had never once taken advantage of my offer to explain them in detail any time he wished. It seemed that Mr. Pearson was far too busy with matters abroad which interested him, and which no doubt were important, but not to the extent that Canadian interests . . . could be put to one side. I spoke of the "gaps and deficiencies" of the first draft statement that had been prepared . . . for the Prime Minister to make to the President at White Sulphur Springs and how I had learned of this very late but had managed to persuade those concerned that such

a pronouncement would end all possibility of maintaining our rights under Article II of the Treaty of 1909.[126]

The phrase "gaps and deficiencies" which had been used by External Affairs in referring to the Canadian case had been included, McNaughton thought, for this reason: there had been some "statement of resentment," it seemed, "made either by the President to the Prime Minister or, more likely . . . by Dulles to Pearson asserting that such proposals [of diversion] could not be tolerated by the U.S. and that to persist would endanger Canadian-United States relations."[127]

The next major move, as shown in the chronology, was the visit to Washington by a Canadian delegation headed by Jean Lesage. McNaughton "spoke of the instructions drafted for Mr. Lesage . . . and how these had been modified just before he left Ottawa as a result of warnings which I had expressed and which had found support from representatives of Finance, Trade and Commerce, The Treasury. . . . In the result it seemed that the so-called diplomatic discussions in which Mr. Lesage had represented Canada in Washington had been quite inconclusive. . . . Mr. Lesage had, in fact, made it clear that Canada would insist that all arrangements would need to be within the Treaty of 1909 including most particularly the maintenance of Article II. The United States representatives had not offered objection."[128] In other words, Lesage, as he had assured McNaughton, had surrendered not "one iota" of Canada's strong bargaining position. Thereafter, as was reported, the "ill-starred attempt of Mr. St. Laurent to transfer discussion . . . from the International Joint Commission into diplomatic channels" was "as dead as a dodo."[129]

It was 1957. McNaughton was seventy years old, still straight and alert but with more white than dark in his hair and closely trimmed moustache. His eyesight was not as good as it had been but he had lost none of his mental vigour; "though he peers through bi-focals at the framed maps he assesses potential dam sites with skill and pithy comment. 'That's no shucks,' he would say, or 'that little tit of a thing!' "[130] At the general election in June of that year the Liberal strength of 168 out of 265 seats in the Commons was reduced to 105; the Conservatives won 111 but, with 48 seats going to the other parties and Independents, the Tories had no majority. On 17 June, however, the Liberals turned over the administration to the jubilant Tories and on the 21st John G. Diefenbaker became Prime Minister. In January, 1958, Lester Pearson succeeded Louis St. Laurent as leader of the Liberal Party and on 31 March his party suffered an unprecedented defeat; the Conservatives won 208 of the Commons seats at the polls that day, to achieve the most powerful majority in Canadian political history.

"On its accession to office," the Toronto *Globe and Mail* announced, "the Diefenbaker Government at once set to work on the [Columbia River] problem. It was spurred on by the action of Premier Bennett of British Columbia in collaboration with the Swedish Wenner-Gren* interests in a plan to develop power on the Peace River."[131] Negotiations were handed back to the International Joint Commission after a meeting between Leon Ladner† and McNaughton in July. Ladner stated that he had "mentioned the part McNaughton had played in the protection of Canada's water resources and had told the Prime Minister he should seek his Counsel." This pleased McNaughton. The time, he said, was opportune. He had heard confidentially from Washington that Jordan was to be replaced. Ladner asked if the new United States chairman would be Neuberger; it was obvious that Neuberger had been out to have Jordan dismissed. "Not Neuberger," McNaughton said, "but McKay‡ who had been Secretary of the Interior. . . . The suggestion of McKay's name for the I.J.C. meant that the U.S. would seek to make this the forum of discussion; that this confirmed the attitude expressed by various U.S. officials that the diplomatic conversations [had not been] useful."

> McKay's appointment to the I.J.C., if made, would be very significant. I suggested that McKay might be being sent to the I.J.C. to give him an opportunity to bring about a settlement of the international aspects . . . that the need of concessions to Canada was now being widely recognized . . . that there were forthright statements which showed a recognition of the mounting needs of the U.S. for Canadian storage and an acceptance of the fact that it had to be paid for.
>
> I said that if McKay were in fact appointed to the I.J.C. I thought he

---

* Financier Dr. Axel Wenner-Gren.

† A Vancouver lawyer who raised funds for the Conservative Party on the west coast.

‡ Douglas McKay of Oregon.

would come "bearing an olive branch" and that, realizing that the U.S. public had been prepared for concessions, we could expect an offer which might have considerable importance.[132]

In October, 1957, the Speech from the Throne made reference to the Columbia: "My Ministers are pressing for a favourable settlement of international problems in connection with the Columbia River to clear the way for a joint programme with the Province of British Columbia to develop the immense power in the waters of this river." Before 1957 was out Governor Jordan had retired. His place was taken by McKay, who was, as McNaughton had said, a former Secretary of the Interior in Eisenhower's cabinet. When talks resumed within the commission in 1958 not only had the changed political situation in Canada strengthened McNaughton's hand; so had the situation in the United States, where the Republicans were under strong attack from the Democrats for not making a bargain with Canada. The climate was favourable for a settlement. "Things are back in the hands of the International Joint Commission," one newspaper commented, "where the experienced hand of General McNaughton can look after the Canadian side and the U.S. Commissioners, headed by . . . Mr. McKay, have an equal grasp of realities."[133] "General McNaughton awaits the outcome with confidence," said another. "So do we."[134] A third report said: "Canada has need of the power, she has the water, she has the legal right and, we believe, she has the right man with the right team to ensure that her natural resources are put to her own use. Too often we have seen our precious heritage of natural wealth given away for fast money or politicians' good neighbourly whims. General McNaughton has a grave responsibility but we may be sure that Canada's interests are in good hands when he enters the new round of discussion. We wish him well."[135]

Throughout 1958 although the bargaining was realistic and tough there was a spirit of co-operation that had not been there before. But the United States, it seemed, did not intend to commit itself until the report of the International Columbia River Engineering Board came in; that was to be delivered to the commission in Chicago in March, 1959. Meanwhile a group of Canadian economists had been commissioned to report on what scheme of development was most favourable in the interests of Canada.*[136]

During 1958 a threat emerged in British Columbia that forced the commission's hand. Premier Bennett, who had been frustrated in his attempt to strike a "quick-profit" deal with Kaiser and other interests in 1954, had turned his attention to the Peace River in B.C.'s north. The Peace was not an international river and thus could be developed as Bennett saw fit, without interference from the federal authority. Bennett therefore moved in behind a group of private concerns led by Dr. Axel Wenner-Gren and his powerful B.C. Development Company. By damming the Peace it was estimated that four million kilowatts at two sites could be produced, with further possibilities at other sites. Already, in 1958, it was clear that the Peace–not the Columbia–was to be given priority in development by the B.C. government.

Bennett's object, which was laudable enough, was to spur development of northern B.C. But his assumption that secondary industry would move into the Peace River country was hardly valid. "Industry," as James Ripley pointed out in the *Engineering and Contract Record*,† "prefers to locate near its markets and bring power in." The scheme had three strikes against it, in McNaughton's view: transmission from the dam site on the Peace to Vancouver (the only market in need of large quantities of power) would be six hundred miles, two hundred miles farther than Mica Creek, and that meant costly power; the development was to be carried out by predominantly private interests, and, because the cost of financing privately developed power is higher than public power, this again would make for more expensive power; and thirdly, the Peace project, when combined with the Columbia (which McNaughton had no intention of abandoning) would saddle British Columbia with more than five times as much power as could then be used. Did B.C. need so much power, he asked himself, and was it needed now? He regarded the Peace River scheme, though a magnificent project at the right time, as disastrously premature. To proceed with it entailed a demand from Bennett for

* It was at about this time that McNaughton first saw Sir John Dill's statue at Arlington National Cemetery. Some years after Dill's death an equestrian statue had been unveiled at Arlington by President Truman, and Mrs. McNaughton, who accompanied the General on his IJC trips, saw it one day during a tour of the cemetery; she brought her husband along. "Yes," he said, "that's Jack Dill." He contemplated the figure for a long time.

† September, 1962.

long-term, large-scale exports of power, which Canada had opposed for more than thirty years as being against the public interest; once American industries had become dependent on Canadian power it would be impossible to get the power back when it was needed, as had been seen in the case of Ontario power during the First World War. The Columbia would meet all British Columbia's needs and more, and with downstream benefits—a return of power and cash for flood control—the cost of the power would be very low.

Nevertheless, there is little doubt that the Peace River threat to the development of the Columbia helped the Americans to see things McNaughton's way. It was clear to the Americans that British Columbia had an alternative to Columbia River power while they did not. It also accounts for Ottawa's anxiety to come to an agreement with the Americans on the Columbia and so win the race between the rival schemes.

Towards the end of 1958 McNaughton and McKay had reached a measure of agreement and reported to their respective governments.[137] Early in December Howard Green, Secretary of State for External Affairs, issued a statement, after first checking it by telephone with Diefenbaker, who was in Australia. Green spoke of the "co-operative development of the [Columbia] river, based on co-ordinated operations, but retaining Canadian ownership and control of projects built on the river in Canada." He went on to state the Canadian view that the sections of the International Joint Commission should be instructed "to work out methods for determining and apportioning both the downstream benefits from water stored upstream and the value of such storage for flood control." In other words, the principle of downstream benefits, after an eight-year battle, had at last been accepted by the United States.[138]

Many observers, both in Ottawa and Washington, concluded from this announcement by Green that McNaughton's scheme for diversion to the Fraser was off; that Canada would gain only payments for flood control and a negotiated fraction of the power produced in the American plants on the Lower Columbia. But that was not the case. Canada had made it clear that she would preserve her right to carry out the diversion of the Columbia to the Fraser when she was ready to do so. Fraser power was not yet needed. So, for a precise term (probably twenty-five to thirty-five years) the Canadian government was ready "to negotiate an arrangement with the U.S. for co-operative development of the Columbia." This would, moreover, give time for science to reconcile the use of the Fraser for power and fish; the salmon industry had raised vociferous protests against damming the Fraser and the influence of that on the B.C. government had been powerful.[139]

This took the negotiations a long step forward. We have seen that the value of hydroelectric power, which would be most useful for peaking in the future, was likely to increase. The Americans, aware of this, wished to obtain a final agreement on the sharing of the resources of the Columbia but this McNaughton was not prepared to give. He had no intention of entering into any agreement to dispose in perpetuity of the resources of the Columbia.* And Green's statement made it clear that federal policy was opposed to the plans of Premier Bennett for the development of the Peace River before the Columbia. "The postponement of such development," Green said, referring to the Columbia, "in favour of some other power scheme could well result in making the development of the Columbia River in Canada impossible."[140]

The tempo of the negotiations quickened in 1959. At the end of January the IJC received requests from both governments recommending the principles that should be applied to determine first, "the benefits which will result from co-operative use of storage of waters"—in other words, the plan of development; and, second, "apportionment between the two countries of such benefit, more particularly in regard to electrical generation and flood control"—the downstream benefits.[141] In March the International Columbia River Engineering Board submitted its report. The board, as has been noted, had considered three possible plans of development, none of which included the diversion of the Columbia to the Fraser: a non-diversion plan, which the Americans favoured, and which included Libby Dam; the Dorr diversion, otherwise known as the "McNaughton Plan," which made Libby impossible; and a compromise between the two, which allowed for a limited diversion of

* His reasoning, as given before the External Affairs Committee on 12 December, 1957, was this: "If we do these developments ourselves, we put ourselves in a position that when the Peace River Power has come in . . . we have something to integrate with in our own country. If we do not hold on to our Columbia, Fraser and Kootenay, and do these developments ourselves at this time, it means that when these further powers of the north are available, these may well slip out of our hands also into the service and into the economy of another nation."

the Kootenay (at Canal Flats), a smaller storage at Mica Creek than McNaughton had envisaged, and Libby Dam. The report showed not only that the McNaughton Plan was technically possible; it was "the most economical use of the water, the total investment costs being estimated at about eight per cent below the next alternative plan."[142] By this time the Canadian economists had also reported. The maximum diversion plan—the Dorr diversion plus the diversion of the Columbia to the Fraser—was the best in the interests of Canada. Then, at the end of the year, the IJC submitted its report, "Principles for Determining and Apportioning Benefits. . . ." to the governments of Canada and the United States.

With regard to the plan of development to be followed, the principles contained one recommendation that, "to provide optimum benefits to each country," the facilities for storage and downstream power production should be installed "in the order of the most favourable benefit–cost ratio." To implement this, McNaughton put forward for consideration that water should be stored at the highest possible elevation consistent with supply. This would maximize the potential energy stored for use by both countries through which regulated water would pass.[143] Thus Mica, an enormous reservoir sited as far north as it was possible to get, would be placed far ahead of any other project especially if the supply to fill it was augmented by Kootenay flows. There would be a fifty-fifty split on the downstream benefits; that is, Canada would get fifty per cent of all the power gained by the United States. The principle that Canada should be paid for flood control—half the value of the damage thus prevented—was also recognized.

The Canadian press was jubilant: "Gen. Andrew McNaughton has the stuff that makes great Canadians. He deserves official recognition for his outstanding victory in getting Canada a fair share of Columbia downstream benefits. This feat will stand as a great achievement in the history of Canadian international bargaining. . . . Gen. McNaughton endured much uninformed and unfair criticism on both sides of the border. But he knew his objective. He ignored all the irrelevancies. He developed a long-headed strategy. He stuck to it—to Canada's lasting benefit. . . ."[144] Though it was reported that the United States would return "value-holding electricity, not inflation vulnerable dollars,"[145] that was no longer quite the case. McNaughton's determined statement that the return of downstream benefits must be in power, not cash, had been qualified by Diefenbaker in a Vancouver speech: for a "strictly limited period," the Prime Minister said, cash might be accepted and credited to the cost of building the Canadian dams.[146]

The Winnipeg *Free Press* talked of a fifty-fifty split of downstream benefits. Canada's share would "easily supply all British Columbia's power needs for years ahead," it suggested on 17 August, 1959. "In the beginning Canada will take its share in expanding annual installments and will not develop any power on the Columbia north of the border. When that share is fully used it will be time to install turbines on the Canadian dam sites" and in the future to develop the Fraser by diversion. "Altogether," the report concluded, "this seems to be an admirable settlement, the result very largely of General McNaughton's stubborn struggle for Canada's interests. . . ."

The Canadian case for the best development of the Columbia Basin was based on geography, engineering, economics and the law—and it was strong. McNaughton, the "old soldier" who had been reported as being "on the verge of winning a great legal-diplomatic victory," it would seem had almost done so.[147] But a treaty had still to be negotiated. The Americans, fearing that McNaughton would be a member of the negotiating team, pressed for a strong replacement for Douglas McKay, who had died in July, 1959:

In picking a successor for Douglas McKay, the State Department . . . should take serious cognizance of the great immediacy and importance of matters being handled today by the commission. These problems . . . may have great diplomatic and legal significance, but above all they are engineering problems.

It is appropriate to note here that Canada's three-man half of IJC is headed by an engineer and administrator of great prestige, vigor and ability—A. G. L. McNaughton. General McNaughton represents Canada's interests in the IJC with unquestionable competence.

. . . there are certain problems before the IJC that can't possibly be settled except through negotiations. And in negotiating these matters we need a champion like Canada's in our corner.

In the best interests of our nation President Eisenhower should name a worthy counterpart for General McNaughton and fairly soon. And it is to be hoped that a man of proven technical competence in the water resources field will be chosen—ideally, one who knows the Columbia Basin water situation, as this is a question of prime importance before IJC.[148]

A Holiday in Bermuda, March, 1957
(BERMUDA NEWS BUREAU)

An examination of the South Saskatchewan River project with Gordon Mackenzie of the Prairie Farm Rehabilitation Administration
(THE DIRECTOR, PRAIRIE FARM REHABILITATION ADMINISTRATION, CANADA DEPARTMNENT OF AGRICULTURE)

Nelson Daily News

Why the Celebration?

Yesterday, Premier W. A. C. Bennett and Dr. Hugh Keenleyside were celebrating the construction of High Arrow Dam. We wonder why.

At the time, Mr. Bennett said that "signing Columbia River Treaty was the happiest day of his life." But now he must be so disappointed.

And Dr. Keenleyside said in April, 1964, that the successful sale of all Canadian flood control and power benefits for 30 years to the U.S.A. for $344 million would pay for:

1. All capital costs and operational expenses of dams and reservoirs at Duncan, Arrow and Mica.

2. Pay half the capital costs of generators at Mica.

However, according to B.C. Public Accounts of March 31, 1968, when the big dam was barely started, $344,628,253 had **already** been spent on the three dams and that that "final costs cannot be determined."

And according to PROSPECTUS, a Government of B.C. publication of Jan. 11, 1967, Arrow's costs would be $229 million. That is $100 million more than the Treaty's estimate of $129.5.

While celebrating, let us remember and appreciate that:

1. We have surrendered control of Columbia's flows to the dictates of Bonneville Power Authority, forever.

2. We have sold and have been paid for our estimated half of the increase in downstream U.S. capacity of 2,700,000 KW. It is now reported that Grand Coulee's capacity is to be increased from 2 to 9 million KW. But Canada does rot share in this increase, nor in the resulting expansion of United States business and employment.

The High Arrow, alone, destroys 40,000 acres of valuable timber and agricultural land.

Population in the wide, flat valley between Revelstoke and Arrowhead is wiped out. The area is a wasteland of thousands of stumps which will be navigational hazards.

2,200 people have been moved.

The recreational value of the shores of the river and the lakes between Castlegar and Revelstoke is sadly diminished by the loss of the natural beaches.

It is odd that Mr. Bennett should now be celebrating the High Arrow when we recall that in May, 1967 he said, "The Treaty missed out. The thing that's at fault in the Treaty is that it was signed too quickly."

General McNaughton would have agreed with Mr. Bennett;

And so do we,

THE MOURNERS OF THE ARROW LAKES

Non-Partisan Association

Notice in the Nelson *Daily News*, 10 June, 1969 on the occasion of the dedication of Keenleyside (High Arrow) Dam

Conservationist, 1962-1966

(NATIONAL FILM BOARD)

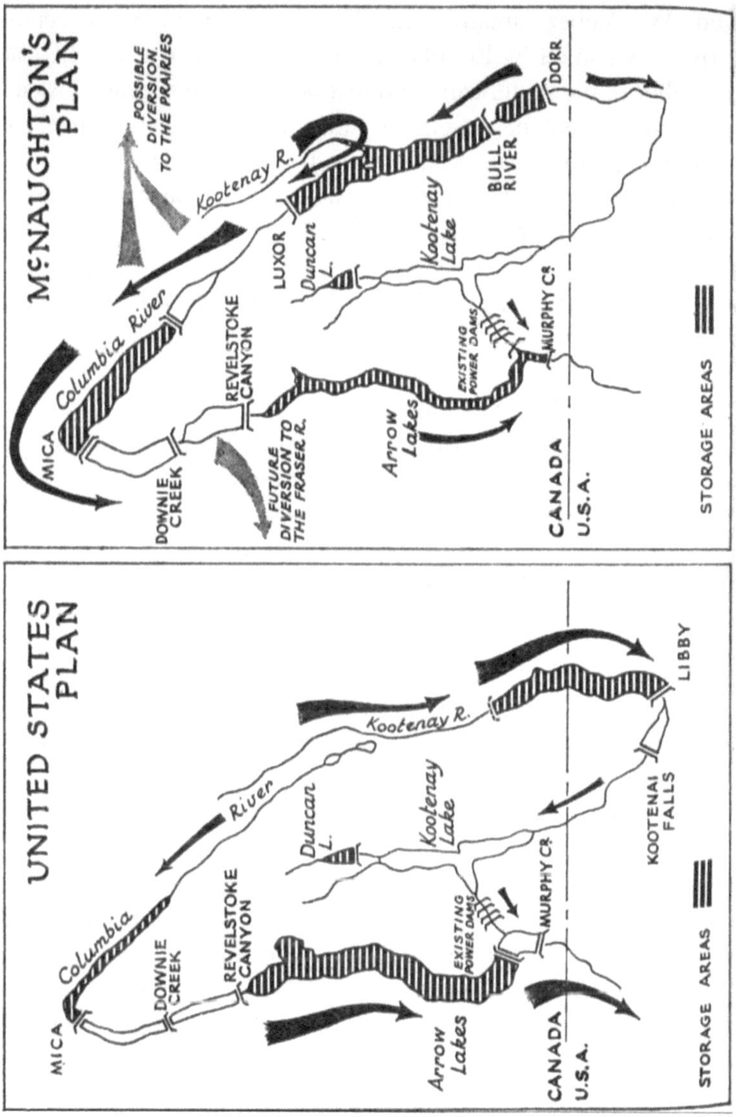

The sizes of pools at Libby and Arrow Lakes show diagrammatically the approximate amounts of water to be stored there.

Eugene W. Weber, already a member of the United States section of the IJC, was named by Eisenhower as acting chairman. But he was not a member of the American team that negotiated the treaty; instead Itschner, now Lieutenant-General, was named, a nominee of unquestioned competence. And on the Canadian side, for reasons that have never been satisfactorily explained, General McNaughton took no part in the negotiations.

The negotiating teams, as announced by Diefenbaker in January, 1960, were as follows:

Canada:	E. Davie Fulton (Minister of Justice), *Chairman* Gordon Robertson (Deputy Minister, Northern Affairs and National Resources) A. E. Ritchie (Assistant Under-Secretary, Department of External Affairs) E. W. Bassett (Deputy Minister, Lands and Forests, British Columbia).
United States:	E. F. Bennett (Under-Secretary, Department of Interior), *Chairman* I. B. White (Assistant Secretary, Department of State) Lieutenant-General E. C. Itschner (Chief, U.S. Army Corps of Engineers).[149]

Itschner had had long experience on the International Columbia River Engineering Board, having been appointed a member in October, 1949. He was chairman from July, 1955, until the board reported in 1959, and was the architect of the American plan: stacked against him as a negotiator was Bassett who had merely been a junior member of the Canadian section of the Engineering Committee since 1954. "It would perhaps be naive to ask," commented Larratt Higgins,* "*where* General McNaughton was in this picture." Leon Ladner, who had been appointed member of a Royal Commission on Energy, pressed for the inclusion of McNaughton on the negotiating team. As he told the General, "Your tact and sound judgment will be

---

* *Saturday Night*, 26 May, 1962.

invaluable at certain moments of the negotiations. . . . What a magnificent work you have done and are doing. If the Government does not erect a statue in your honour, I will!!!"[150] But McNaughton was not brought in.

"It has become the direct responsibility of others," he said, "to take the framework of definition of benefits, of methods for their measurement and allocation, and of procedure of co-operation in their use, and to apply these to the specific circumstances of the Columbia River Basin. You will appreciate that this is an extraordinarily important assignment which has been given to the two teams of negotiators. . . . We now move into a stage where the Congress of the United States will become the forum of debate, and continued vigilance will be required to present the rightfulness of the Canadian position and the reasonableness, the fairness and the equity to both parties of the solutions we offer."[151]

The negotiations began in February with the Canadian team fully aware that Canada held the aces. The right to divert the Columbia into the Fraser at some future date had neither been abandoned nor weakened. "The economic values of such diversion for B.C.'s long-term development have been seriously assessed by the 'economists panel,' " said the *Financial Times*. "If Canada should bargain them away–which now seems unlikely–at least it will be in full knowledge of what they are worth."[152] Bassett, a provincial civil servant, presumably would protect the interests of British Columbia.

At first things went smoothly. Conditions were right for a favourable settlement. Though General Itschner fought hard for the inclusion of Libby, the United States recognized that the McNaughton Plan for the Dorr diversion would give most benefits to both countries at least cost. In flood control and power the value resulting from the McNaughton Plan would be slightly greater to the United States than that resulting from Libby and the non-diversion scheme. The McNaughton Plan, however, would not only be much less costly ($200,000,000 as opposed to $324,000,000 for Libby); it would bring 400,000 kilowatt-years of energy annually at-site in Canada, which would be totally lacking under the other scheme. This very satisfactory situation prevailed, as McNaughton reported, until the middle of June:

> . . . after the matter passed from the Commission to the negotiators, until mid-June 1960, the record shows there was acceptance by the U.S. of the basic proposals I had made, namely, that the development of the Kootenai in the U.S.A. including Libby . . . and of High Arrow in Canada would be dropped and the much more advantageous proposal for Dorr, Bull River-Luxor adopted together with the early machining of Mica. . . . This arrangement represents a *reduction* in the total investment for both countries of more than $250 million and an *increase* of 450 megawatts of Firm Energy net at load centres.[153]

High Arrow, it may be noted here, was a proposed dam on the Columbia not far north of the international border, that would raise the waters of the Arrow Lakes. It would drown out twenty-seven thousand acres of ground, as well as established communities and communications, simply to store some seven million acre-feet of water for the United States. There would be no benefit whatsoever to Canada in the way of power production.

Such, then, was the situation in June. The McNaughton Plan obviously was in the bag but to make absolutely sure that the American understanding of the clauses for diversion were the same as explained to him by McNaughton, Fulton held a secret meeting with the United States negotiating team at the luxurious Seigniory Club on the Ottawa River, east of the city.

> He found that they were, Elmer Bennett, Chief U.S. negotiator, was reported . . . as saying that the U.S. had offered to give up Libby dam and permit total Kootenay diversion to get an agreement. Neither Fulton nor his opposites . . . had any reason to prolong . . . [the] final conference beyond formalities of signature.[154]

Tom Gould, the respected Ottawa correspondent of the *Victoria Times*, explained what happened shortly afterwards in Ottawa:

> Behind a door covered with sound-dampening green felt, an assorted group of engineers, economists and politicians met to discuss the fate of the Columbia River Basin. The place was Ottawa. The time, early spring of this year. One of the British Columbia representatives suddenly blurted out: "B.C. has no intention of allowing the construction of dams in the East Kootenay Valley."
>
> A hush fell over the room. Justice Minister Davie Fulton, Canada's senior negotiator, immediately called a recess. The senior U.S. negotiator, Elmer Bennett, wore a puzzled look while elation shone

on the face of Gen. Itschner, chief of the U.S. Army Corps of Engineers. . . .[155]

Premier Bennett, still infatuated with his vision of Peace River power, had deliberately sabotaged the Columbia negotiations for, without flooding the East Kootenay, the Dorr and Bull River-Luxor pools could not be formed. All Bennett wanted from the Columbia was quick money to pour into the other scheme. Thus, during negotiations with the United States, he "undercut Canada's bargaining position" and "unilaterally vetoed" the McNaughton Plan. "This was the decisive political act which set the stage for the content of the draft treaty. . . . This act gave birth to Premier Bennett's two-river policy. In the most simple terms this policy provides for *storage* on the Columbia to meet U.S. needs. Cash received for this 'service' will go to finance at-site power development on the Peace. It means, 'Sacrifice the power potential in Canada on the Columbia in order to cut down the cost of developing the Peace.' A very short-term perspective indeed!"[156]

Fulton was dumbfounded, as he had every reason to be. The British Columbia government had been kept informed of the negotiations every step of the way; "its few objections had been met, no demands for changes had been expected from that quarter. Had B.C. wanted changes, national interest required that they be raised before final negotiations and in private with Ottawa representatives. None had been asked."[157]

B.C. was satisfied, and Fulton confidently but incautiously permitted a full delegation of the B.C. government . . . to be present at the formal session—where they had no right to be.[158]

Then the devastating blow had fallen and, with that, Fulton's bargaining power was stripped away.

The question remains, had British Columbia the right to act as it did? Bennett's stand not only disregarded the national interest; it also ignored the inter-provincial interest, for the McNaughton Plan was of potential future benefit to the Prairie Provinces. Was not B.C.'s conduct a usurpation of federal rights in international treaty making? And was it not an act which B.C., or any other province, could logically extend later to any kind of treaty which affected it?[159] Fulton (and he was Minister of Justice) explained that ". . . while Canada has the treaty-making power and the jurisdiction reflected in the International Rivers Improvement Act, the Province of British Columbia has the broad and extensive jurisdiction given by the Constitution over the use to be made of property and resources situate in the Province."[160] But three members of the External Affairs Committee, which approved the final treaty and protocol, in expressing their dissent from the majority opinion said this: One of the "most disturbing features" was the "willingness of successive government representatives of Canada to accept a second-best approach . . . because of the dictation, or veto, of the representative of one province–British Columbia. We protest," they continued, "that this is contrary to the Canadian Constitution which clearly vests legislative power in respect to the development of international rivers in the Parliament of Canada. . . ."[161] And C. B. Bourne of the Faculty of Law, University of British Columbia, said: "Provincial authorities could not frustrate a determined unilateral attempt to develop an international river by the federal government."[162]

Be that as it may, the Canadian government accepted Bennett's veto. Two possibilities were open: to drop the negotiations, with the prospect of reopening them, after the differences with B.C. had been settled, from a position of strength; or to continue them from a position that had now been undermined in the hopes of salvaging at least something from the wreck. The Prime Minister was said to have badly wanted to sign some sort of treaty with Eisenhower on his first visit to the White House and he ordered the negotiations to continue. Fulton, therefore, was under instructions to get a treaty for the Chief to sign.[163] And there is nothing inherently improbable in this. By 1960 the Diefenbaker honeymoon was over; the country was almost ready to divorce itself from him. The magic with which he had dazzled the eyes of the public had washed away and the Prime Minister was revealed as an indecisive and discredited man. His reputation had sunk both at home and abroad, but to put through an international treaty, on the scale the Columbia was likely to be, might help restore his flagging prestige.

Canada, as must have been obvious to the Americans, was a house divided. But in a family quarrel about the car, the husband does not, as a rule, go out and give the car away.

The giveaway started with Libby. The McNaughton Plan, by

diverting the flood waters of the East Kootenay, had guaranteed flood control on the Kootenai, which was what the Americans wanted. Now there would be no Dorr diversion, and the Americans, not unnaturally, asked what was to happen about flood control in these changed circumstances? So the Canadians *pressed* the Americans to include Libby Dam by offering great inducements. Of this Fulton said: "When British Columbia decided that it would not permit this use of its resources because of the tremendous flooding involved, this had to be abandoned. Since the only solution of the flood problem downstream . . . is provided by the Libby Dam, it became necessary to contemplate Libby if any Treaty were to be concluded."[164] We have seen that when the United States applied to the IJC for permission to build Libby Dam in 1951, they had offered seven million dollars in compensation for the flooding that would occur in Canada. Now Canada would make available not only the land, and prepare it for flooding at a cost then estimated to be between ten and fourteen million dollars, but also to bear the cost*[165] "in consideration of the benefits accruing to it." These benefits would be those of regulated flows to Canadian plants at Trail on the West Kootenay (after the Kootenai has re-entered Canada), which in 1951 had been considered negligible. Now, since Canada according to the treaty was to have no say in how these flows should be regulated, the benefits might well be nil.† In fact, if the Americans were to use these flows for peaking, as was most likely, the benefits would be nil and could affect adversely the generation of electricity in the Trail area.‡

---

* For the record, McNaughton advised "that under no circumstances should the U.S. be allowed to pay directly for any property or construction in Canada for if we did so we would be open to the claim that an encroachment on our sovereignty had been 'bought and paid for' and this, it could be expected, would be very troublesome later on." He did, however, expect Canada to claim downstream benefits.

† According to the treaty, if a variation in the operation of storage is considered to be of advantage to Canada, the U.S. shall consult with Canada and if the variation would not be to its disadvantage it shall vary the operation accordingly. The U.S. is under no obligation to grant this favour. It is very likely that the U.S. will "over-regulate" Libby (as they do their Hungry Horse Dam), to produce high flows on the Kootenai in the winter and cut off the flow altogether in the summer. Under the treaty they can also do this with Duncan Lake, another treaty project, and so the benefits to firm power at the West Kootenay plants would be minimal. The only recourse left to the Trail plants would then be to sign an interconnection agreement with the United States on terms very advantageous to that country, as has been done on the Pend Oreille in order to obtain more capacity at Waneta.

‡ "The trouble with Libby," McNaughton said, "is that U.S. becomes the upstream state on the Kootenai with absolute unrestricted control of Libby storage and refill

"It is only fair to the U.S. negotiators," McNaughton commented, "to say that their draft contained no mention of 'in consideration of the benefits accruing to Canada'; they were entirely concerned with benefits to be obtained by the U.S. to raise the overall benefit-cost ratio of an uneconomic project to the extent that it would be made acceptable to the Congress of the U.S.A. *The words added were at the instance of the Canadian negotiators* * to make it seem that there was some *quid pro quo* of substance."[166] And Canada would reap no downstream benefits from Libby Dam either for flood control or power, despite the fact that forty-two miles of her territory would be flooded at her own expense; "all benefits which occur in either country from the construction and operation of the storage accrue to the country in which the benefits occur"—a lopsided arrangement, to say the least. "In the fine print, like some improvident mortgage," said McNaughton, "the equitable basis of the equal division proposed in the I.J.C. Principles has been vitiated."[167]

Apart from these appalling concessions, the option to construct Libby Dam which the Canadians gave the Americans had much more far-reaching consequences. *Libby was the key to U.S. mastery of the Columbia.*† It competed, and was incompatible with, the McNaughton Plan to maximize the generation of power in Canada. The Columbia-to-Fraser diversion would be lost to Canada forever and, with Libby built, and thus a vested interest to which the waters of the Kootenay had been "dedicated," Canadian diversion of the Kootenay to the Columbia would not prove possible either. The treaty included clauses for diversion. We can divert, the supporters of the treaty argued, starting in a small way twenty years hence and concluding it eighty years from then,[168] and these rights were claimed by the Department of External Affairs "as a benefit from the Columbia River Treaty."[169] As Canada already had the unrestricted right to divert (under Article 2 of the Boundary Waters Treaty of 1909), to *forego* those rights, even for twenty years, under the terms of the new treaty

---

which can be used to drown out West Kootenay plants or cut off the supply for extended periods."

* Italics supplied.

† John Krutilla, an American who has studied the Columbia River Treaty, contends that the United States built Libby Dam for political reasons. Its construction was a net loss to the United States and an overall net loss so far as concerned the most efficient use of the basin as a whole. See his *The Columbia River Treaty* (Baltimore: The Johns Hopkins Press, 1967).

seems a strange benefit to claim; but it is typical of the External Affairs apology for a shocking treaty. With ratification of the Columbia Treaty by Canada the right to divert would be forfeited altogether for two decades, limited for four decades and then it would be handed back for another four decades but–and this is important–*subject to litigation concerning downstream vested interests on the Kootenai that will have grown up.*[170] Under those conditions, we may be sure, Canada will be unable to afford the Kootenay diversion. And can it be thought that a British Columbia government, which had vetoed the Kootenay flooding in 1960, would be any more willing to sanction drowning out the Kootenay Valley twenty years afterwards, when a different sequence (including the flooding of Arrow Lakes) had been established, the population had grown, and the ground to be flooded had vastly increased in value? Even that would be a naive assumption.

With Libby, the key project from the United States point of view, "handed over," as McNaughton put it, together with "a bonus to make it seem the U.S. were doing Canada a favour and without restriction of any sort on the use of storage . . . I do not believe an outcome equitable to Canada could be secured. Mr. Fulton and his colleagues *had surrendered and sold the pass* and all that remained thereafter for the U.S. negotiators to do was a 'mopping up' operation which, ably abetted by the B.C. representative on the *Canadian* negotiating team, they performed with great technical skill."[171]

Libby, in fact, set up a chain reaction in Canada's Columbia Basin that worked to the advantage of the United States. The treaty provided for three projects in addition to Libby Dam: High Arrow, Duncan Lake and Mica Creek. High Arrow we have examined briefly. This was a large storage project useful to Canada only for the return of downstream benefits but of great value to American downstream plants. It has been reported that, at Bennett's insistence, High Arrow had been accepted by the Canadian negotiators even before B.C. sabotaged Canada's bargaining position in June.[172] As a high dam at the mouth of the Arrow Lakes would cause extensive flooding it may be wondered why Bennett should insist on this; but in the light of subsequent developments it is not illogical from Bennett's point of view. When vetoing the McNaughton Plan he could say, "Look, the Arrow Lakes area will already be flooded; British Columbia cannot tolerate the flooding of the East Kootenay as well." What is illogical is that Bennett's action would cause a more valuable part of B.C.'s territory to be put under water than would have been the case under the McNaughton Plan; and it would displace more people—2,250 by Libby and High Arrow but only 1,000 by Bull River-Luxor and Dorr.

Bennett's arguments were that valuable land would be flooded in the Kootenay and that an east-west transportation barrier would be created. "These two arguments," wrote an engineer who made a thorough study, "are obviously specious. Experts agree that flooding would actually improve the land in the valley by providing a controlled water supply [thus wiping out floods and ensuring flows in the dry season] and new chances of irrigation. As to a transportation barrier, one already exists—in the form of rivers and mountains. The reservoir would add very little to the Kootenay's width."[173] And, as the agricultural land in that area gets better from north to south, it is certainly odd that Bennett should forbid the flooding by Luxor of ground said to be "almost worthless" while permitting Libby to flood an area which contained "the most valuable land of all."[174]

Fulton's version of the inclusion of High Arrow is this: "With Bull River-Dorr [the McNaughton Plan], it might have been possible to create a pool of storage, including Mica . . . without Arrow Lakes. But when British Columbia took the firm position that they would not permit the flooding in the East Kootenay necessary for Bull River-Dorr storage, then High Arrow became an absolute necessity. . . ."[175] High Arrow, then, came in as a treaty project that would flood forty thousand* acres of Canadian ground. And, unlike flooding in the East Kootenay, the loss of agricultural land there would be permanent. "Because of the steep-sided valley, no new development at a higher level is possible . . . agriculture would be wiped out permanently by High Arrow Dam."[176] The three members of the External Affairs Committee, who dissented from the majority position that the treaty and protocol should be approved, said this of Arrow Lakes:

The Treaty unnecessarily includes in its projects the High Arrow Dam which will cause disaster to long-established Canadian communities, the unnecessary destruction of natural resources and sixty years of public and private investment in public facilities in agriculture and

---

* The External Affairs *Presentation* gives a figure of 27,000 acres; but the ICREB Report (Appendix I, p. 80) states: "The reservoir area at full pool level would be 130,200 acres, which is 40,000 greater than the area of Upper and Lower Arrow Lakes."

industry *merely to sell benefits to the United States which are temporary in character and of diminishing value.*[177] *

McNaughton saw the dislocation of established communities as the most serious effect of High Arrow. But he queried its usefulness on technical grounds also. External Affairs claimed that "This reservoir at Arrow Lakes made it possible to operate the upstream projects [Mica and associated sites] to generate power in Canada for Canadian needs and then to re-regulate the resulting river flows so that they crossed the border to the United States in a flow pattern suitable for downstream benefit generation and consistent with our Treaty commitments."[178]

Thus High Arrow was to be used to re-regulate flows from Mica to match United States requirements for power after Mica had been machined and was being regulated specifically for at-site power in Canada. "Under these circumstances," McNaughton observed, "it is a little odd to call for the construction of High Arrow to be completed before Mica itself, and long before machines are to be installed at Mica." Until Mica was machined, its storage could easily "provide the regulation for downstream benefits in the U.S. and similarly for flood control."[179] And as the U.S. requirements for the regulation of flows along the Columbia would prevail–under the treaty they would have the right to order how Canadian storages would be operated, as will be seen–there was grave doubt in McNaughton's mind that "it would be economically practicable for Canada to proceed with the installation of Mica generation."[180] In that case, without at-site power generation at Mica, he saw no need for High Arrow (included in the treaty as a re-regulator) at all. Finally, he queried the cost of the dam–some seventy-two million dollars–which made the cost-benefit ratio appear favourable. This figure had been used by the negotiators but, according to McNaughton, there would be difficulties with the foundations of the dam resulting in a cost of between a hundred and a hundred fifty million dollars before construction was through. "It is," he said, "a matter of great importance in making a decision to know what the additional costs would be and what would remain in the way of net benefits, if any, even without taking proper account of the misery which the construction will cause to a large, well-established group of our people."[181]

* Italics supplied.

A dam at Duncan Lake was another treaty project. This had been included in the earlier Canadian proposals for this reason: the Kootenay diversion, called for in the McNaughton Plan, would deprive the Consolidated Mining and Smelting Company at Trail of some Kootenay water; Duncan Lake Dam, therefore, was to be constructed to make up for the loss of this water at the West Kootenay plants. That was all. Its usefulness to Canada disappeared with the scrapping of the McNaughton Plan.[182] Nevertheless, the Americans wanted it; it would be another useful storage for them. Fulton did not say much in justification of the addition of Duncan Lake except that Arrow and Duncan Lakes would provide "together 8.5 million acre-feet of storage."[183]

And so we come to Mica, the remaining project, which had been the heart of the McNaughton Plan. Well, the defenders of the treaty have said, Mica is included; there can be no grounds for criticism there. "Mica," Davie Fulton said, "makes possible the generation in Canada of tremendous blocks of power. It is the key project for the development of power on the Columbia River in Canada, not only at the site itself but at the downstream projects of Downie Creek and Revelstoke Canyon. Mica alone, when machined, is capable of producing more than 6¾ billion kilowatt hours of energy per annum, or about the same amount as is returned in the form of downstream benefits for all these Treaty projects."[184]

But the Mica project of the treaty was not the Mica that McNaughton had planned. This would be an emasculated Mica—"Little Mica," it has been called—a storage deprived of almost six million acre-feet of regulated flow of water because the Kootenay had not been diverted. Mica, high on the Columbia, would have a large reservoir capacity but, without the inflow of Kootenay waters, it would refill very slowly after drawdown, and "Drawdown to meet Treaty obligations to the U.S.," commented an American observer, "would take precedence over drawdown to meet at-site power needs."[185] Thus, without the full pool that diversion would have given—enhancing any at-site power operations—Mica's storage would probably go downstream to meet American requirements for peaking, which would not provide the steady flows required for the generation of firm power in Canada.

The kind and amount of regulation [of flows and refill on the

Canadian Columbia] which it will be open to the United States to impose in its own interests is unlimited and could be ruinous to Canada. . . . This implies that anything and everything Canada may ever want to do must be considered in the light of an unpredictable exercise of then American rights—a risk which may make unfeasible any developments beyond those compelled by the treaty.[186]

So, although the inclusion of Mica looks good on paper, its use as a Canadian power-producer will be greatly impaired by the treaty terms.* And the same applies to Downie Creek and Revelstoke Canyon, both downstream from Mica. By the time Mica could be ready for the installation of generators, the U.S. requirements for regulation would be for "peaking," Canada's for "firm power." The two are incompatible. With the rights of regulation defined in the treaty all in favour of the United States, the United States desire would prevail. McNaughton's bleak summary reads as follows:

It is, in fact, *very doubtful if, in these circumstances, it would be economically practicable for Canada to proceed with the installation of Mica generation,*† a situation which would no doubt suit the U.S. Government very well because their interest in the Columbia River basin is in storage upstream for use for power and flood control benefits in the U.S. unimpeded by any other consideration whatsoever. That is, their interest in the Canadian Columbia is as a simple "storer of water" or provider of "reservoir space."[187]

It is doubtful if Fulton understood this. In a letter published in the September, 1962, *Engineering and Contract Record* he said: "Critics now advance the view that under the Treaty the United States has been endowed with the right to direct the operation of the Canadian storage. This is simply not the case. Neither the responsibility for, nor the control of, the operation of the Canadian storage dams for power will rest with the United States." As to that, McNaughton said:

Mr. Fulton may be expressing a naive hope but the strict legal interpretation of the Treaty, which it is evident is also the view of the U.S. authorities, is the direct opposite. I note that it is only in respect

---

* The average headpond level for Mica is significantly lower for United States regulation studies than it is for Canadian load. This can be seen diagrammatically in reports of the External Affairs Committee hearings, April, 1964, p. 618.
† Italics supplied.

to power that Mr. Fulton claims that jurisdiction remains with Canada. So presumably he accepts the fact that in regard to flood control, U.S. authority to direct is spelled out precisely in the relevant clauses of the treaty . . . .[188]

What McNaughton was getting at, of course, is that Mica, the only Canadian site in the treaty for the generation of power, should, in the Canadian interests, release its water in a flow pattern that would result in the steady generation of firm power in Canada. Now if the United States called for the holding back of the water because of flood risk downstream, Canada would have to comply, to the detriment of her steady flows and power generation. And the United States had the right, vested in the treaty, to decide when this should be done. Thus flood control could affect adversely power production in Canada. Worse, so could power generation in the United States.

And not only *shall* Canada [McNaughton went on] operate the Canadian storage on the call of the U.S. but Article IV (5) of the treaty goes on to prescribe that any additional water resource development *shall not* be operated to adversely affect stream flow so as to reduce flood control and *hydroelectric* benefits which the *operating plans* which the U.S. *will submit* will otherwise provide. "Submit" is a deceptively mild word but it does not hide the fact that *Canada must carry out these plans* which can be varied by the U.S. with no requirement other than to *consult* with Canada in advance. Consult does not mean that the U.S. need pay any attention whatever to the view of Canada.[189]

McNaughton predicted that there would be conflict, for this reason. The United States would, in all likelihood, use the Canadian storage for peaking. We have seen that, in any mixed-power system, hydroelectric power can most economically be used for this. The American requirement was for a steady inflow at their Grand Coulee Dam from which releases would be made to meet peaking needs. They would obtain it by over-regulating Mica to compensate for unregulated flows elsewhere in the system. Over-regulation means that the outflow is cut off, perhaps completely, during the high-flow period in order to impound as much water as possible behind the dam for use later in the year; and, in that later period, large flows would be released, very probably far in excess of Canadian requirements. Therefore, the over-regulation of Mica—starving the turbines at one period and flooding them the next—could only have adverse effects on the production of Canadian power. And, without any Canadian generation at all initially, the first operating plans would be drawn up by the United States in the interests of their own country.

Thus the American interest in peaking would be vested, right from the start: and the American operating plan an existing plan that could not be changed unless the United States agreed—highly unlikely because of the resultant great loss to them. But when Canadian generation had been installed, although in opposition to the interests of the United States, it would be important to redirect the plan to firm power, calling for steady flows at Mica.

Under the Treaty "failure to agree" on operating plans would constitute a *dispute* which, by Article XVI (1) *et seq.*, either party may refer to the I.J.C. or other tribunal for *decision.* The I.J.C. or other tribunal operates under the provisions of the Treaty which are specific, and . . . of great importance . . . is the objective of the regulation defined in Annex A, para 7 as *optimum power generation* in the basin as a whole. The Canadian installed capacity will be small compared to the U.S., so in optimum generation *it will be the U.S. interest in the regulation of the out-flows from Canadian storage which will predominate* and no doubt the I.J.C. or other tribunal which is required to apply the Treaty provisions strictly, must so direct.

In consequence it will be plans for peaking which Canada will be required to implement under pain of "appropriate compensation"; and, since equity is not at issue under the Treaty as it would have been under the I.J.C. Principles, *there will be no recourse but to carry out a plan which while designed to give optimum benefits in the basin as a whole . . . gives the U.S. the kind and amount of benefits designed by them, and, as regards Canada, does great harm to firm power production which is the Canadian need.*[190]

This was recognized by at least one American authority, though evidently not by Fulton. Professor M. E. Marts, writing in *Water Power*, February, 1961, said:

Another problem involved is the conflict between on-site power at Mica Creek and releases of Mica storage to meet the Canadian commitment to the United States. Releases of storage during the winter downstream season mean, of course, loss of head for the remainder of the season. Although this problem is inherent in all storage-power dams, it is aggravated because of Mica's size and cost, its location at the upper end of the system where inflow is small relative to the size of the reservoir and transmission is costly, and particularly because releases from Mica storage will be for a foreign power system, not for a British Columbia system of which it is a key project.

To ameliorate the operating conflict at Mica, Marts suggested, Canada might give up any thought of producing power at Mica: "An on-site power plant might not be constructed at Mica, but many government statements have encouraged the Canadian public to expect not only power but low-cost power from Mica."

So much, then, for Fulton's claims of "more than 6¾ billion kilowatt hours of energy per annum" to be produced in Canada by Mica. It is too early to know if McNaughton's fears will be justified but if they are the Stars and Stripes can be run up over every one of the treaty projects: Libby, High Arrow, Duncan Lake *and* Mica. "The basic plan of this treaty," Stewart Udall, Secretary of the Interior, told Congress, "is for storage of water in Canada" for power generation and flood protection in the United States.[191]

"It was," McNaughton said, "in the control of storages and the financial benefits that the United States outnegotiated Canada."[192] It would be worthwhile to set out each clause of the treaty side by side with the General's comments, but space does not permit.* We have seen something of the control of storages directly or indirectly by the United States. A striking feature of the clauses relating to control is this: although Canada is the downstream state from Libby Dam, she has no say whatever in the regulation of the storage there; the United States, downstream from the Canadian storages on the Columbia, can control them very well.

In the financial aspects of the treaty Canada was just as thoroughly outmanoeuvred. Under the treaty, as we have seen, Canada was required to provide storages at High Arrow, Duncan Lake and Mica Creek—under a strict timetable and with penalties for lateness—at a cost to Canada of about four hundred and sixty million dollars.† These storages were to be used for flood control and to smooth out the flow at generating plants in the United States. To provide flood

---

* A detailed analysis of the Columbia River Treaty and Protocol will be made in a book by an expert in the subject, which is in the course of preparation.

† This figure, as McNaughton predicted in the case of High Arrow, proved totally unrealistic.

control in the United States *forever*, the United States would make a single payment of some sixty-four million dollars ($52,100,000 for High Arrow, $11,100,000 for Duncan Lake and $1,200,000 for Mica), together with four small payments for special calls under Article 6 of the treaty. These total payments would not even pay for High Arrow Dam. American anxiety to have the High Arrow storage is revealed in these figures, as are the penalties: $192,000 a month for delay in completing High Arrow, but only $4,500 a month in the case of Mica. Dr. John (Jack) Davis, the Member for Coast-Capilano and Liberal critic of the treaty, said this of the sixty-four-million-dollar flood control payment: "We are prepared to make a lump sum, all-out and forever settlement." But, as he pointed out, flood control would become more valuable as time progressed, as the United States developed more industries and its population increased. "So why," he went on, "make a flat, all-out lump-sum settlement, even if it is $64 million, when this great advantage increases with time?" It would cost the United States, even in 1960, five hundred million dollars to make alternative arrangements.

Consequently one can see that $64 million, while it is a substantial sum in the eyes of people who look only a few years ahead, is a very low price for the United States to pay for this service in perpetuity; forever, holding back these billion tons of water each year and letting it down in an orderly fashion so as to protect real estate in the United States and to generate more electrical energy. . . . And yet, that was the way the treaty was drawn.[193]

That American rights to operate the Canadian storages for flood control would give them the ability "to generate more electrical energy" was incontestable. As F. J. Bartholomew, dean of British Columbia consulting hydro-engineers, remarked: "We are compelled to obey U.S. orders in accordance with U.S. flood control. This gives the U.S. the right to limit *their* * storage for use for flood protection and conserve it for optimum power production, and call on Canadian storage first for flood protection. . . . Generally this may not cause serious loss to Canada but in certain years it could necessitate a major waste of Canadian water."[194] Though the supporters of the treaty comfortingly suggested that the United States would never exercise its

* Italics supplied.

full rights solely for its own benefit in a manner ruinous to Canada, then why, McNaughton asked, should the treaty include those rights?* And why perpetual rights?

The IJC Principles had firmly suggested a fifty-fifty split of the additional power produced in the United States as a result of the Canadian storage. Even that, through Canadian neglect, was compromised. Davis, speaking before the House (in Committee of Supply) said:

> And as for the fifty-fifty sharing arrangements for downstream benefits, the Canadian negotiators . . . overlooked the benefit that Canadian flows, timed to supplement American secondary generation and so combine with it to produce firm power, would have. As a result Canada gets approximately 40% of the additional firm, reliable power, produced in the United States as a result of the treaty.[195]

This, McNaughton said, would result in an annual loss of $7,900,000, "a very large sum to lose in a conversation between the chief Canadian and U.S. negotiators. When I heard about it afterwards," he went on, "and protested, Mr. Fulton told me he had *committed himself.*"[196]

Not only were the projects to be developed much more favourable to the United States; in the financial aspects the United States would have by far the better of the bargain. "The terms of the proposed treaty," wrote Dr. L. Austin Wright, a former general secretary of the Engineering Institute of Canada, in the 23 September, 1963, issue of the Vancouver *Sun*, "are a disgrace to Canada's political negotiations and an expensive indignity to Canadians." From the time of Bennett's sabotage in June, Fulton had been severely handicapped but McNaughton did not spare him. "The result, of course," he commented, "with B.C. playing in with the U.S. in order to secure cash for Peace River development, has been deplorable and this treaty, which Mr. Fulton negotiated, represents a major calamity for Canada and equally so for B.C."[197]

Dr. Hugh L. Keenleyside, formerly of External Affairs and now an employee of the B.C. Hydro and Power Commission† and a consistent supporter of Bennett's "two-river" policy, has said that

* The IJC had proposed that the American rights should not extend to *any* flood control purposes, but only to the forecast onset of floods of exceptional magnitude.

† Bennett took over B.C. Electric and formed the B.C. Power Commission to market expensive Peace River power.

McNaughton did not oppose the treaty terms. Speaking at Saanichton on 12 May, 1962, he hinted not only that McNaughton was past his prime ("Because I hold him in such high regard, both as a friend and as a public servant, it is to me a cause of profound regret to see him, as he approaches the end of his remarkable career, entering into a controversy in which he is opposed by all informed opinion except his own. . . .") but also said this: "I recall that at the last meeting of the Canadian officials before the signing of the Treaty, the Honourable Davie Fulton . . . went around the table and asked each person if he had any reason to oppose the signing. General McNaughton did *not* object."[198]

This sounds as though Keenleyside was speaking of something he witnessed at first hand–that he was there, which he was not. Keenleyside was never a member of the Cabinet Committee. Fulton, according to a record McNaughton made at the time, asked each person for his view. McNaughton replied that his position was well known and invited Fulton's attention to public statements he had made to committees of the House of Commons; the technical evidence he had given on those occasions was correct and he saw no reason to vary the conclusions he had expressed. What the Canadian negotiators reported to the Privy Council was their affair but he himself did not recommend what was contained in the report. He added, "It was for them alone to decide whether or not to recommend the text *they* had agreed on with their U.S. colleagues to the Government of Canada." But, let it be admitted, he thought the government of Canada would never accept those proposals for, as he said at the meeting in response to a question by Howard Green, "new conditions will arise in future and we must then take appropriate decisions." He had made it absolutely clear, he said, that he "did *not* associate himself" with the proposals and that he did not believe they "would survive."[199]

Nevertheless, they did. The Columbia River Treaty was drawn up and signed in Washington by Prime Minister Diefenbaker and President Eisenhower on 17 January, 1961.

Michael Barkway walked behind two of the men who had negotiated the treaty for Canada, Fulton and Ritchie, who, with Arnold Heeney (the Canadian ambassador), were on the way to the signing ceremony. It was a brisk winter's morning and the three swung jauntily along. Barkway described it for the General later. "I can see it,"

McNaughton said, turning to Barkway with a half-snort, half-chuckle. "Three Rhodes Scholars—three blind mice!"[200]

The United States government quickly ratified the Columbia River Treaty. As Jack Davis said, "The Senate of the United States passed this treaty with a vote of 99 to 1,* and in record time."[201] It was ratified, McNaughton said, "in March 1961 after a brief, non-contentious, session during the course of which it received strong expressions of approval. Appreciation was recorded for the efforts of the U.S. negotiators in securing very large, important and long-lasting benefits for their country."[202]

Fulton—having drawn up the treaty and thus bound to defend his work—in a speech on 7 February, 1961, praised it highly. "It will bring to our province almost unbounded power," he exulted. "The Treaty makes possible, at long last, the development of this richly endowed province's greatest natural resource, and the benefits that will accrue to residents of British Columbia over the next quarter century should bring them a growth and industrial expansion that could hardly have crossed the imagination just a few short years ago."[203] He praised McNaughton: "A theory was developed and put forward by General McNaughton . . . which was of the greatest significance and for which the greatest tribute should be paid to General McNaughton by Canada. This is the theory of downstream benefits. . . ."

In December 1959, as a result of further intensive studies for which again the greatest credit must be given to General McNaughton, the International Joint Commission made a unanimous report back to the two Governments containing its recommendations as to the principles which should be applied. Again, it was a very great victory for the Canadian point of view. . . . Canada is entitled to receive 50% of the extra power production made possible at downstream plants as a result of projects constructed on the Canadian side, and 50% of the value of the flood protection provided.[204]

We have seen how valid these claims by Fulton are. There is no doubt that McNaughton had put millions of dollars in Canada's

---

* The single opponent of the treaty represented coal interests and objected to the *cheap* electric power that the United States would gain.

pocket that, without him, might never have been claimed. *[205] But Canada had bargained millions away. No matter how much the downstream benefits amounted to, McNaughton would have agreed with L. Austin Wright: "The losses due to the inefficient plan that is now proposed make the downstream benefits trivial by comparison."[206]

Canada, despite such speeches as this one by Fulton, did not ratify the treaty in 1961. Ratification was blocked by Premier Bennett of British Columbia who, in an effort to force his own plan for the simultaneous development of the Peace and Columbia Rivers, wanted to finance the Peace scheme (which would cost about one billion dollars) with revenues from the Columbia and, to accomplish this, proposed to sell Canada's share of the downstream benefits to the United States. By July Bennett had gone further. He now proposed that only the two projects that were to be built first in Canada–thus producing an earlier return of downstream benefits–should be proceeded with and Mica left out of the treaty altogether.

Fulton, on 25 July, 1961, in Vancouver, opposed these proposals–now publicly announced by Premier Bennett–strongly. He said the government of Canada rejected this new and "shocking" suggestion.

The whole purpose of the negotiations has been to secure for British Columbia and for Canada an abundant source of low-cost power. This was agreed to by the Canadian and British Columbia negotiators, and by their governments, throughout. British Columbia needs low-cost power. The proposition that we should now allow the United States to retain this entire source of power, for cash, would be to betray our own best interests and would be counter to the whole concept of the Columbia River negotiations.

To build only the High Arrow and Duncan Lake dams would be close to the very thing that some critics of the Treaty have erroneously accused the negotiators of contemplating. That is, *to see Canadian territory flooded for no other purpose than to permit the United States to generate extra power* † with no power generation ensured on the Columbia in Canada. This, especially coupled with Mr. Bennett's own suggestion that all downstream benefits be sold in the United States, would be *the greatest dedication of Canadian resources to the*

---

* McNaughton's long struggle was recognized by the University of Toronto when an honorary Doctor of Laws degree was conferred on him that year. He was described as "a soldier-scientist whose mind had been used, whose health had been taxed, and whose heart had been given in the service of his country."

† Italics supplied.

*benefit of another country and the greatest betrayal of Canadian interests that it would be possible to contemplate.** For this reason alone the present Government of Canada will veto such a proposition. . . . For all these reasons, neither of these courses can be permitted by the Government of Canada.[207]

Less than a year later Diefenbaker cut the ground from under Fulton's feet. The Speech from the Throne in 1962 talked of large-scale, long-term exports of power that should be encouraged, but though the government fell before the matter could be debated, the incoming Liberal government, as will be seen, let Bennett have his way. The Diefenbaker government's change of stand made Fulton's position impossible. A man of integrity, he left the cabinet in November, 1962, to lead the Conservative Party in British Columbia against the Social Credit government of W. A. C. Bennett.

Fulton's statements of July, 1961 (and others in November of that year in which he described Bennett's plan for the sale of power as offering the Americans "the greatest windfall since the purchase of Manhattan Island") are interesting in the light of the subsequent sanction of the export of power by the federal government.† With that, only Mica, on Fulton's own admission, saved the treaty from being the "greatest dedication of Canadian resources" to the benefit of another country and the "greatest betrayal of Canadian interests" that could be contemplated. Mica remained as a treaty project. And yet the "Little" Mica of the treaty–so different from the earlier mammoth concept–may have its efficiency as a Canadian generator reduced still further by the flow-pattern which the Americans have the power to establish; we have seen General McNaughton's fears about this. How much, then, is the Columbia River Treaty worth to Canada?

The most that could be said of the treaty was that it was second-best. Another member of Diefenbaker's cabinet—Douglas Harkness, who, as Minister of Northern Affairs and National Resources, had been a member of the Columbia Committee–admitted that. Asked if Canada could have obtained a better deal, he replied, "Yes, if British

---

* Italics supplied.

† The federal government had little choice. In August, 1961, Premier Bennett had expropriated the private B.C. Electric Company and by that made himself the sole major purchaser of electricity in the province. If Bennett refused to buy Columbia power, what was to be done with it? Bennett was in a strong position to enforce export to the U.S.A.

Columbia had been prepared to go along with the plan that we had finally arrived at, I think we would have got a better deal as far as Canada is concerned than the plan we finally got." Harkness favoured the Kootenay diversion and "Big" Mica; not High Arrow and certainly not Libby. To sanction the sale of electricity to the United States he considered a serious mistake. "To accept so much cash payment in place of fifty per cent of the downstream power increment I think was a very serious long-term error . . . the value over the [term] of the agreement of fifty per cent of the increased hydro-electric power delivered into Canada would have done us at least five times as much good as the cash payment which was finally accepted."[208]

McNaughton made no bones about the fact that he considered the treaty, and Bennett's further proposals, "disastrous" and a "sell-out of our vital rights and interests in perpetuity," and he opposed the government. As yet the treaty was not ratified and the door was still open for re-negotiation. Diefenbaker had had his day in Washington; he obviously considered that half a loaf was better than no bread at all and viewed McNaughton's opposition with increasing concern. Early in 1962 he decided to remove him. Thus, for his pains and persistence in the interests of his country, and after a lifetime of devoted service, McNaughton was retired in April, 1962. He was seventy-five on 25 February, 1962, and the reason Diefenbaker gave was that he had reached the retirement age for judges under the Judges Act! McNaughton had been many things, but never a judge.[209] He himself said bluntly that he was "fired."

The manner of his going is not without interest. The government asked McNaughton to stay on after his seventy-fifth birthday to attend the fiftieth anniversary meeting of the IJC, to be held in Washington from 3 to 5 April, 1962. The request was made by Howard Green, the Secretary of State for External Affairs, and McNaughton agreed on two conditions: that he be allowed to use that meeting as a last chance to get better terms from the Americans than those contained in the draft Columbia Treaty; and that his retirement should be kept secret until after the meeting as, if it were known, his hand would be weakened when pressing for re-negotiated terms. Green agreed to both conditions. McNaughton arrived in Washington on the evening of 2 April where he was met by Ambassador Arnold Heeney. Next day Heeney passed to him a telegram from Diefenbaker, which contained the text of a press release about his retirement timed for ten o'clock on the 3rd. "He concluded without any hesitation or difficulty at all," said Michael Barkway, "that Diefenbaker had deliberately broken the agreement made by Howard Green in order to destroy his authority to re-negotiate the treaty even before the meeting started."[210] McNaughton attended the meetings with his wife. They received the gifts and the tributes. But the Columbia Treaty terms were not discussed; there was no point in that now.

McNaughton arrived back in Ottawa and "blew up that day." He had "built up a head of steam" all the way home and when the reporters met him he let it go. He called Diefenbaker "a black dictator" and "a few choice terms like that" and, all in all, he caused a stir.[211] Lester B. Pearson, the leader of the opposition, was on his way to Welland and happened to be at the airport at the same time. On hearing from reporters of McNaughton's attack on Diefenbaker, Pearson smiled and said, "Give the General my warmest regards."[212]

The Liberal attitude is now important to our story, for Diefenbaker did not continue long in office. On June 18, 1962,* the Conservatives lost ninety-two seats, leaving Diefenbaker still Prime Minister but without power. On 8 April of the following year the Liberals formed a minority government under Lester Pearson. And Pearson, it seemed, was in McNaughton's corner.

Pearson went on to Welland from the Ottawa airport in April, 1962, and there he referred to McNaughton's statement that "Canada had been sold down the Columbia River." He was quoted by reporters as saying that the treaty had been signed "in unwarranted haste by Prime Minister Diefenbaker for reasons of political expediency," which surely was the case. Pearson said "the treaty should be *re-negotiated*† at once by a Canadian government which would pay sufficient regard to Canada's interests."[213]

During December, 1962, as we have seen, Jack Davis, who had been research director for B.C. Electric prior to his election as Liberal M.P. for Coast-Capilano, attacked the draft treaty very strongly in the House.‡

---

* McNaughton was approached at an embassy dinner before the election and asked not to attack the Conservatives on the Columbia Treaty issue. He replied that his stand was based on a "strictly national attitude" and because he considered the Columbia so important it "must be above party politics."

† Italics supplied.

‡ His address was not reported in B.C. at all save for six lines in the Vancouver *Province* that "he wants to scrap the treaty." The newspapers in British Columbia made a shoddy job of informing the public.

As to whether the present treaty, signed nearly two years ago, is a good treaty or not, let me say this: I personally believe that the treaty must be *re-negotiated.** I believe many hon. members hold the same view. It must be re-negotiated for several reasons. . . . [214]

Davis had then torn apart several of the treaty's clauses and ended by saying that it was not too late to study seriously "alternative plans for carrying out an immense development which has tremendous long-term consequences for the country."[215] And Pearson, signing himself "Mike," had thanked McNaughton "for the help he had given Davis and other British Columbia members "on this important question."[216] McNaughton could hardly be blamed for drawing encouragement from the attitude of the Liberals.

The General, though no longer a public servant, had decided to fight the Columbia battle to the end. Prime Minister Diefenbaker had given the House an undertaking, in the face of protests and outcries against the sellout, that the treaty would be referred to the External Affairs Committee of the House of Commons before it was ratified and McNaughton knew he would be called upon to testify before the committee. Though free at last to speak in public, and, in fact, pressed to do so, he felt the matter was not a proper subject to argue in public speeches; that the treaty should go before the committee, as Diefenbaker (and, later, Pearson) had promised. "What is needed," he said, "is for people with the right, duty and authority, to go into it thoroughly and take evidence under oath."[217] McNaughton had faith in that committee. But though he got his day in court, it was a mere formality. The decision had been taken.

That decision, apparently, was made by Pearson before he came to office. On 29 April, 1962, President John F. Kennedy gave a dinner for Nobel Prize winners in the Western Hemisphere and Pearson, who had won the Nobel Peace Prize for setting up the United Nations Emergency Force in 1956, was present. Kennedy's strong antipathy towards Prime Minister Diefenbaker was no secret; at any rate, Pearson was well received at the White House on this occasion.[218] Pearson made a statement, in Washington, about the Columbia Treaty which was as vigorous *for* the treaty as his previous one at Welland had been *against* it. On this occasion he said that the

---

* Italics supplied.

purpose of the negotiation he had proposed "would be to add protocols to the treaty as it now stands." A protocol would not have to be referred to the United States Senate for ratification provided it made no substantive changes in the treaty. The only way to change the treaty to give optimum benefits from Columbia development to both countries would be to eradicate Libby Dam, as we have seen. However, as for Libby, Pearson continued, "the option given the United States to proceed with the Libby Dam within five years is an integral part of the treaty and would be unaffected by the addition of any such protocols."[219] And that was the way it ultimately developed. Surely it is significant that only three weeks after his statement at Welland, without having had sufficient time to study even a fraction of the engineering data available regarding such a complex matter as the Columbia River Treaty, even if he had wanted to, Pearson in Washington had made up his mind to accept the treaty's most unfavourable clause for Canada. All McNaughton's efforts to correct the treaty during the next two years, reasoned as they were, amounted to no more than an exercise in futility after Pearson's trip to the White House in April, 1962. McNaughton was not aware of this. Liberal statements in the House, notably that by Davis as late as December, 1962, called for re-negotiations and appeared (from the letter Pearson had sent the General) to have had the backing of the leader of the opposition.

For the first time in more than fifty years McNaughton was unemployed. Even so he kept him self fully occupied. Though he no longer had a staff he managed to get things done "by volunteers, friends, who rallied around and helped him. He had the ability to make people want to help him. . . ."[220] He received assistance in research, typing and duplication. He prepared testimony, ready for the External Affairs Committee, made technical speeches before the Engineering Institute of Canada and in 1963 published an article in the spring issue of *The International Journal* in which he gave his views on the Columbia River Treaty. This important article—perhaps too technical for general consumption—was included as part of his testimony before the External Affairs Committee. McNaughton took the time to analyse in detail the treaty terms, and the more he went into them, reported an economist who worked with him,

. . . the more alarmed he became, because upon closer inspection, the language of the Treaty does not mean at all what it seems to mean on first reading. One would suppose that the logical way to read a contract before it becomes binding is with great scepticism. In other words, at this stage, the onus is surely to prove that the language is not capable of multiple construction or construction detrimental to Canada. Read in this manner, the Columbia River Treaty is a frightening document.[221]

The Liberals came to power in 1963, and that seemed to give Canada a second chance to make something worthwhile of the Columbia Treaty; McNaughton's hopes were high. But one of Pearson's first acts was to visit President Kennedy at Hyannisport and there, it would appear, he entered into definite commitments with the President regarding the treaty. On 18 June McNaughton had a talk with Paul Martin, Secretary of State for External Affairs in Pearson's government, who had been put in charge of the Columbia negotiations. By this time there had been discussions with Premier Bennett of B.C., which led to the Canada-British Columbia agreement of 8 July, 1963; this gave Bennett the power to deal directly with the United States, especially on the sale of power. Martin and McNaughton discussed the draft treaty but it is obvious they were widely apart. To McNaughton the agreement was a disaster that he would do his best to set right. He mentioned the seaway and Mackenzie King's urging that he should become chairman of the IJC Canadian section "to build the St. Lawrence without a treaty." The seaway had been built and McNaughton said he had "no feeling that, in this, I had given away any Canadian right" and he hoped "our arrangements in the new *régime* would not be allowed to slip." Martin made it clear that he was "under obligation to support the proposed [Columbia] treaty"; that there were "commitments from Hyannisport and in the discussions recently with Bennett." The impression McNaughton received was that the purpose of the meeting had been to let him know "that Pearson at Hyannisport and in meetings with Bennett had agreed to ratification of the treaty with only minor changes."[222] A month later a draft protocol, incorporating these changes, had been drawn up (together with a draft of Canada-United States notes that were to be exchanged) for review by the representatives of British Columbia on 24 and 25 July.[223]

Nevertheless, Martin asked for McNaughton's views, which at intervals for about a year, the General gave him. One of the main arguments that developed was the method of conducting engineering studies. "The engineering consultants appointed by the British Columbia Government," McNaughton noted in a letter of 22 August, "appear to have been given terms of reference strictly confined to the Treaty projects only"; their instructions embraced no alternatives. He considered that "an extremely unsatisfactory position for the responsible Government [Federal] on the eve of decision."[224]

Martin could "see little advantage in calling for further studies on a matter which has been decided by the responsible government [provincial]."

"I do *not* agree," McNaughton replied heatedly, when he realized what Martin meant, "that the government of B.C. is the government responsible for final selection [of projects] by which I understand you mean the ultimate decision. The Columbia and the Kootenay are rivers which flow out of Canada and, under the [British North America] Act, Canada, by the International River Improvements Act, has asserted jurisdiction. The Government of Canada is therefore *the final authority* and is responsible, at the least, that harm is not done Canada. The water . . . all of which is of Canadian origin," must be "physically as well as jurisdictionally under the sovereign control of Canada, to regulate and to divert as Canada's interests and those of her provinces determine."[225]

But the tactics adopted with regard to engineering studies continued. These amounted to the parcelling out of individual projects to consulting firms for study; and the projects in themselves were sound enough from the engineering point of view. In other words, as Austin Wright put it, the engineers "planned what the politicians told them to," without looking at alternatives.[226] It was a better sequence of projects, and a consideration of alternatives, that McNaughton looked for, which would maximize the benefits of Columbia development for Canada. Thus it was a dialogue of the deaf. Martin completely missed McNaughton's point and accused him of being "the only engineer with an intimate knowledge of this subject who seriously questions the conclusions reached by other engineers and engineering firms," and McNaughton finally gave up all hope that any other plans would be studied:

It appears that when the present Government took over you and your associates accepted this document [the draft treaty] in principle, and subject only to such minor changes as would *not* require the approval of the Senate of the United States.

These considerations have subsequently dominated the arguments which you have advanced and consequently also have conditioned fundamentally the reports made by the technical officers and the consultants you have engaged to advise you. . . .

In the result, while I have taken the greatest care in the answers I have given to your many comments and inquiries, it is evident that we have been working in diverse frames of reference. I do not believe therefore that our correspondence has resulted in an effective discussion of the fundamental questions at issue.[227]

There had been some plain speaking:

I say to you, Mr. Martin, as Secretary of State for External Affairs, Canada, with the greatest seriousness [McNaughton wrote on one occasion] that if this proposed Columbia River Treaty is ratified, Libby will be built by the U.S.* and for all time thereafter, this action, made possible by yourself and colleagues in the Government of Canada, will have deprived Canada of the beneficial use and control over the waters of Canadian origin in the East Kootenay. The only benefit we will receive will be what may come to us as a by-product, of little account, of the regulation of Libby which is vested in the U.S. to be carried out without restraint other than the minor requirement presented in the I.J.C. Kootenay Lake Order regarding levels.

May I say also that even if the treaty or protocol should remove the right of the U.S. to claim damages for our East Kootenay diversion the U.S., having invested some hundreds of millions of dollars in the construction of Libby and Kootenay Falls downstream, can be expected to exert the greatest political, economic, and moral pressure to *persuade* Canada to forego any plans for diversion.

My counsel to you, as an old friend of very long standing, is to withdraw from this dangerous imbroglio, while yet you may, for the sake of Canada.[228]

But the exchange of correspondence altered nothing. It was, however, useful to the government, for the Prime Minister could advise anxious inquirers, as he did, that "the Minister of External Affairs, the Honourable Paul Martin, and others with whom I have entrusted the task of negotiating a new agreement with the United States, will continue to seek the advice of General McNaughton. . . ."[229]

* Construction started in 1966.

In another letter written on the same day (25 July, 1963), to George B. Cady, the defeated Liberal candidate from Trail, B.C., Pearson said:

Thank you for your recent letter concerning Columbia River Treaty. As you know its terms are to be changed in several important respects. These changes as you are probably aware will be incorporated in a Protocol to the Treaty the contents of which have been determined in Ottawa and will soon be under negotiation with the United States.

While the financial and other terms embodied in the present draft Treaty will be improved, the physical plan will not be altered to any great extent. *We are not, in other words, adopting the McNaughton Plan.** Nor is it our intention to eliminate the High Arrow Project at this stage. Assuming that the United States is willing to pay a sufficiently high price to finance all of the construction in Canada the Treaty *Storage* † Projects, i.e. *Mica*,‡ Duncan, the High Arrow, will go ahead as presently scheduled in Article IV of the Treaty. . . .

Were it not for the fact that the Protocol to the Treaty is now under negotiation, I would outline the terms and conditions which it contains. Please rest assured that *these have been designed to meet most of the criticism which General McNaughton and others have raised* § concerning the Treaty in its published forms.

Prior to drafting the Protocol the Honourable Paul Martin had discussions with many people. He has sought the advice of General McNaughton and various other authorities who have made an extensive study of the Columbia River and its great potentialities for the future. They have, therefore, all had a voice in influencing the outcome. . . .[230]

This letter, which was published in a British Columbia newspaper on 2 August, 1963, and the duplicity that it revealed, came as a shock to McNaughton. He was being asked for advice by Martin when, as was clear from Pearson's letter to Cady, the Prime Minister and his colleagues had already taken a decision on the terms of the protocol and were discussing them with the United States. He objected most to the impression the Prime Minister was giving that his views had been considered and that the treaty had been revised in accordance with

---

* Italics supplied.
† Italics supplied.
‡ Italics supplied.
§ Italics supplied.

McNaughton turns out with the Red Chevrons on ANZAC Day, 1963
(COURTESY MAJOR C. C. J. BOND)

McNaughton, Colonel Commandant, The Corps of Royal Canadian Electrical and Mechanical Engineers, in the RCEME mess with Captain W. L. Sheldon, chairman of the National RCEME Corps Publicity Committee
(COURTESY THE RCEME CORPS PUBLICITY COMMITTEE)

GOVERNMENT HOUSE
OTTAWA

His Excellency the Governor-General's statement on the occasion of the death of General the Hon. A.G.L. McNaughton, 11th July 1966.

With the death of General Andrew McNaughton, Canada has lost, not only one of our most gallant soldiers but, as well, a distinguished citizen in many fields of endeavour. There is hardly a Canadian who does not know of the contribution he made both in war and in peace.

He was a soldier to the end of his life, fighting for the causes in which he believed, with the integrity, candour and tenacity for which he became famous. He was never a man to be intimidated by authority and strove for what he considered to be right with a complete self-giving of all his powers. He was one of those all too rare human beings who combine a brilliant intellect with a dauntless heart.

In our century, the right of commanding is no longer an inheritance. It is a reward won only by hard work, courage and compassion, by constant striving for, and ceaseless dedication to, a cause greater than oneself. These qualities alone make a man worthy to command others. They alone inspire the finest efforts from those who follow. Such qualities were the hallmark of Andrew McNaughton and they are the reason why his example continued to enkindle pride and high resolution in our armed forces long after he had left active service command.

In General McNaughton's death I lose a very dear and life-long friend for whom I had deep affection and the highest regard. He will be greatly missed by countless friends and admirers. With his passing, Canada has lost one of its finest sons and most courageous soldiers.

As with all great men, a large part of General McNaughton's success must be credited to the unfailing devotion and constant support given to him by his wife at every stage of his career. To her my wife and I extend our most affectionate sympathy.

General Vanier's tribute

The presentation of McNaughton's sword and decorations to the National Research Council, June, 1967

them—which was a travesty of the truth. He was now perfectly aware of what was going on: the correspondence was being conducted solely "to develop any weak points in my argument, *not* to gain information from me on which to base a realistic and constructive approach to a renegotiation. . . . In other words it seems the Liberal Party have already surrendered to the U.S. and all that is going on is a cover up to make the result more palatable to the public. I have described the conversations with me and the subsequent letters as an attempt at an *anodyne*." But he continued the correspondence. "While I deplore the shortsightedness of Pearson *et al* I welcome the opportunity . . . to explain the reasons for my support of Plan ix(a)* and my rejection of High Arrow and *Libby* . . . all this both verbally and in writing so in due course a record of the warnings will be available."[231]

McNaughton saw the protocol for the first time in March, 1964. On the 5th he had read Martin's "statement on the Columbia River Treaty and Protocol" with "care and close attention." He regretted that he was unable to find in it "evidence of correction to the many points in the Treaty giving rise to anxiety as regards the protection of the vital interests of Canada."

I note that in a number of places you have expressed similar anxieties to those which I have brought to your attention, but when I have come to the executive clauses of the Protocol as now drafted I find that *no* corrections have in fact been made to limit the extravagant powers which were to be vested in the U.S. by the Treaty. In fact, in a number of cases, by the use of imprecise language it appears that the damaging effects on Canadian interests have been enhanced.

I am very disappointed that despite my best endeavours over many months in our correspondence I have not been able to make these dangers clear to you.[232]

The External Committee's hearings, however, had not yet been held and support for McNaughton's position was coming in from many quarters, most powerfully perhaps from the provincial government of Saskatchewan. In a brief to the federal government Saskatchewan protested that the treaty and protocol did not seem to protect Canada's right to make future diversions from the Columbia-Kootenay system to the prairies. If the economy of the prairies was not

* The McNaughton Plan.

to wither for lack of water for irrigation, domestic consumption, industrial use and power production, Columbia water would be needed. And Saskatchewan thought that Columbia water would not be available unless the terms of the treaty and protocol were altered.

Firstly, it argues that these documents may turn out to prohibit any diversion whatever. . . .

Secondly, it argues that even if diversion out of the basin is possible the qualification that such diversions must be for *consumptive uses only* will block the transportation of Columbia water into the South Saskatchewan [River] because, once in the latter, it would unavoidably be used for power production at plants already established. Finally, it agrees with other critics of the treaty . . . who have pointed out that if developments using Columbia (or Kootenay) water take place downstream in the U.S. it will be impossible for Canada to "repatriate" that water, whatever the treaty may say.[233]

"Canada's negotiating teams, led by Hon. Davie Fulton on behalf of the Diefenbaker government and by Hon. Paul Martin in the discussions regarding the protocol, appear to have given little thought to anything except finding a settlement acceptable to the Americans and to Premier Bennett of B.C. Yet . . . the Columbia is not of interest to B.C. alone. It may, in fact, be of life-and-death importance to the Prairies."[234] This prompted a typical sally from Bennett of B.C.: "Tell Saskatchewan to keep its cotton-picking hands off our resources!"[235]

David Cass-Beggs, the general manager of the Saskatchewan Power Corporation, supported the brief very strongly. Engineering studies in the province established that diversions from the Columbia over the Rockies offered the cheapest and best way of bringing more water to the arid prairie regions.[236] E. W. Bowness of Calgary, a veteran of more than half a century of engineering work on power developments throughout the continent, described the treaty as being totally unfair to Canada. "This doesn't read as a treaty between friendly countries," he said, "but like one forced upon a conquered nation by the victors."[237] Favourable reports of the effect the treaty would have on Canada he termed "misleading propaganda and absolute lies."[238] Dr. L. Austin Wright addressed the External Affairs Committee:

Beyond any shadow of a doubt I consider the Treaty very disadvantageous to British Columbia and to Canada. I am unable to understand how any Federal Government can give approval to any such one-sided arrangement. Certainly from the engineering and economic considerations there is no possible justification for such an uncompensated give away of our heritage. There may be political considerations which are to be all-powerful but if such be the case it will be to the everlasting disgrace of the Government. . . .

The proposed treaty is wrong from almost every point of view of Canadians. . . . It is a preposterous proposal. . . . Surely this travesty of justice will not be approved by your Committee. Surely we can get back to reason and economics. Surely politics, in the most miserable sense of the word, is not going to be substituted for reason, honesty and fair play. I feel certain that if the present Government approves of this fantastic treaty the people will rise up and smite them and I will be glad to help them.[239]

F. J. Bartholomew of Vancouver was "convinced the Governments at Ottawa and Victoria were determined to secure a Treaty with the U.S. purely for the political credit and standing which it would generate for the parties concerned . . . regardless of the Treaty terms,"[240] and R. Deane, of the Consolidated Mining and Smelting Company, agreed with him. These were leading engineers in Canada whose attention had not been confined to one project, one facet of the whole, as, on government instructions, that of the consultants had been. They saw the complete scheme of development, as McNaughton did, and they were appalled. Nevertheless, the treaty, with its protocol and sales agreement, was ratified by Canada on 16 September, 1964.

Ratification followed hearings by the External Affairs Committee which have been described as "a farce from start to finish."

In the first place the Minister for External affairs told the committee it could make no changes whatsoever. It could only recommend approval or rejection without any amendments. Under such terms of reference it is not surprising that most committee members did not even bother to read the briefs that were presented to them. At the hearings themselves the committee proceeded usually with a bare quorum present, 11 out of 35. . . .

Despite the irrefutable arguments put forth by the outstanding Canadian statesman-engineer-scientist, General McNaughton, and the vigorous objections of the Government of Saskatchewan, a large number of trade unions and distinguished *independent* engineers, the committee dutifully went through the motions and voted to recommend acceptance.[241]

Three members of the committee, who dissented from its findings, agreed with that assessment:

In our view, the whole procedure in referring this treaty to the committee after final negotiation of the Protocol constituted a mockery of parliament. While evidence was invited and the committee was asked to scrutinize the treaty, it was made clear by Mr. Martin, the Secretary of State for External Affairs, from the outset *that no recommended change in the treaty would be accepted. . . .*[242]

They reminded the government of the words of Jack Davis, head of the Liberal Power Committee, who had described the treaty in the House as "nothing short of a fiasco"; that it was "inconceivable that the government of this country should ever enter into any agreement, let alone a comprehensive treaty with another country to last at least 50 years, which failed to procure any demonstrable advantage for the people of Canada. No man, no government, no party, can be proud of the achievement." They would, therefore, when the resolution to approve the treaty came up in the House, urge Parliament to reject this fiasco, this sellout of Canadian resources, this embarrassing treaty.[243]

McNaughton said the same. "Moreover, when the Treaty and the Protocol, which I believe adds nothing, came to be placed before the External Affairs Committee all that was asked for was approval of a political decision already taken by the Government. In the result, under the weight of opinion which you marshalled in support of your particular position there was, likewise, no effective discussion of the alternatives which had been advanced by myself and others; these were brushed to one side and ignored. . . ."[244]

But, of course, treaty and protocol easily cleared the House of Commons as, with both major parties bearing equal responsibility for their production, they were bound to do. Davis, who was by now Pearson's Parliamentary Assistant, was understandably absent when the vote to ratify was taken.

Perhaps, as an American writer suggests, McNaughton weakened his case in the eyes of some waverers, by expressing a strong "go-it-alone-Canada" philosophy which reflected

a deeply felt nationalism, as well as a distrust of the basic fairness of the United States. This view was contrary to the prevailing Canadian attitude which tended to idealize the existing peaceful relationship between the two countries, and sought to encourage joint enterprises and international cooperation[245]

Viewed in the most philosophical light, the treaty, "plus, of course, the face-lifting protocol," must be accepted, Michael Barkway said, as the best of "the miserable choices" that were left. But, he added, "We need not pretend it is more than a shabby second best." Downstream benefits had been won and so we should "swallow our pride and thank God and General McNaughton that it is not as bad as Mr. Bennett, M. St. Laurent, and Mr. Diefenbaker in turn might have made it."[246]

But the treaty is worse than Barkway could have realized in 1964. This may perhaps be gathered from one small example. Grand Coulee Dam in the State of Washington had a capacity for the generation of 1.944 million kilowatts. It was assumed during the treaty negotiations that Canadian storage would add 1.728 million kilowatts and it was for a half-share of this that Bennett settled, not in power, but in cash. It was announced in the House of Commons on 18 May, 1967, that the increased output at Grand Coulee will not be 1.728 million; it will be 7.2 million kilowatts, over four times as much. But it is far too late for Bennett to get any share of that. This was predicted by McNaughton in his *International Journal* article and it gives, as an economist has pointed out, "some measure of the magnitude of error in assessing the real benefits from Canadian storage, and consequently of Canada's half share of them."[247]

But these figures were only a guide and played little part in determining the cash payment to Canada as Dr. Hugh Keenleyside admitted:

The essence of the idea . . . was the sale by Canada and the purchase of the United States . . . not of a certain number of kilowatts hours but of a *service*–the controlled flow of Columbia River water across the boundary in accordance with an agreed plan of operation. British Columbia contended that the *payment* by the United States for this *service* should *cover at least the cost of the three storage dams. . . .**[248]

Bennett sold the downstream benefits, for thirty years, upon ratification of the treaty. He got, according to the official figures,

---

* Italics supplied.

$501.1 million dollars in U.S. funds for the power (and for flood control on completion of the three dams that must be built at Arrow Lakes, Duncan Lake and Mica Creek). The Canadian government, in hailing the event, stated that by 1973 there would be a surplus of 53.4 million dollars after the dams were built—enough to pay for half the machining at Mica (if the U.S. regulation of flows ever makes that worthwhile). Keenleyside indicated, however, that the costs of operating the storages had not been included when arriving at this figure and these would reduce the surplus to 6.2 million dollars over costs of 447.7 million dollars for the same.[249]

But worse was to come. On 12 May, 1967, it was reported in the House of Commons that even the revised surplus of 6.2 million dollars had now evaporated. Through increased construction costs for the dams—as McNaughton had warned—British Columbia will require *an extra two hundred million dollars to finish the job.* Not only did Canada give the river away; she will pay some two hundred million dollars for the privilege. Asked about this in the House by H. W. Herridge, Member for Kootenay West, the Minister of Energy, Mines and Resources (Jean-Luc Pépin) replied:

> This matter has now become the exclusive responsibility of the government of British Columbia. Only last week I was assured that the government could take care of the situation without help from anybody. I would add that because of the fact that some of these dams will be completed earlier than the date foreseen in the agreement, there will be added benefits which will help compensate for some of the increased costs.[250]

Premier Bennett, speaking a day earlier in Vancouver on the same matter, blamed the federal authorities for the whole mess; they had negotiated a poor treaty: "The treaty missed out. The thing that's at fault is that it was signed too quickly." He went on to claim that the arrangements he had been ready to put into force with Kaiser Aluminum for a storage at Arrow Lakes in 1954 had been "three times better than the treaty."[251]

In these words of Bennett we have the true assessment of the Columbia River Treaty, no matter what the External Affairs apology of 1964 may say. And Bennett, though he now disowned it, was in fact the chief architect of the treaty. At least in 1954, Kaiser would have built the dam at its own expense and would then have returned to

British Columbia twenty per cent of the downstream power free of charge; whereas now Bennett, having sold the downstream benefits, must construct dams out of *his* pocket to provide more than five times as much water (drowning out established communities and land in the process) and after all that he will end up two hundred million dollars poorer. Warned by a critic of the stature and knowledge of McNaughton, the Liberal government of 1954 had laughed at Bennett and his scandalously short-sighted deal; but two successive federal governments had managed, by their bungling ineptitude, to commit Canada to an arrangement that was three times worse.

And, as a disastrous and frequently overlooked by-product of the treaty, industries that might have been attracted to Canada by cheap power have gone instead to the United States. McNaughton predicted that to the External Affairs Committee in 1953. Professor Harry V. Warren, of the University of British Columbia has estimated that, because of this, British Columbia will be deprived annually of about five hundred million dollars in capital investment and three hundred million dollars in new production and foreign exchange in the aluminum industry alone. He contended that "cheap power is the decisive factor in determining the location of aluminum plants," that "fresh supplies of cheap power brought several aluminum plants to the U.S. Pacific Northwest," and that "British Columbia could have attracted these plants had it not dealt the United States an abundance of cheap power against which its power could not compete."[252] The Americans had never expected Canada to be so naive. As early as 1924 Francis E. Shenahon had pointed out to the American Society of Civil Engineers:

It will be an economic question for Canada to decide whether she patiently awaits the inevitable coming of industries to her own territory, with the inevitable building up of her commercial importance, or will transmit this power, with losses, to available markets in the United States. Questions of customs duties will enter but the chances are that, in the end, the Mahomet of industry will be compelled to go to the Canadian mountain of power.[253]

Furthermore the Americans gained, from the Canadian storages, the consumptive use of water in the western United States and that, McNaughton contended, was what they were really after:

They were trying to get their principle values—water for consumptive purposes—as a by-product of power development and kept the argument on power development and skinned us at every turn. . . .

Let me give you some figures. . . . The total of the three storages that we're to build average 15.5 million acre feet (an acre foot equals 325,851 gallons). The Americans have the right to the full use of that amount, to release it as they wish. . . . Now what's that water worth? The present value of water at the headgates of irrigation works on the southeastern slope of the Rocky mountains in the United States varies from about $2.35 to $10 per acre foot.

With the knowledge that water is going up in value, 10 or 15 cents a year . . . the average over the 30 years (the first stage of the treaty) can certainly be taken at $10 an acre foot. To show you that's not unreasonable, the value of water today in a citrus field is about $70 an acre foot. So 15.5 million acre feet made available is $155 million a year. . . . The value for power purposes was put at $501 million for 30 years, whereas if you take $155 million a year as the figure, even allowing the Americans a good discount because they are paying in advance, you'll find it works out to $2.65 billion over the 30 years.

In other words, the consumptive uses of the water, which are not provided for in any way in the treaty, are worth five times more than the power benefits paid Canada.[254]

There is a song being sung in British Columbia today which is called "The Dam Song";* it might equally well be termed "The Epitaph for a Mighty River" and it goes like this:

There's seven wonders in this world,
Their names are known quite well.
From Babylon gardens to the Sphinx
The ancient stories tell.
But now the greatest wonder
Is the folly in our land,
They've sold us down the river,
With the new High Arrow Dam.

Four miles up from Castlegar
The Columbia will be dammed,
And the water that they store there
Controlled by Uncle Sam.
This sets my mind to wonderin'

* Reproduced by permission of Joe E. Irving and Earl R. Fraser; and Charles H. Buchanan, President, Interior Sound and Promotion, Nelson, B.C.

Just what the fate will be
Of the big Columbia River
From the Rockies of B.C.

We could divert the Kootenay
Into Columbia Lake
By way of Thompson River
And on down through Hell's Gate.
We could build a mighty complex
Of factories all our own
And manufacture Canada's ore
While foreign countries groan.

It's not that we're ungrateful
We've never been before.
Been trying to stand up by ourselves
Since Eighteen Sixty Four.
But now upon this issue
We all should take a stand;
Canada for Canadians!
It's our God-given land.
Is this land *your* land—or is it *their* land?
This land was made for you and me!

Chapter 7

McNaughton's Last Campaign

McNaughton was not apt to repine. With the ratification of the Columbia River Treaty a battle had been lost but not, as yet, the whole campaign. He was never one to look back nostalgically at the past and therefore the history of the Columbia was of interest to him only in the lessons it provided for future action. Viewed in that light, it was important. If Canada were going to use that treaty as the prototype for future "deals," she would quickly lose her vast resources.

It was obvious to McNaughton that the real issues of the Columbia Treaty had not reached the Canadian public. The problems had been mainly technical and extremely complex. While much had been made in the press of whether the downstream benefits should be sold for cash or not, this was not the paramount issue. And whether electricity should be exported, with technological advances particularly in transmission, was no longer a question of the greatest importance. He had fought his battle to have the treaty re-negotiated fundamentally; the real issue had been the terms of the treaty itself and who was going to control the river. Also, because he felt that his main task lay with the External Affairs Committee, McNaughton had turned down many invitations to speak in public when he could have explained what was at stake. The battle had been lost but it had taught McNaughton, at least, one lesson: that he must draw the attention of the people to the vital need for the conservation and proper development of fresh water.

Of all the resources that a nation has, water is the most important. It is the indispensable resource without which people cannot survive. If there is insufficient water there can be no farming; crops wither and animals die. Factories depend on it and so it is vital to an industrial economy for power and for manufactures. Of the electrical energy generated in Canada in 1960, ninety-three per cent was generated by hydroelectric plants; water power has been behind Canada's rapid industrial development. John Root, a member of the Ontario Water Resources Board, pointed to the growing need for water in the industries themselves: it takes from 9,000 to 19,000 gallons to produce a ton of finished steel; from 5,000 to 65,000 to make a ton of paper (depending on type and process); 60,000 gallons to produce 1,000 pounds of textiles; 300 gallons to convert a hide into leather; and 300 gallons to manufacture one gallon of beer.

Canada is fortunate at present in that it possesses an abundance of water; enough, certainly, to sustain a steady growth in population and increased industrialization. But consumption increases more sharply than growth in population indicates. Domestic requirements grow with modern life, especially as societies become more and more urbanized. Not many years ago, for example, a few buckets of water from the pump would meet a family's daily needs; today the average city dweller uses much more than a hundred gallons a day for household purposes. Water pours through toilets, washing machines, dishwashers, baths and showers. It sprinkles and squirts lawns, streets and cars. How much water does a facility such as a "car wash" use? And factories grow greedier. The Ontario pulp and paper industry uses 550,000,000 gallons of water every day and the industry is expanding.

In agriculture there have been warning signs. The dust-bowl years of the thirties were one. Of the land devoted to crops in Canada, only 1.4 per cent is irrigated and yet, as population grows, more irrigation will prove essential. Wheat requires fifteen hundred tons of water per ton of crop. Though in Canada as a whole there is no water shortage, the prairies will need increased supplies before much longer. These could have come from the Kootenay and Columbia. Instead, the Americans gained *consumptive* use of Canadian flows (without making any payment), with California doing its best to outflank Oregon for the tremendous benefits which had been gained.*[1] But

* "Regardless of which American jurisdiction wins ultimate control of the water," McNaughton said, "it is apparent there is no longer any pretence of Canadian

there are still the Peace and Athabasca. In addition it may be possible to divert water from some of the northern rivers which flow, virtually unused, to the Arctic Ocean and Hudson Bay.

In the United States, where population and the degree of industrialization are far greater than in Canada, there are shortages of water in certain regions. In the southwestern states the water table is steadily sinking at rates in excess of ten feet a year through heavy demand on underground water. Wells run dry and have to be sunk deeper and deeper. Sometimes, as in the thickly populated Los Angeles area, seawater seeps in to replace the drastically lowered fresh supplies, so that wells become useless. In dry states, such as Texas and Arizona, underground reservoirs of "fossil" water, which have been there since the Ice Age, have been discovered and tapped to support fruit and lettuce farms. But these supplies, once exhausted, cannot be replaced. Elsewhere huge cities—New York is one—have suffered drought. Under these conditions it is hardly surprising that American eyes turned enviously towards Canada and that plans took shape in the United States to divert Canadian supplies of water to the south.

It should never be forgotten, however, that the land of our southern neighbour is drained by one of the most generous water systems in the world, adequate, with proper care, to meet the nation's needs for all time to come. The problem is that the system is, by and large, polluted: "It is a dismal fact that we now have seriously contaminated and despoiled almost every creek, river, lake and bay in the entire United States."[2]

Streams and rivers have been filled with sewage, rubbish and industrial wastes. A person wading into the once lovely Potomac, "rotting at Washington's doorstep," would run the same risk "as swimming in a cesspool"; Niagara is tainted "with a stench like that of rotten eggs"; the Hudson "repeatedly is contaminated by raw sewage before it flows past New York City, where it receives another colossal discharge of the stuff, more than 400 million gallons every day"; the Mississippi is described as "the colon of mid-America"; Lake Erie is dying and may soon "become a vast greenish sump"; and even Lake Tahoe in the Sierra Nevada Mountains—a lake that was clear and unspoiled a few years ago—is already green and murky from being

---

ownership. It is only prudent to forecast that any future attempts by Canada to use portions of Columbia water on the Prairies or in the Fraser will be met by the most unyielding opposition by the United States."

dosed with sewage. Sewage-treatment plants are expensive; they produce no profits; "they are for the benefit of 'the other fellow' who lives down-stream." And yet water can be reclaimed, as has been demonstrated in Los Angeles. A plant there reclaims water at a cost of between eight to nine dollars an acre-foot, whereas water brought in from the Colorado River meant paying fifteen dollars per acre-foot.[3] The high cost of Colorado water, no doubt, provided the incentive for this development. Had it been known in 1962, when the plant was built, that Columbia water for consumptive purposes would shortly become available from Canada free of charge, it might not have been thought worthwhile.

McNaughton, while he was aware of the deteriorating situation south of the line, did not anticipate the full-scale attack that the Americans contrived. He expected a gradual encroachment on Canadian resources, stimulated by the Columbia victory, and thought that the first move would be made against the Yukon.

The Alaska boundary dispute had been settled in 1903, under American pressure, with the United States getting the frontier it desired. The American "Panhandle" effectively shuts off the Canadian Yukon and northern territories of British Columbia from the Pacific. But the headwaters of the Yukon River were left under Canadian control and these waters are one of the greatest untapped sources of power in the world. The Americans now wanted to have the river dammed and the headwaters turned to flow through the Panhandle to Taiya, where a great aluminum industry was to be built in American territory. McNaughton thought the Americans would press for this diversion first; their Federal Power Commission, in the National Power Survey of 1964, had indicated the development, which McNaughton thought would have "far-reaching potential for vast loss to Canada."[4] But the Americans had far more grandiose ideas than that.

They saw in the Columbia River Treaty a signal that all Canadian water was for sale at bargain rates and they can hardly be blamed for that. "Even so," Larratt Higgins, the economist, said, "the speed and enormity of the response from the United States was breathtaking."

About three weeks after the signing of the Protocol to the Columbia River Treaty on January 22, 1964, news media across the length of the continent received a bundle of paper weighing about two pounds. This was the introduction by the Ralph M. Parsons Company (an engineering firm in Los Angeles, a water-short locality) of a "concept" labelled the North American Water and Power Alliance or NAWAPA for short. It is defined by its promoters as "a concept . . . for collecting excess water from the Northwestern part of the American continent [Canada] and distributing it to the water-short areas of Canada, the United States and Mexico."[5]

Under NAWAPA the Americans would get their Rampart Dam on the Yukon River; and much else besides. They planned to divert the Peace, the Yukon, the Mackenzie, the Fraser, the Columbia and the Kootenay Rivers; in the Rocky Mountain Trench a five-hundred-mile lake was to be created, mostly in Canada. From this lake the water would pass to thirty-three states and to portions of Canada and Mexico. As well as redistributing what the Americans termed "surplus" water, NAWAPA provided for a waterway from Vancouver, on the Pacific coast, to Duluth, on Lake Superior, thus increasing the flow through the Great Lakes-St. Lawrence system sufficiently to increase Niagara Falls power generation by fifty per cent. The programme would cost an estimated hundred billion dollars and take thirty years for the construction of canals, tunnels, reservoirs, pumping stations and everything else needed to change around the flow of the many rivers.

Steps were taken, the New York *Times* reported, "to familiarize industrialists and Government agencies in the United States with the broad outlines of the proposal . . . and the initial reception has been generally good. George S. Moore, President of the First National City Bank of New York, seemed to be the consensus . . . when he said recently, 'There was general agreement that the NAWAPA project was essential if our economic growth and standard of living is [*sic*] to be sustained.' The skids are being well greased . . . it is conceivable that virtually overnight NAWAPA could be on the road to reality."[6] "Reaction to the plan in the United States," Senator Frank Moss of Utah reported, "has been enthusiastic." It was the same in Mexico but, Moss conceded, "forces in the Dominion of Canada" were "somewhat divided."[7]

Well might the Americans be enthusiastic:

NAWAPA will deliver 78* million acre-feet [of water] annually for industrial, municipal and agricultural use in the United States. . . . Irrigated land in the United States would be increased by 40 million acres. The national income from agriculture, livestock, mining and manufacturing would be increased by $30 billion a year. Recreation, wild life and fish assets would be improved as far east as the plains States. Flood control would be an incidental but by no means unimportant gain.[8]

And to obtain these benefits all the United States would have to do was to persuade Canada to look beyond a narrow, national interest; to convince her that water is a continental resource to be shared by the peoples of Canada, the United States and Mexico. Of course, with the Columbia Treaty in mind, the transaction would be facilitated by making Canada's "get-rich-quick" politicians believe that they were capitalizing on resources. "I think this is the context of mutual benefits," Secretary of the Interior Stewart Udall said. "We got the Columbia River Treaty through . . . with the Canadians because we talked in terms of mutual benefits and I think this is the only way."[9]

The NAWAPA scheme was removed from the realm of science fiction by Senators Frank E. Moss of Utah, Joseph S. Clark of Pennsylvania and Lee Metcalf of Montana who, in 1965, laid a bill before Congress to establish a Department of Natural Resources together with "Senate Concurrent Resolution 55," which was described as expressing "the sense of Congress that the U.S. government refer the U.S. so-called NAWAPA project to the International Joint Commission for study and an engineering survey, and that the Canadian Government be invited to do likewise;" also that "the Canadian and United States Governments be informed of the results of the study by December 31, 1966."[10]

McNaughton studied the NAWAPA project early in 1964, and its proposal for the integration of Canada's water with that of the United States—for the benefit of the United States—deeply shocked him. Out of the hundred and twenty million acre-feet of water to be diverted annually from the Rocky Mountain Trench, practically all of it of Canadian origin, the Americans had the colossal nerve to suggest that only twenty-two million acre-feet (eighteen per cent of the total) should

* Actually NAWAPA would divert 120 million acre-feet, 98 million to the United States.

remain in Canada! And all this had sprung from the disastrous Columbia River Treaty and Protocol which, an economist remarked, "now stand as the basic blueprint for the development of all Canadian resources in the name of continentalism. . . . The Columbia River Treaty and propositions such as that which calls itself the North American Water and Power Alliance follow very closely the classic lines of imperialistic expansion: the exploitation of the natural resources of the underdeveloped colony (Canada) by the great industrial motherland (the United States)." And, more pointedly, "The Columbia River Treaty and these schemes are a denial of the possibility of Canadian development to the limit of Canadian resources; they are a means of allowing the United States to develop beyond the limit of its resources on the basis of those belonging to another country."[11]

McNaughton determined to fight NAWAPA to the limit of his powers. He had two objectives in mind: to make Canadians aware of their natural resources; and to expose the dangers of "continentalism" to Canada. The two were closely interlocked.

McNaughton's whole life had been a calculated and relentless pursuit of Canadian interests. From his youth he had identified himself with the idea of a strong, autonomous Canadian nation and he had worked consistently towards that goal. He was the arch-apostle of Canada as a strong and independent country and that concept moved more deeply in him than any other. It had provided the purpose of his life; it was the kernel of his belief; and it was the mainspring of his attitude towards politics and politicians. He was a man of straightforward feeling and judgment, forthright and resolute, who was quick to praise, quick to condemn, policies that would affect Canada's destiny in a good or adverse sense. It was for Canada that he would fight, never for himself; and that was the secret of his spell. The "magic" and "mystery" of McNaughton are thus explained, for it was this simple concept, and his absolute dedication to it, that made him so compelling.

In the sixties Mackenzie King's old fears of "domination from Downing Street" were laughable. The bogy of interference by Britain in Canada's affairs had long been laid. After two world wars Canada was sovereign and independent, linked with Britain in the

Commonwealth but no more than that. Canada was linked by her own choosing not only with Britain but also with the NATO powers in a collective defensive pact; she was, too, a partner with the United States in the defence of North America. None of these encroached upon her sovereignty. But, as the ancient ties with Britain weakened into sentiment, especially among the old, a new generation with other ideas had grown up. This generation, reinforced by European immigration, had little sympathy for "the British connection"; and it had little fear of the United States. Television programmes, movies and magazines were flooding across the border from the United States; travel between the countries was easy. All this was conducive to a trend to think in terms of North America—of continentalism—and to be more concerned with attaining a high living-standard and material things than with retaining Canadian nationality. "Don't knock the Americans," a new and powerful voice seemed to say, "you'll be one soon." And why not? Canadians and Americans, externally, are much alike; most of them speak a common language, drive the same cars and live similar lives. Continentalism, it was thought in Canada, would bring more opportunities, higher wages and cheaper cars.

Not all Canadians felt like this. Some were becoming worried. "Travel where you will in Canada," a prominent Canadian told James Reston of the New York *Times*, "and you will find some evidence of your influence in our society." He cited American-owned factories, theatres showing American movies, American-made military equipment used by the Canadian forces, shops filled with American products or those of Canadian subsidiaries, and American flags flown by motel owners in the hope of luring American dollars. "American cultural, economic and political influences so pervade our way of life," said another, "that we have begun to wonder if one relatively small nation can retain its independence in the face of the strong pressures generated by our giant neighbour to the South."[12]

There was good reason for alarm. In 1965 Americans controlled sixty per cent of Canada's petroleum industry, fifty-two per cent of its mining and smelting, thirty-five per cent of pulp and paper, forty-five per cent of manufacturing, fifty per cent of chemicals, electrical and farm machinery, and almost a hundred per cent of its automobile and rubber industries.[13] In all, over sixty-two per cent of Canada's economy is controlled in the United States. And this continentalism, this merging of the Canadian economy with that of the United States, is very dangerous. First, there will be economic unity with the United States; then will come political union with the United States. History has shown that political control inevitably follows economic control. The activities of the American United Fruit Company in Central and South America are one example; the old British East India Company is another.

McNaughton was firmly opposed to economic and eventual political union. His attitude is best summed up in these words of T. C. Douglas, leader of the New Democratic Party, which the General noted with approval: "Many people are won over by the myth that if we get economic and . . . political union with the United States we'll be as wealthy as the Americans." Those people, Douglas pointed out, saw only New York, Chicago or Los Angeles. They ignored such states as Montana, the Dakotas, the Carolinas, and regions such as Appalachia and the Mississippi, where conditions were far different, and great industrial vortices were busily sucking in the resources and manpower of the peripheral areas and growing rich at their expense.

I ask you what would happen to Canada in such a situation. Do you think for instance the automobile industry would stay in Canada or would they simply close their American subsidiaries in Canada and expand their plants in the United States? The simple fact is that Canada would become part of the periphery of the American economy. It would be an area to supply what the Americans desperately need–raw materials–gas, oil, minerals, timber, pulp, water, fuels of all kinds. We would be the hewers of wood and the drawers of water. We would supply the raw materials to feed a great insatiable industrial machine.

For every two man-hours Canada gets out of extracting iron ore from the ground, somebody else gets twenty man-hours for making it into iron, two hundred out of processing it into steel, and two thousand for the manufacture of finished steel products.

You don't develop an economy by getting the two man-hours extracting raw materials from the ground; you develop a great economy by processing that raw material either into semi-finished or finished products. There would be no assurance whatsoever that in an economic union very much of that processing . . . would take place in Canada.

So let no one be deceived by the myth that economic union with the United States would suddenly make us all as wealthy as the people who live in Beverley Hills. We would be more likely to look like the hillbillies in North Carolina . . . the maximum economic development, the greatest opportunities for jobs, the greatest possibility for a higher standard of living lie in developing a Canadian economy rather than in allowing our economy to be absorbed and become part of the American economy.[14]

The position in Canada was already bad enough with regard to industries, farm products, and resources. In Diefenbaker's day Canadian trucks, cars and parts could not be sold to mainland China, though export had been approved, because the manufacturers in Canada were the subsidiaries of American companies and they would not fill the order. It was American policy, not Canadian policy, that was carried out, and it was the same with flour. The Pearson government negotiated sales of wheat and flour to Russia, with some of the flour to go to Cuba. The millers in Canada, again subsidiaries of American companies, refused to mill the wheat because it was American policy not to trade with Cuba. Thus the loss of the economy means the loss of the power to make decisions. It is the same with the development of the resources which the Americans own. Though Canada might wish to develop resources and expand the economy, it is American interests that prevail; and Americans often prefer to use up resources in more unsettled parts of the world while they may, and leave Canadian resources idle. To quote Douglas again:

The old line parties have come to accept this continentalism which is developing so rapidly in Canada. They accepted it in the matter of the cars and trucks for mainland China. . . . They accepted it more recently in the matter of the flour milling contract for Cuba. They accepted it most of all [when] both of them voted for the Columbia River Treaty [and] turned the entire control of the fifth largest river on the North American continent over to the United States. . . . The Americans got a very good deal on power, but it wasn't power they wanted, it was water. That Columbia River water is going to be piped 1,200 miles down into places like Oklahoma and southern California because water is becoming the scarcest and most precious commodity on this continent. And this is only the beginning. . . . The United States, of course, has now succeeded in polluting almost every river they have. They are now desperately looking for fresh, clean water. The process is on to sell water as we have already sold a large part of our economy.[15]

Thus McNaughton made his decision: he would fight continentalism and his vehicle would be water. Canadians must decide: either to retain the resources of an independent country and develop them; or to integrate the economy with the United States and, the sooner the better, join them and get it over. He chose that Canada should be a strong and independent country and his actions were governed by that. He knew, in the prevailing climate with regard to water*[16] and continentalism in general, that he would in all probability face a chorus of name-calling from his fellow Canadians. He would be termed "a narrow nationalist, a protectionist, an isolationist, an anti-American," whereas "whenever the United States moves to protect its own economic interests at the expense of Canada the general Canadian reaction is to recognize such moves as legitimate ones" for the protection of vital United States interests. "By these strange Canadian standards, it is acceptable for an American to be nationalistic, and it is deplorable for a Canadian to want his country to survive."[17]

McNaughton waged his campaign essentially in the form of three speeches though, had he lived, there is no doubt that others would have followed. His first was aimed at the engineers of the country and was presented at the annual general meeting of the Engineering Institute of Canada in May, 1965; his second was delivered to the business community and others at the Canadian Club in Montreal during October, 1965; and his last was given to Canada's intellectuals, the Fellows of The Royal Society at Sherbrooke, Quebec, in June, 1966, a month before he died. In addition, he reached the broad masses of the Canadian population through the pages of *Star Weekly*. Most important, however, were his speeches, especially the last. As they followed each other, one could almost sense the turn of the tide.

At the Engineering Institute, McNaughton had a deeply interested audience. These were the men who had been effectively silenced by

---

* The report of the Western Canadian-American Assembly, held in August, 1964, had included this sentiment: "Canada and the United States are moving in the direction of a new and significant policy for the development of energy resources, particularly water power, on a continental scale. Recent technological advances which have made the border increasingly irrelevant have brought about in both countries a willingness to consider an encouraging degree of integration."

governmental policies whereby the Columbia projects had been parcelled out for study, "each given a part of the recipe, like the Benedictine Monks,"[18] and each precluded by professional ethics from expressing views about the merits of the scheme as a whole. Before any commitment was made by the government to study NAWAPA, McNaughton wanted to reach these men. They must be enlisted in the fight against NAWAPA, and in the larger campaign to direct the use of water to the future needs of all Canadians. What was needed, in fact, was a national water policy, which, it will be recalled, Lesage had enunciated in 1955 and St. Laurent had abandoned in favour of direct diplomatic negotiations shortly afterwards. Canada, it was clear, had had no water policy since 1956. For that reason, McNaughton said, "The Columbia River Treaty was negotiated in a policy vacuum": it was now time "for a new understanding of our water situation and a policy to guide future development of this crucial resource."[19]*

The General's thesis, a simple one, was

> that Canada's fresh water is fast becoming a scarce commodity of steadily increasing value which we hold in trust for future generations of Canadians. Since our future needs are not yet known, but are likely to be greater than we now expect, and since water supply once sold to a foreign user can never be reclaimed, it follows that if Canada is to prosper as a nation, our water cannot be sold, or export possibilities even considered, until we know a great deal more about our situation than we do at present. This means that we must steadfastly reject any proposal for the use of our waters, even for a supposed term of limited duration, which could in any way prejudice our full control over these waters when it will become in our interests to use them at home.

There was room, of course, for co-operation with the United States in the development and use of waters along the boundaries. That was properly a matter for the International Joint Commission. But national waters were an entirely different question. Any American proposal that the IJC should study these must be rejected out of hand. This was a matter for Canada, and Canadian engineers; it had nothing to do with the United States. And when studying Canada's national water resources it was important to bear certain things in mind:

---

* For the full address, see *The Engineering Journal*, June, 1965.

. . . although the overall national water supply appears adequate, at least on paper, there are major regions, such as the Prairies, where present supplies are inadequate. Second, while the national total seems to indicate a surplus that could be exported, much of this apparent surplus is in remote regions. The specific sources that could physically provide exportable water are generally those that are best located for our immediate needs as well. Finally, severe pollution is by no means unknown in Canada and in fact, in several highly developed areas, it has reached crisis proportions in this country as well as in the United States.

Few, before these words of General McNaughton, appreciated the significance of the proposal by Senator Moss that NAWAPA should be referred to the IJC. After all, NAWAPA was an international scheme and the IJC was an international body set up to study international water problems. But the waters that were to be studied were Canadian waters. To let them be considered international property, and to let their future be decided by an international body, was to lose the battle before it started, as McNaughton knew. To accept NAWAPA as a proper matter for the IJC would really set the momentum rolling. As one engineer, who heard McNaughton, put it:

NAWAPA was a Buck Rogers' scheme and it was laughed at. But then it became talked about more and more seriously and you have engineering papers at very august societies being presented on it. So it goes beyond the laughing stage and reaches the serious discussion stage and the next step, if the trend continues, is to launch a formal engineering study saying, "Sure, this is a fantastic scheme, but let's study." So you set up an engineering programme and spend ten million dollars studying and you run into the exact situation that you hit on the Columbia: that the plan you've studied is the only one you know anything about, you've already invested a great deal of time and effort, you have a bunch of people who are expert in it, and the momentum is almost impossible to stop.

The politicians get up and say, "Well, we'll be good neighbours. We'll study this joint property." But this is not joint property, this is our property, we must decide what we are going to do with it.[20]

And, as McNaughton said, although there *appears* to be plenty of fresh water in Canada, the future needs of the prairies must be considered, as well the growing incidence of pollution that will reduce supplies. His second point amounted to this: that a great percentage of

Canadian water is not usable by anybody. "It's in small glacier lakes, scattered all over the place, under a haphazard drainage system and therefore useless for export.* The exportable water is from the rivers—the St. Lawrence, the Columbia, the Nelson—and these are waters that are vital to Canada's own development. Right away, there is an immediate or foreseeable future for these waters."[21]

As for any American proposal that half these revenues (from water and power) would go back to Canada—"a good talking point to counter any nationalistic resentment that might develop in the Dominion against the concept of shared resources," McNaughton took issue with that:

> . . . if the United States regards a 50-50 split of purely Canadian resources as equitable, they are dead wrong! Shared benefits are quite proper on an international river like the Columbia or on a boundary river like the St. Lawrence. But the suggestion of splitting a resource owned entirely in Canada smacks more of plunder than co-operative development. . . . When water is sold as a commodity, the owner is entitled to the full market price.[22]

But it must not come to that.

> We must resist any temptation in the foreseeable future to sell water of national origin to the United States, regardless of how attractive current prices may appear. At the same time, we must make determined efforts to increase our knowledge about water and the ways in which it can best contribute to our national development. . . . We should reject with the greatest firmness any proposals for United States use of Canadian waters that relegates us to the role of "hewers of wood and drawers of water". . . .
>
> In conclusion, I venture to suggest that, as members of the engineering profession, we bear heavy responsibility both individually and collectively to contribute towards the formulation of water policy which will be of maximum long term benefit to Canada.
>
> Normally, an engineer's function is to design and build in accordance with an objective that has already been defined. But in the case of water development, this objective for Canada has not as yet been given, and part of our task must be to help in shaping a national purpose to fill the present policy vacuum. Only then can we get on

* And yet Northern Development Minister Jean Chrétien on a tour of the Arctic is reported by *Time* Magazine, 2 May, 1969, as having said: "Within 25 years we will be exporting water."

with the building job, secure in the knowledge that we will be building a better Canada.[23]

McNaughton had struck home. He received a standing ovation from the assembled engineers, who later awarded him their gold medal for the best paper on an engineering subject in 1965.

It was his address on October 4 to the Canadian Club at Montreal, however, which brought the subject of a water policy for Canada to the attention of the people and made it an issue during the federal election of 1965. On that occasion McNaughton pointed out that water is the fundamental resource in any civilization. And there is never too much. The amount merely determines the upper limit of development.

In Canada, the upper limit of development is high in great part because of our water resources. This fact should be an inducement for us to speed up our progress rather than to relinquish the fundamental asset that makes it possible. It is therefore nonsense to talk about a surplus and it is dangerous folly even to contemplate selling water. All our water can be translated into growth somewhere. Let this growth take place in Canada.[24]

There must, then, be no export of Canadian water to the United States, where there were abundant natural supplies of water. The problem in the United States was the shortage of clean water.

A no-export policy for clean water would impose no hardship upon our American friends. They have it within their own power to make available vast quantities of *clean* water and the unit cost is stated to be much less than for new supplies. All that is necessary is for them to process their abundant supplies of *dirty* water. . . . If our friends to the south have a hundred billion dollars to spend on the problem, let them apply it where a permanent solution would result for a fraction of the cost of the extravagant scheme now being touted under the name NAWAPA, which would dispossess Canada of her water.[25]

What was the point, he asked, of "adding pure Canadian water to the sea of pollution" in the hopes of diluting sewage? "It seems to me that all they will accomplish will be to enlarge the sea of pollution into a continental ocean, and if this should happen, we will still be confronted by the problem we have today but on an infinitely larger scale. We will have to clean up a continental cesspool or perish. . . The superficial attraction of adding clean Canadian water to existing polluted supplies probably lies in the assertion that it seems easier to do so from a political point of view."

Above all, he demanded a clearly enunciated Canadian policy. There had been some "pious sentiments" but no more than that:

Let us insist that the politicians tell us of their policies in these matters while they are appealing to us for our votes.

Let us hear what they propose to do about cleaning up our own pollution in Canada, and let them tell us what they are going to do about polluted boundary waters in which we share the blame with the United States. What are the *Canadian* plans for the development of our northern waters; we have already heard American plans for some of these waters which are so abundant that there seems even enough for our American friends to contemplate offering some of our water to their Mexican friends perhaps to make amends for some of the damage they have done by their pollution of Mexican water with salt.[26]

The American plans and proposals, he went on, were being made in the name of continental development.

Let me suggest to you that there are two meanings to the words "continental development." One of these, which is commonly being used in the United States and in certain quarters in Canada today, simply means the concentration of growth in already existing centres of power, and by this I mean the concentration of growth and power in the United States which is to be made possible by feeding Canadian resources to the south.* But there is another meaning to the words "continental development." This other meaning denotes balanced development throughout the continent, and particularly development of Canada's north. . . .

Up until very recently, we did not have the technology to exploit our north, but now it appears to be in sight. Development of our hinterland will never take place if its resources are drained off before the conditions of sustained growth can be created, and Canada will never be more than a one dimensional nation until such time as the north does develop.

---

* These words, which are so reminiscent of Douglas's (which have been quoted) were used a month before Douglas spoke.

If our north is developed by Canadians, the result will be to add not only economic strength, but a depth arising from the diversity of cultures and political traditions that will enrich as well as strengthen the whole of North America.* This to my mind is a true concept of continental development.[27]

There *must* be a policy, McNaughton concluded, but, "if we cannot decide what to do immediately, then I suggest that we should hang onto our resources until we have definite plans. By doing so, we will protect our future. In this way we can make no mistake because the value of the commodity rises day by day. . . . Let us always remember that, with water available, industry and population will be attracted to Canada."[28]

This address by McNaughton aroused considerable interest in the country–through the press, which supported and quoted him extensively–and among the political parties. It was said in the House that "General Andrew McNaughton, who is respected as one of the most colourful and dedicated patriots in Canadian history . . . has never hesitated to direct his fire at wavering politicians when a national issue was at stake and his current campaign to preserve Canadian water for future generations of Canadians may well rank as the most important of his many contributions to our country. Without a doubt, Mr. Speaker, his dramatic introduction of this issue into the . . . election put some significance into what we considered an unnecessary election campaign."[29]

It came at an opportune time, shortly after the dangerous Merchant-Heeney Report, which was drawn up by Arnold Heeney (the former ambassador to the United States and McNaughton's successor at the IJC) and Livingston T. Merchant (former American ambassador to Canada). One of their recommendations, which was favoured by continentalists, was this:

---

* Here we see very clearly McNaughton's advocacy of Canada as a sovereign nation. Assimilation into the United States would drag Canada's many nationalities and cultures into the American "melting pot." The "mosaic" would be destroyed. It can hardly be supposed that the French-Canadian people, for example, would retain their language, their culture, their traditions or their system of civil law. To preserve the "mosaic" would make, according to McNaughton, for a richer society with richer values. And, politically, Canada, as a middle power not tied either to Britain on the one hand or to the United States on the other, could still make her voice heard for peace in the councils of the nations.

It is in the abiding interest of both countries that, whenever possible, divergent views between the two governments should be expressed and if possible resolved in private, through diplomatic channels.[30]

In effect, Canada and the United States should settle their differences quietly, behind closed doors. McNaughton had feared those closed doors at the time of the Columbian negotiations and he feared them still; but his public exposure of NAWAPA would make it very difficult for the Secretary of State for External Affairs "to present to this House and to a committee another treaty and say: Here it is; you can say all you like about it but you cannot change a damn word."[31]

McNaughton's address, in fact, coming in the middle of a federal election campaign, caught the old parties in disarray. He had challenged them to state their views on a water policy for Canada and, indeed, they did their best. On 15 October L. B. Pearson, the Prime Minister, said:

We have lots of water. . . . We will be discussing this with the U.S., who are very anxious to work out arrangements by which some of our water resources are moved down south. This can be one of the most important developments in our history.[32]

Three days later Arthur Laing, the Minister of Northern Affairs and National Resources, said:

Diversion of Canadian water to the U.S. is not negotiable. There is no such thing as a continental resource. We own it."[33]

Pearson, having had second thoughts, no doubt, after Laing's forthright statement, shifted ground. He was quoted in the Toronto *Star* of 26 October as having said:

... in the long-range future a surplus of water might permit export, but certainly that will not be in the immediate future. Export can be considered only after we have worked out our own national water policy.[34]

And Laing, having put his head together with Pearson, on 28 October said much the same:

If [the U.S. should become starved for water] we certainly would not sit idly by when we have 40 per cent of the world's fresh water. It is not our intention to dispose of water without first investigating all alternative uses.[35]

Meanwhile the Tories had displayed an equal ambivalence. John Diefenbaker, leader of the opposition, said on 16 October:

The government appears willing to pull the plug on one of our most vital resources. The Conservative party will not agree to a continental diversion plan. This would mean the chipping away of Canadian sovereignty.[36]

But eight days earlier, his lieutenant (Alvin Hamilton, a former Minister of Northern Affairs and National Resources) had come out in support of NAWAPA:

Canada should agree to a U.S. plan to build a series of canals to carry fresh water from northern Canada to parched areas of the U.S. In return Canada should be guaranteed a 15 per cent annual growth rate for the steel industry, which would eventually give Canada half the North American output.[37]

Then, according to the Toronto *Globe and Mail* of 25 October,

The Conservative party's national water policy calls for the diverting of northern rivers to supply southern Canada. Mr. Diefenbaker made this clear Saturday night. He did not explain how, if northern waters are diverted to the Great Lakes, the U.S. could be prevented from tapping the lakes.[38]

And, from the same newspaper, it seemed that Hamilton, too, had thought again:

Instead of selling water, Mr. Hamilton proposed that the industries in the U.S. that use extensive supplies of water and energy should plan their future expansion in Canada where the water already exists.[39]

This would be funny if it were not tragic. It is clear that neither of the two major parties had the faintest inkling of what a national water policy should be. And yet the government of the day, on the lines of the Merchant-Heeney Report, might well have embarked on a round of negotiations with the Americans—for which they were patently ill-equipped—that would have moved Canada's water resources to the south, if McNaughton had not foreseen the danger and sounded his blunt warnings.

McNaughton warned the politicians and the newspapers warned the people. Press accounts, such as this one in the Victoria *Colonist* of 19 October, were common:

It is a rather strange spectacle when the leaders of the two largest political parties in Canada are not only at odds with each other on the question of water export, but at variance also with colleagues of their own. Thus we are left in doubt as to the policy of either party, which serves only to emphasize the recent warnings of General McNaughton that Canada is proceeding in a political vacuum. His point was that programs shaped by special interest groups can get out of control for lack of a firm national purpose and that Canada should look to its future development before starting international studies. The warning is surely worth heeding.

The danger was not yet over. After the election the Pearson government remained in power and its policy of "creeping-continentalism," as the leader of the New Democratic Party termed it, continued.[40] The pressure on Canada to share her waters—already strong—would grow stronger, and besides this there were the obvious quick-dollar temptations: a desire to help to balance payments by selling water; and to have the United States share, or even pay, the costs of bringing the waters of the northern rivers into the centre of the continent. Two men in the Liberal ranks had spoken very firmly against water exports. One was Laing, whose department was responsible for water resources, and the other was his Parliamentary Secretary, John Turner. Turner, in Washington at the end of 1965, made it clear that he had firm notions. He warned the Americans against glib talk about a "continental water policy" and said that, before Canadian water could ever be considered as continental water, "We must determine Canadian needs—now and for as long as we can conceivably predict."[41]

It may be of interest to note that Pearson, in his post-election cabinet shake-up, removed both Laing and Turner from having anything to do with water. Water would now come under a new Department of Energy, Mines and Resources, headed by Jean-Luc

Pépin. This was indicative, one observer noted, "of a thaw in water-export policy." Pépin, "a reputedly bright political novice," showed "none of Laing's unyielding approach." In fact, he said that if, in using water resources "in our own interest," the country could also use them "in the American interest," then "we shall do that." Moreover, he did not believe "that Canada should reject all American approaches until every last gallon of Canadian water is measured and allocated to some future Canadian use."[42] McNaughton agreed with the Member of Kootenay West who said in the House, "We fear the direction in which we are going."[43]

McNaughton opposed, as Kenneth McNaught put it, "the bankrupt policy underlying the old parties' fear of issues involving nationalism."[44] He strongly condemned the sort of political expediency, perhaps the expediency of a single individual, that would barter away tomorrow's values for the empty approbation of today. The campaign he was fighting was in the long-term interests of his country, and would determine "within two or three generations which parts of this continent will keep on growing and developing and which will shrivel for lack of water."[45] It might determine, in fact, whether Canada would survive as a nation at all. The outcome was vital and McNaughton laid careful plans for the next battle, which proved to be his last.

On 8 June, 1966, General McNaughton confronted Senator Frank Moss of Utah, NAWAPA's most powerful proponent, before The Royal Society of Canada at Sherbrooke, Quebec. He left his audience in no doubt concerning the importance of the issue.

McNaughton, it was reported by Professor E. Dolman, had arrived early, "in fine fettle and obviously ready for the fray." Moss, on the other hand, "arrived a little late, having been misdirected to the city of Quebec."

The audience was worthy of the occasion–representatives of all sections of the Society, expectant, and keenly attentive. The President, Dr. W. Kaye Lamb, acted as chairman of this Plenary Session and appropriately introduced each speaker in turn. The two protagonists knew each other's viewpoint from previous encounters, and did not attempt to mask their differences. But whereas the Senator's smile hardly deserted him, even when volley after volley of grape-shot was fired across his bow, the General's joviality vanished soon after he mounted the platform, and his facial expression ranged from frown to scowl.[46]

Moss, who spoke first, strongly advocated the adoption by both countries of the NAWAPA scheme. In the United States it would permit increased flow in the Upper Missouri and Upper Mississippi during low flow periods and provide ample supplies of clean water for all the arid western states, including supplies for the restoration of ground-water where that had been depleted. There would be enough left over for Mexico in amounts many times greater than that gained by the Egyptians from the construction of the High Aswan Dam. Canada would benefit enormously. NAWAPA meant the creation of a vast power-generation system across Canada, supplying new industrial and agricultural water, and water transportation to the Prairie Provinces. It would stabilize Columbia and St. Lawrence flows and bring water to the Great Lakes to stabilize their levels. British Columbia would gain in storage, power and navigation facilities, and have a huge lake on the west side of the Rockies. "Both the U.S. and Canada must determine what we should do—and determine it fairly soon."

In perhaps the most devastating attack of his life, McNaughton tore NAWAPA apart. The scheme was "unacceptable" and, "despite some temporizing pronouncements which had been issued by distracted politicians," he would set forth the position that should be taken by Canada and the provinces*[47] which, he believed, represented the views of technical men and administrators who were best qualified to judge.

The "NAWAPA propagandists" talked of great quantities of Canadian water spilling unused into the Arctic Ocean. "But the major sources for the scheme are hundreds of miles from the Arctic Ocean. They are, in fact, the rivers of the Canadian cordillera which provide a great series of prime power sites, rivers which form the basis of one of the world's great concentrations of the forest product industry, rivers which provide some of the finest salmon runs in the world."

There are detailed plans on the Canadian drawing boards; there are projects now under construction to harness these flows. The associated mineral and forest resources are already staked out, and the required human and financial resources are being attracted to the region. The NAWAPA promoters would move all of this out of Canada, the people, the industry, the water. It can only be described

* Premier Bennett (who had learned much from the Columbia) now had a different attitude towards water exports. He said that British Columbia would sell electricity but *not* water. "Even to talk about selling it is ridiculous. You do not sell your heritage."

as madness to believe that Canada has surplus water in an area that is so obviously earmarked for major resource development and where so much is already taking place.[48]

But NAWAPA, McNaughton went on, had nothing to do with the maximum development of rivers or resources in Canada:

Its purpose is to flood the valleys in Canada and to drain off the water in regulated flow for beneficial use in the United States. But the valleys themselves are of vital importance to British Columbia, because they contain the level land which is so vitally needed for roads and railways, for industries, for people and for agriculture. Whitehorse and Prince George would be submerged, and their land with them, as would countless miles of railway and highway. These irreplaceable assets would be destroyed in the name of trans-mountain navigation.[49]

What was the point, he asked, of bringing some of the waters of the Canadian cordillera to the Great Lakes?* The Canadian West–not the United States Northwest or Southwest or Midwest–was its logical beneficiary.

The Canadian Prairie region can look forward to maximum development of its agricultural potential made possible by water for irrigation. It can also look forward to major developments in mineral, fossil and forest resources. The logical consequence of such development will be a major petrochemical industry, metal producing industries, pulp and paper industries–and these all call for large supplies from the annual flow available.

The northward course of our Prairie empire is already being staked out by prospectors and timber cruisers. It has already reached the great tributaries of the Mackenzie system–the Peace and the Athabasca. Massive lead-zinc deposits are being mined at Pine Point on Great Slave Lake, one of the huge natural balancing reservoirs of the Mackenzie system. . . . Mineral resources have already been staked out in the region of another great Mackenzie reservoir, Great Bear Lake.[50]

---

* In any case, McNaughton's experience at the IJC had taught him that the Great Lakes were extremely sensitive to augmented supplies of water. The most serious problems were caused by flooding, not low water, and Canada was about to be sued by the United States for damages due to high water alleged to have been caused by Canada in 1952. To bring in more water to the Great Lakes would aggravate these risks.

The NAWAPA scheme was "a mammoth inroad into Canada's lawful rights and interests," and, more than that, it had been described by Dr. Trevor Lloyd, professor of geography at McGill, as "an exercise in sophomore civil engineering which has received far greater attention than it deserves." It was a concept,

> . . . monstrous not only in terms of physical magnitude, but monstrous also in another and more sinister sense in that the promoters would displace Canadian sovereignty over the National Waters of Canada, and substitute therefor a diabolic thesis that all waters of North America become a shared resource of which most will be drawn off for the benefit of the United States midwest and southwest regions where existing desert areas will be made to bloom at the expense of Canada.[51]

And on what did this concept rest? On the premise that there were large quantities of surplus water in Canada and, certainly, there were hundreds of thousands of lakes. But, said McNaughton, "this is water in inventory; if it is removed it will not be replaced; moreover its removal could reduce the amount of rainfall and therefore reduce the supply of running water." And its removal might also disastrously affect the climate, for "the large surface of these lakes has a profound effect in modifying our climate from . . . harsh extremes." It could well be the same if there were any tampering with the flow of rivers in Canada's North, for effects upon climate and vegetation had been noted by scientists as the result of even minor changes. It was only running water that could be regarded as being available for use, for this was the only water replenished by rain and snowmelt. And the *total* stream flow in Canada, including the Arctic flows, was not limitless, but "of a similar order of magnitude to that of the United States."

Referring again to the proposed dilution of polluted American water with clean Canadian water, McNaughton continued:

> I am inclined to believe that in the long run perhaps one of the most unfriendly acts that Canada could do to the United States would be to offer our water at bargain prices. Because if we did, the attack on pollution might be postponed until the problem was no longer soluble and, at the very least, the cost would be astronomical.[52]

The government, then, should not commit Canada's waters in any hasty or ill-advised fashion. Canadians had everything to lose by that. "We have everything to gain," McNaughton concluded, "by waiting until the essential information is available upon which we can make our *own* assessment of the subject of sharing resources and our *own* plans as to the course of action we will adopt.".

Let no one believe that Canada could sell water to the United States on a temporary basis, to relieve the shortage, and then cease supplies when the need at home was present. The Americans wanted the water permanently and NAWAPA was a step towards permanent integration, not only of water but of the two economies. McNaughton ended by quoting the comments of William S. Foster, editor of *The American City*, on the implications and risks of water export:

> There's a great deal of risk. Offhand, at first glance, you would say that the risk is great for the southwest because the water belongs to Canada and, presumably–if at Canada's wish–it could simply turn off the valve and the water would not be there any more. But there is a more subtle threat than that. I wonder just what would happen to the sovereignty of Canada if a nation like the United States . . . would suddenly say, "You no longer have the right to deny us the water even though you need it in your own industrial and agricultural development"? I've often thought about that and wondered what the really long range risks are to Canada, and I think they're considerable.[53]

Senator Moss, in his speech, had made it equally clear that Canadian water, once committed, was committed for all time. "I want it understood," he said, "and I'd like to underline this–I don't think Canada could manage this on a turn-on, turn-off basis, to say this year we are going to give you . . . water and that year we turn it off, because it just won't work that way. That's the reason it has to be studied and projected because, depending on water, cities grow and people move in and industries establish, and they just can't shut down and move away for a year. We are talking about a long range and continuous thing. There is no denying that."[54]

At Sherbrooke, McNaughton carried his audience with him. Dolman has drawn attention to "the obvious sympathy aroused in this ostensibly hard-headed, scholarly audience by a great patriot's fervent admonitions. All our largely latent Canadianism was inspired by the vigorous forthrightness of this engineer, who was also scientist, soldier, statesman, and conservationist–surely a lesson," Dolman shrewdly commented, "for more 'successful' politicians of so much lesser stature."[55]

On the disposal of Canada's water depends the survival of Canada as a nation; it is the foundation of all the future prosperity of this country. That was the message McNaughton wished to impart. These issues became his cause and he served them until his death in July, 1966. The issues did not serve him; it was with complete selflessness that he presented his case and there are strong indications that he was successful.

At the end of 1965 the New Democratic Party adopted this resolution:

> Resolved that we support the principles of General McNaughton for water control in Canada, which include insistence on Canadian autonomy in development and control of water resources, with Canadian needs receiving priority.[56]

Arthur Laing, although water resources were no longer his responsibility, continued to keep a close eye on the subject from the Liberal front bench. In January, 1966, he gave faint hints of "pressure" and "a deal":

> I have said that of all the resources of Canada, fresh water is the greatest. We have in this nation over 30 per cent of the fresh water on the face of the earth. Now, with exploding population, with declining resources of other kinds in other countries, it seems to me that if we are not hasty, if we reserve as long as we possibly can a final commitment of them, then this nation is going to experience a tremendous national capital gain in the improved value of these resources.[57]

But, two months after McNaughton's Sherbrooke speech, his views had hardened. Speaking to the American Bar Association he said: "Canadian water is not now negotiable, and I am not certain that it ever will be."[58]

Davie Fulton, who was a candidate for the leadership of the Conservative Party in 1967 (and who, it was reported, had been a "tortured soul" since his capitulation to "expediency" at the time of the Columbia negotiations),[59] now strongly advocated a water policy for Canada. In February, 1967, he said the government should be inviting population and industry to Canada to make use of domestic water supplies rather than thinking of water export.[60] And there were other signs. The Canadian Chamber of Commerce, in a brief to the cabinet during December, 1966, urged Ottawa and the provinces "to make an exhaustive inventory of Canada's water resources and a long-range forecast of requirements before considering water exports to the United States."[61]

Most hopeful, perhaps, is the attitude of the press in keeping McNaughton's struggle alive, and in keeping the Canadian people abreast of new developments. "His greatest service to Canada," wrote Patrick Nicholson in July, 1966, "may be yet to come. And it will come if we ponder and act upon his advice. . . . [A] great chain of industrial cities . . . half a century from now, will have taken shape in the northern Prairies. Our first great socio-industrial complex, reaching from Montreal to Hamilton [will be] saturated; our second is developing between Edmonton and the Peace River, which will ultimately be the home of 100,000,000 Canadians. Our water in the west will all be required there; we must not divert it to the U.S.A. . . . If industry needs that water, let it come here to use it, rather than divert our riches south to the U.S.A. Canadians should never forget the last great words of that great Canadian. . . ."[62]

And more than a year later an editorial in the Kingston *Whig-Standard* * gave evidence of unrelaxed vigilance. In reply to further overtures by Senator Moss the newspaper spoke out firmly:

> Clearly, Canada doesn't need any more deals like the Columbia River Treaty of 1964. . . . The Dominion Government should mince no words in informing the United States that our water resources are not negotiable and may never be made available to the U.S. while we retain our sovereignty. Our water belongs to us and we should decide when and how it should be used. . . .

As General McNaughton often pointed out, Canadian water should be guarded with the utmost care, especially against the "continental policy" envisaged by Senator Moss and N.A.W.A.P.A. For a developing nation such as Canada, the continental concept in the use

* 22 August, 1967.

of such natural resources is a most dangerous one.

The campaign continues. The Americans will exert the utmost pressure to achieve their ends. On Saturday, 3 September, 1966, at Summersville, West Virginia, President Johnson said the world will soon face a choice between war or ample stocks of pure water. "We are in a race with disaster, "he said, "and people denied it will fight for it."[63] Three days later Davie Fulton, in the unlikely role of water champion, called on the government to make a policy statement that there will be no continental sharing of Canada's water until the federal government's water policy is clearly stated and has been debated by Parliament. Fulton too, it seemed, feared the behind-closed-doors approach, as well he might.[64] But we need not experience alarm nor need we feel neighbourly concern. The United States has plenty of water. As Raymond Nace of the United States Water Resources Branch admitted, "We in the United States are not running out of water but we are running out of ways to avoid paying the bill for adequate water works and rational management."[65] Thus, as General McNaughton pointed out in Montreal, the American initiative springs from political motives, for they know that to enforce adequate anti-pollution measures would not make for popularity at the polls.

If Canadian politicians, with that in mind, will turn from expediency and look to the kind of Canada that McNaughton envisaged—strong, thriving, happy and completely independent—he will have won not only his last, but his life's campaign.

Chapter 8

The Final Years

In April, 1962, McNaughton left the public service; he never returned. This, it may be thought, was a time for leisure; a time to look back and sit gossiping with friends. He had earned it. But it was the future that interested him, not the past (save for the lessons the past might teach), and in the future he recognized that he still had a duty to his country. As a private citizen he continued to fight the Columbia River Treaty until its ratification in September, 1964, when he took up in earnest his battle against the North American Water and Power Alliance.

He was vigorous, his mind as clear as ever, and he was used to hard work. McNaughton would have agreed with Plutarch who said, "God forbid that I should ever be at leisure." In retrospect, he described the Second World War years in Britain as a "dog's life," and certainly the hours had been long, frustrations had accumulated and tempers had grown shorter. But there is no doubt that McNaughton had done what he had to do with gusto and enthusiasm. After the war, when he was weighed down with United Nations, Permanent Joint Board and International Joint Commission responsibilities, he allowed himself no respite; he drove brain and body relentlessly forward by his indomitable will.

Among the many important campaigns he had waged throughout his life, he considered the safeguarding of resources the most important. And so that he could concentrate on it he shed, one by one, certain other tasks which had always interested him. Since 1946 he had been directly associated with The Corps of Royal Canadian Electrical and Mechanical Engineers, first as Honorary Colonel, then as Honorary Colonel Commandant and finally as Colonel Commandant from 1962. In March, 1964, he said farewell to the Corps he had done so much to found. Three years later a drumhead service at Camp Barriefield barracks marked the twenty-third birthday of the Corps; that day the name was changed to McNaughton Barracks.

In 1965 he gave up the chairmanship of the Associate Committee, National Aviation Museum, which he had held since 1954. This was the committee of the National Research Council whose dual purpose was to foster the establishment of a National Aviation Museum and to support aviation research. The museum, a first step towards a national science museum envisaged in the report of the Massey Commission, opened on Dominion Day, 1961. McNaughton became honorary chairman of the National Aviation Museum Association, "a fitting recognition," (in the words of E. A. Côté of Northern Affairs and National Resources, under which department the museum came), "of the contribution that you have made to the creation and growth of the National Aviation Museum."[1] In 1965 Dr. Guy Ballard, the president of NRC, "reluctantly honoured his wish to resign."[2]

In an age of compromise to have a cause is singular. McNaughton devoted himself to his and made sure he had time enough to serve it properly. But he kept in touch with other things. His vast knowledge was strengthened by an extensive correspondence with old friends, eminent in various walks of life. People called on him. Official visitors from Britain or from Washington would come to Ottawa and take an hour or so to chat with him. He kept himself well-informed. He turned out with the Red Chevrons once a year, attended the opening of Parliament, and went to meetings which interested him; he still visited NRC as a man who "loved laboratories" and who came "to see what the boys were doing."[3] One such visit was in his seventy-ninth year, when he went to see how survey research had developed from its beginnings. Dr. L. E. Howlett has described it:

I took him around our photogrammetric section where we do have some rather sophisticated modern developments. There was a certain tendency on the part of some of my colleagues to treat him as they would normally treat a man of his years and distance away from actual scientific activity, and to give an easy answer. But they quickly found that the General came back with another question; that he knew perfectly well what he was asking, and that he wanted a real answer to it. The General was very interesting in this way. Most of the people in the section had had no dealings with him in a personal sense before, but after he had spent an hour in their laboratory they had a terrific respect for the amount this man, who as far as they were concerned had been dealing with water resources for the past ten or fifteen years, knew; and that he could come in and ask intelligent and very penetrating questions. The admiration he stirred among quite a different generation was remarkable and it impressed me that the General could still throw the same kind of spell that he threw on me about thirty years ago.

There was always that quick interest to take in something new. He burst in with all his old vigour, hair a bit tousled—his mental keenness and desire to know were still there.[4]

James Ripley, a young engineer who called at McNaughton's home to talk about the Columbia, remembered their first meeting and described the impression the General made on a member of a younger generation:

I was a bit leary of going. I'd heard about him all my life and admired him, but he was such an important man, such a big name, I wasn't even sure he would talk to me. After five minutes, I felt I'd known him all my life. I've talked to many similar people, Prime Ministers and whatnot, and most of them go out of their way to let you know that they are big men, but he never did. He was always himself, and he was a gregarious and sort of warm-hearted person anyway, and he was interested in everybody he met.[5]

Walter Stewart, who interviewed McNaughton in February, 1966, said much the same:

The General met me at the door of his grey-stuccoed house, set among trees in a quiet, dead-end street in central Rockcliffe. He is a six-footer, erect and handsome. He carries his 79 years with a swagger; his step is quick, his hand-clasp firm, his manner almost regal. A shock of neatly brushed white hair tops a square face marked by grey-brown eyes behind brown-rimmed glasses, a sturdy nose and long upper lip mounted by a rectangle of aggressive white moustache. . . .

At first I asked what I hoped were pointed questions—"You're retired, general, why don't you stop making a fuss?"—but I soon abandoned any attempt to control the interview and gave myself over to a struggle to keep up with his incredibly agile mind. . . .

I went to see him under the suspicion that this obdurate old soldier was a crank, an odd-man-out against the natural course of Canada's development; I came away convinced that unless we listen to McNaughton, we may barter away control of our most precious resource.[6]

New generations had come up but, as at Sherbrooke when he addressed The Royal Society in 1966, McNaughton held the foreground still. That year the Professional Institute of the Public Service of Canada gave him a special gold medal, the first in the institute's history, for a lifetime of service to Canada and to his fellow man. He was still serving. Fifty years before he had been a vital figure, searching for better artillery methods on the Western Front. Now he stood athwart the modern scene looking, as he always did, farther into the future than his hearers. Men saw his indomitable figure on the stage and knew that his patriotism burned as bright as ever.

With his wife he spent the weekends at their cottage at Montebello on the Ottawa River. He loved the river and the distant hills; moreover it was close to the Seigniory Club where he could entertain his friends. He planted thousands of trees at Montebello, by the house and down towards the river, and he enjoyed walking through his plantations. But trees brought their problems. The branches had to be trimmed and his wife, ever solicitous about his comfort and his welfare, forbade him to climb the trees. He developed a method of cutting them which worked very well. He made a crossbow and an arrow with a heavy head. He attached a line to the arrow and then he would shoot it over the branch he wanted to cut down. He used the line to pull a moving chain-saw into the tree. Finally he guided the saw with a rope from the ground until it cut off the branch he wanted. There was much else to keep him busy; watering the gardens and weeding, and making things for the cottage in a workshop which was a vital part of the house. He was not far off eighty. There was a new flag in Canada; the world he had known was changing but he kept pace with it. He was still fighting—and fulfilled.

Sometimes McNaughton was forced to look back and his comments on draft histories are valuable. He was asked to describe his role and that of the Army in the Second World War but, until his biography was started, he would not do so:

Rightly or wrongly, I have not the least doubt that the policies of the

Canadian Army, which I initiated and drove forward, were in fact correct and in the best interests not only of Canada, but also in our contribution to the defeat of our enemies and to success in our joint endeavours with our associates. It would of course be too much to expect that this would be recognized as yet, more particularly because those who are now giving voice [Churchill and Alanbrooke particularly] are mostly the very persons who sought for various reasons of their own to exploit Canada and Canadians for their own ends, and whom it was my privilege to oppose.[7]

Only at the very end did he seek to put the record straight. By then the familiar stage was growing empty—Churchill, Alanbrooke, Roosevelt, Mackenzie King, Ralston, even Crerar, all were gone—though Dill, on his horse at Arlington, remained in a sort of way.

He had seen many of his dreams for Canada realized. Great ships passed up the St. Lawrence Seaway not far from his summer home; Air Canada planes, on the Ottawa-Montreal run, flew overhead. The nation had full autonomy; the armed forces were independent. Canada was, as Bruce Hutchison described it, "a nation constitutionally British, geographically American, but wholly Canadian," and McNaughton was fighting to keep it so. Canada to him *was* the promised land, as described to the Children of Israel:

For the Lord thy God bringeth thee into a good land, a land of brooks and water, of fountains and depths that spring out of valleys and hills. A land of wheat and barley and vines. . . . A land wherein thou shalt eat bread without scarceness, thou shalt not lack anything in it; a land whose stones are iron, and out of whose hills thou mayest dig brass.

All that, and more, was Canada. It was an outstanding heritage to be passed, without mortgage, to generations of Canadians still unborn. McNaughton had been Canada's champion all his life. No wonder an ex-gunner, who had known him in the First World War, wrote: "I am so proud that I have lived a little in your time. Canada can ill afford the day when you—my dear, honest, gallant sir—cross the bar."[8]

On Friday evening, 8 July, 1966, the McNaughtons left Ottawa to spend the weekend at their cottage. Mrs. McNaughton drove the grey Buick. On Sunday one of the General's old aides, Lieutenant-Colonel "Chou" Price and his wife Betty, reached Montebello. They all dined at the Seigniory Club and returned to the cottage later in the evening. The talk turned to Canada's water and McNaughton reviewed the situation in an animated way: it was moving in the right direction, he was convinced of that. By this time the sun was setting behind the hills beyond the bend in the river. Thunder was in the air over Ottawa and the western horizon was black with clouds shot through with orange and tinged with pink. McNaughton drew his wife and guests outside and moved them this way and that to see the changing effects to the best advantage. He liked sunsets. Years before, at the funeral of Colonel Elroy Forde, the Chief of Signals, he had written down and preserved these words:

> So be my passing
> My task accomplished and the long day done,
> My wages taken. And in my heart
> Some late lark singing.
> Let me be gathered to the quiet West,
> The sundown splendid and serene.[9]

But he had no premonition; he was in the middle of a fight which was not yet over, and he had the bone-deep will to see it through. Nevertheless, in the early hours of the next morning, 11 July, he died peacefully in his sleep. Not long afterwards the sun they had all watched sink in moody splendour, such a short number of hours before, rose again on a Canada that was without McNaughton.

The eleventh of July, and the two days that followed, were no more than a blur of impressions to me. There was the McNaughtons' Rockcliffe home, brilliantly lit though it was a sunny day, and relatives and friends assembling like the gathering of a clan. Governor-General and Madame Vanier arrived in the afternoon. The telephone rang constantly. The military funeral was arranged in accordance with careful planning. Above all, I remember Mrs. McNaughton, bereaved but splendid. The next day passed in preparations for the funeral, which took place on the afternoon of the 13th, a blazing summer's day.

I was an honorary pallbearer, and proud of that. The funeral service at Christ Church Anglican Cathedral was simple, as he would have wished. It was conducted by Bishop J. O. Anderson of Winnipeg, McNaughton's old chaplain, and there were Roman Catholic clergy, as well as Protestant, in attendance. There was no eulogy, just a short scripture reading and two hymns; of these, "Fight the Good Fight" I remember best. Candles shone in the artificial light, reflected from medals and swords generously sprinkled throughout the cathedral, packed with mourners.

In the bright sunlight outside a gun carriage awaited the heavy flag-draped coffin, borne by men of the Royal Canadian Horse Artillery, with the General's cap and sword placed on it. Governor General Vanier, an old friend, in full uniform and Prime Minister Lester Pearson led the ranks of dignitaries and mourners. Slowly the long procession moved off through crowded streets as far as the Supreme Court Building. The Signals band, in bright uniforms; sailors, in white; Royal Canadian Mounted Police in scarlet coats; cadets; gunners; Royal Canadian Electrical and Mechanical Engineers—interminably, it seemed, the contingents slow-marched over the hot asphalt to the beat of muffled drums, as bells tolled solemnly overhead.

The sun glinted on McNaughton's decorations carried carefully on velvet cushions; it shone on sword scabbards and the dark metal of reversed arms; it twinkled back from the silver stirrups, with boots reversed, of a black, riderless horse led by a gunner of the Royal Canadian Horse Artillery. More than the gun carriage with its load, more than anything else, that prancing horse symbolized that McNaughton had gone, and I thought of the words he had used in 1957 on his son's gravestone:

> Here let the slow bell toll;
> On the other side
> The trumpets sound for him.

Bugles sounded later that afternoon in Beechwood Cemetery, their mournful notes echoing over the bright grass, splashed with shadows. Pipers played a lament. Minute-guns boomed from Nepean Point, setting flocks of starlings whirling. Then the coffin, still draped with the flag and bearing McNaughton's cap and sword, sank slowly into the grave.

During the next few days, and even weeks, tribute was paid McNaughton in British, American and Canadian newspapers and in letters from persons who had known him at various times of his life. At the instigation of Senator "Mike" Mansfield of Montana, articles from the New York *Times* and the Washington *Post* on McNaughton's passing, which appeared on 12 July, were read into the *Congressional Record* of the United States. "He was not only a great Canadian," Mansfield said, "but he was a great North American as well."[10] Editors wrote more in pride than sorrow. "General A. G. L. McNaughton. The very roll of his name was like breaking out a flag." He had "a twinkle in his eye at 79, was as at home sloshing about in shin-high overshoes on civvy street in a winter storm as he was a shining figure in full dress uniform, his fine head inches above all the rest and seemingly conducting his reconnaissance of a state banquet or ceremony as though it were an eve-of-battle sortie."[11]

Her Majesty The Queen wrote; so did prime ministers, ministers and ambassadors. Kenneth Bailey, the Australian High Commissioner, said: "You have proud memories of one of the most distinguished lives of service to Canada in this whole generation." "No person of his generation gained a higher place," wrote Arnold Heeney. Louis St. Laurent pointed out that the "lifelong services" to his country of "this great Canadian" had been "an outstanding contribution to the development of Canada and its prestige in world affairs." He had indeed left behind, wrote Hugh John Flemming, "a heritage of accomplishment." Of all great Canadians, said Prime Minister Lester B. Pearson, "Canada has had no greater man to serve her."[12]

Others came, in their letters of condolence, to witness to every phase of McNaughton's life. One had been with him at the Second Battle of Ypres in 1915: "From then on his brilliance and efficiency left us all far behind."[13] Another, who had also known him in World War I, "saw at close range the impression that he made with his original contribution in Counter Battery work. Regular soldiers years his senior came to learn from him."[14] As a gunner he was "an inspiration" and would always be remembered as "the finest gunner that served in the Royal Regiment of Canadian Artillery."[15] His work for survey, and in opening up the North between the wars, was remembered: like Ulysses, "whose purpose was to strive, to seek, to find and not to yield," McNaughton had also believed it "not too late to seek a newer world." His contribution to the development of Canada was "at least as great as anyone in this generation."[16] Remembered also was his unemployment relief scheme and his encouragement of aviation: unemployment relief had "advanced the

Trans-Canada Airway development by more than five years,"[17] and Canadian aviation owed him much.[18] At NRC he had inherited "a small scientific unit suffering from the Depression . . . and moulded it into an important and virile organization."[19]

Britain remembered with gratitude "his services to this country and to the cause of freedom in the Second World War"; nor did she forget "his subsequent endeavours in the cause of freedom in the United Nations."[20] One who had served under him in the First Canadian Army thought of "his great qualities of heart and mind"; the Army had been "unconditionally devoted to him always."[21] His "contribution towards battle worthy weapons was second to none."[22] Cardinal Roy (the Military Vicar) remembered him as "a great scholar" as well as "a great military chief."[23] He was "without peer," Archbishop O'Neill wrote; few "loved Canada more or did more for Canada."[24] And Douglas LePan, who had been his education adviser during the Second World War, saw him as "one of the greatest men and the greatest Canadian I have ever known."[25]*

No one mentioned his term as Minister of National Defence, though he did more for Canada then, perhaps, than at any other time. He was a good Minister, if a poor politician: "A man whose life was guided by the essence of Christianity and truth" and for whom "paltry politics were beneath contempt."[26]

At the United Nations, Philip Jessup, who had been on the American team, praised McNaughton highly: "Canada has contributed a lot of talent to the U.N., but nothing better than Andy. What a vibrant spirit he was!"[27]

So far as atomic energy in Canada went, his interest had been keen for the welfare of his country, "both in scientific achievements and in its natural resources."[28] On the Permanent Joint Board on Defence, an American, who had been associated with him, saw his work as "unselfish devotion to the best interests of his country and of the world."[29] "A man of genius," Canadian Air Marshal Reyno wrote,

---

* Writing in *Globe Magazine* (22 September, 1967) about youth shrugging off responsibilities, Douglas LePan, who was then principal of University College, University of Toronto, said this: "I have known at least a few men whose lives suggested that responsibility, far from being in any way conventional or diminishing, could be instead a means for sounding a very wide range of human possibilities. When I remember, for example, the passionate intelligence and intensity that General McNaughton brought to a long discussion of public responsibilities, the figure of him in my imagination grows large and emblematic and luminous."

"who achieved true greatness."[30] His work on the International Joint Commission earned many tributes. Robert Moses (who had been chairman of the New York Power Authority) remembered what McNaughton had done for the St. Lawrence power development: "Like the Iron Duke, he was four-square to all the winds that blow."[31] The seaway itself was "an outstanding example of what a personality can achieve in directing the course of events."[32] Douglas Stuart, who had been U.S. ambassador at the time of the final seaway negotiations, deplored his loss: "We in this age are in need of great wisdom and sterling character such as he possessed."[33]

Jean Lesage, writing as the leader of the opposition in Quebec, remembered McNaughton's insistence on a national water policy at the time of Kaiser's offer to build the Arrow Dam. "He had one of the greatest minds and he was kind enough to make me profit from his knowledge."[34] A member of the IJC stated: "He was listened to with respect and at times with awe; his legacy is such as to perpetuate his presence among us."[35] So far as the Columbia went, "History will relate that his aims were for Canada and her future generations rather than for the moment, as the politicians would have them."[36] Nevertheless, "his engineering ability will be followed still more closely in the future. He has indeed pointed the way for Canada—a great general, a great engineer, a great Canadian. 'Nature might stand up and say to all the world, *This was a man!*' "[37]

It is clear that no two persons saw McNaughton in quite the same way. They remembered him in a particular sphere and called him great. Surely this greatness in so many ways is the measure of the man? Some, indeed, did see a larger man, one who had "made his spirit live and act in every corner of Canada,"[38] and his deeds in Canada would "always stand as a monument to his memory."[39] But he was more than a national figure. Throughout his life, he had been "absorbed in something significant, something he could fight for, something worthy of his qualities of heart and mind";[40] that could not be forgotten. "One expects people like General McNaughton to be immortal. I think they are, really, because people never forget them and surely that is what immortality is."[41] Never again would his shouts "in the teeth of the tempest" be heard with "an exaltation of the spirit";[42] but at least, if tributes mean anything, he would be remembered.

The general public will no doubt best remember McNaughton as a soldier, for it was during the Second World War that the spotlight of publicity shone on him at its strongest. Good soldier though he was, however, he was much more than that; his eminence grew in the postwar years, while that of other distinguished soldiers faded. "I think if you look at his career after the war," commented Dr. Omond Solandt (Chancellor of the University of Toronto), "and compare it with that of other generals who had a distinguished record during the war, you'll see the difference. He would, in fact, be quite a famous man if he had never done anything until the end of the war. His career since the war has been a series of successes in various complex and difficult jobs, all of them involving a strong background of science and engineering."[43] Soldier, engineer, scientist and statesman—that was McNaughton.

It remains only to emphasize the spring that moved him. Sir Saville Garner, who, as British High Commissioner had known McNaughton well, said of him: "All who came in contact with him were touched by the strength of his character, his utter integrity, and his firm sense of purpose,"[44] and certainly it was purpose that ruled McNaughton's life and led to his great achievements.

Dr. C. J. Mackenzie, who knew him as well as any man, has pointed to purpose as the source of McNaughton's strength and difficulties in the foreword to this book. What he has said was true of McNaughton. He would argue, battle, arouse hostility—but never for self. He employed wit without malice, toughness without arrogance, and all because of purpose—his dedication to the welfare of his country. There are few Canadians, even those who have been irritated by him, who have not experienced at least some satisfaction in being so strongly represented in war, at the negotiating tables of the peacetime world, or in being so steadfastly defended against American encroachment by this man who was the foremost champion of Canadian autonomy. There is no doubt that he was a great man; as that, McNaughton impressed those who knew him closely in widely divergent fields. More arresting is the assertion by Sir Arthur Steel-Maitland, Lord Bessborough, Reyburn, Ignatieff and Reyno that the spark of genius had touched him; and even those five did not know all the facets of this many-sided man. If he was a genius, it was his sense of purpose that had made him so, and his own words to graduates about to embark on life's great journey testify to the importance purpose had for him:

If the world of affairs, to which you are now about to transfer your activities, were governed by the laws of logic there would, of course, at once be a proper niche in which your energies would be absorbed. . . .

But, fortunately or unfortunately, the one characteristic which seems to be most conspicuous for its lack in the world, as we know it, is logic and this is the case whether it be in business, in the Arts and Sciences, in the Public Service, or in Government itself. Everywhere things seem, at least on the surface, to be made up of discontinuities and it is probable that you will find it exceedingly difficult to pick out and identify, let alone to follow, the thread of purpose which, without any doubt, links all things together in a pattern; but it is a pattern in which the harmonies are only revealed by faith and the light of experience; and experience comes only with patience, long continued. . . .

So the path which I predict for you is no smooth open road leading straightly into the future, but one of ups and downs, of trials and tribulations, and obstacles. You would not have it otherwise. Steel is not forged except by fire; trials are to consume the dross; and obstacles are a challenge to be overcome so that, when you shall reach the heights of some accomplished undertakings, you may not only take your satisfaction in the end itself but know that you have become a more useful instrument for the further tasks which always lie ahead.

If I have pictured for you a vista not made up of quiet, and peace, and simple pleasure, there is no need to be pessimistic on that account. For the world, despite its conflicts and uncertainties—or possibly because of them—is a very interesting place. Things everywhere are in a flux of rapid, successive change, and nowhere more forcefully than in Canada, this marvellous country of ours, where opportunity expands before the eye in ever-widening fields for useful endeavour. . . .

And so, with the hope of widest benefits to human happiness on the one hand and danger, real and always incipient, on the other, we must have leaders of purpose and clear vision in the choice of objectives—leaders who will give the most careful forethought in planning, leaders who will be ever alert in the nicest balance to safeguard the nation's welfare and at the same time prevent distractions from disturbing the forward course on which we are embarked.

It seems to me that, after Faith in Divine Providence, Purpose is the great attribute which those who aspire to leadership must strive for.

Purpose, definite and firmly held, and always pressed with vigour. . . .[45]

Biographical Sketch

General the Honourable Andrew George Latta McNaughton, P.C., C.H., C.B., C.M.G., D.S.O., C.D., M.Sc., D.Sc., D.Eng., D.C.L., LL.D., PSC, IDC.

Born: Moosomin, Sask., 25 February, 1887; son of Robert Duncan McNaughton, formerly of Saint John, N.B., and Christina Mary Ann McNaughton (Armour), formerly of Ingersoll, Ontario, only daughter of William Armour of Dorchester, Ontario, and Ann Prentice, his wife.

Educated: Public School, Moosomin, Sask. (1894-1900); Bishop's College School, Lennoxville, P.Q. (1900-1905); McGill University (1905-1910); postgraduate and instructional staff in Electrical Engineering and Mathematics (1910-1914); Royal Staff College, Camberley, England (1921); Imperial Defence College, London, England (1927).

Degrees: B.Sc., McGill, 1910; M.Sc., McGill, 1912; PSC, Royal Staff College, 1921; IDC, Imperial Defence College, 1928.

Married: Mabel Clara Stuart, daughter of Godfrey Weir, Montreal, P.Q., 17 September, 1914, at Valcartier, P.Q.

Children: Christina Pauline Stuart, 10 August, 1915; S/L Andrew Robert Leslie, A.F.C., RCAF, 21 July, 1916; Brigadier-General Edward Murray Dalzell Leslie, D.S.O., C.D., 20 October, 1918; S/L Ian George Armour, RCAF, November 6, 1919 (killed in action over Emden, Germany, June, 1942); Leslie Anita Stuart, 14 September, 1924.

Military and Civilian Record:

1909/ 1910	Special course for university candidates for commission in regular army, McGill University.
4 May, 1909	Provincial Lieutenant, Canadian Militia.
9 May, 1910	Lieutenant, 3rd Battery, CFA.

16 May, 1911	Promoted Captain.
28 May, 1913	Promoted Major to command 3rd Battery, CFA.
22 September, 1914	Major, 4th (7th) Battery, CFA.
21 October, 1915	Major, 21st Howitzer Battery, CFA.
11 March, 1916	Lieutenant-Colonel, 11th Brigade, CFA.
27 January, 1917	Counter-Battery Staff Officer, Canadian Corps Artillery.
October, 1918	Appointed to command Canadian Corps Heavy Artillery.
10 November, 1918	Promoted Brigadier-General.
	Twice wounded; three times mentioned in despatches; awarded Distinguished Service Order.
1919	Appointed member of committee for reorganizing Canadian Militia.
1 January, 1920	Commissioned in Canadian Permanent Force, appointed Director of Military Training and Staff Duties, National Defence Headquarters, Ottawa.
1921	Royal Staff College, Camberley, England.
1922	Resumed appointment, Director of Military Training and Staff Duties.
1 January, 1923– 22 December, 1926	Deputy Chief of the General Staff.
1923	Joint inventor of cathode ray direction finder.
January, 1927– April, 1928	Imperial Defence College, London, England.
1 May, 1928	District Officer Commanding, Military District No. 11, Victoria, B.C.
1 January, 1929	Promoted Major-General, appointed Chief of the General Staff.
1930	Adviser, Canadian delegation to Imperial Conference, London, England.
1932	Adviser, Canadian delegation to Conference for Limitation of Armaments, Geneva, Switzerland; member of committee on Transatlantic Air Service, Imperial Economic

	Conference, Ottawa; chairman, Associate Committee on Survey Research, National Research Council; chairman, Inter-Departmental Committee for Trans-Canada Airways; vice-president, Canadian Geographical Society; responsible under Ministers of Labour and National Defence for organization and administration of works projects for the care of single homeless men.
5 October, 1935	President, National Research Council.
17 October, 1939	GOC, 1st Canadian Division.
10 July, 1940	Lieutenant-General, Commander VII Corps.
25 December, 1940	Appointed to command Canadian Corps.
6 April, 1942–26 December, 1943	General Officer Commanding in-Chief, First Canadian Army Overseas.
27 September, 1944	Promoted to General.
1 November, 1944	Retired from Canadian Army.
2 November, 1944	Created Privy Councillor and appointed Minister of National Defence.
21 August, 1945	Resigned as Minister of National Defence.
1 September, 1945	Appointed chairman, Canadian section, Canada-United States Permanent Joint Board on Defence.
2 April, 1946	Appointed Canadian representative, United Nations Atomic Energy Commission.
26 September, 1946–16 March, 1948	President, Atomic Energy Control Board of Canada.
January, 1948–December, 1949	Permanent delegate of Canada to the United Nations, representative of Canada on the Security Council.
1 January, 1950	Appointed Commissioner, Canadian section, International Joint Commission.
17 July, 1950	Elected chairman, Canadian section, IJC. Retired 15 April, 1962.
30 June, 1959	Resigned chairmanship of Canadian section, Canada-United States Permanent Joint Board on Defence.

Decorations: Distinguished Service Order (1917); Companion of the Order of St. Michael and St. George (1919); Companion of the Order of the Bath (1935); Companion of Honour (1946); Grand Officer of the Order of Leopold [Belgium] (1946); Canadian Forces Decoration (1955).

Honours and Honorary Degrees: LL.D., McGill, 1920; Member of the Engineering Institute of Canada, 1926, Life Member of the Engineering Institute of Canada, 1949, Honorary Member, 1957; D.C.L., Bishop's College, 1937; Honorary Member of the Canadian Institute of Forestry, 1938; Honorary Fellow of the Royal College of Physicians and Surgeons, 1939; awarded Sir John Kennedy Medal of the Engineering Institute of Canada, 1940; awarded Silver Medal of the Royal Society of Arts, London, England, 1941; LL.D., Queen's, 1941; Honorary Fellow of The Royal Society of Canada, 1941; Honorary Member of the Institution of Electrical Engineers, London , 1942; LL.D., Birmingham, 1942; Honorary Member, Royal Canadian Institute, Toronto, 1942; Honorary Member of the American Institute of Electrical Engineers, New York, 1942; Honorary Member of the American Society of Mechanical Engineers, New York, 1943; Honorary Member of the Institution of Royal Engineers, Chatham, 1943; LL.D., Ottawa University, 1943; Honorary Member, McGill Graduates Society, 1943; LL.D., University of Saskatchewan, 1944; Honorary Member, Royal Canadian Engineers Survey Association, Ottawa, 1947; Charter Associate, Arctic Institute of North America, 1948; Honorary Colonel, 12 March, 1946-18 March, 1949, Honorary Colonel Commandant, 19 March, 1949-19 September, 1962, Colonel Commandant, 20 September, 1962-18 March, 1964, of the Corps of Royal Canadian Electrical and Mechanical Engineers; Honorary Life Member, Canadian Artillery Association (1949); Member, Association of Professional Engineers (Electrical) of Ontario (1950); Gold Medal, Association of Professional Engineers of Ontario (1951); Honorary Colonel, McGill University COTC (1951); Life Member, National Geographic Society (1952); LL.D., Michigan State College (1955); D.Sc., University of British Columbia (1960); commissioned

Boundary Dam power plant, 14 May, 1960; Honorary Member, Native Sons of Canada, 1960; LL.D., University of Toronto (1961); Honorary Member, Association of Professional Engineers of Alberta (1961); citation, Engineering Institute of Canada, Vancouver, B.C. (1961); Doctor of Military Science, RMC of Canada (1963); Honorary Fellow, Canadian Aeronautical and Space Institute (1963); Founders Award from the Institute of Electrical and Electronics Engineers "for his inspiring leadership and his personal contributions in the field of electrical engineering and radio communications" (1964); Member, American Institute of Aeronautics and Astronautics (1964); Special Gold Medal, Professional Institute of the Public Service of Canada "for a life-time of service to Canada and his fellow man" (1966); D.Eng., University of Waterloo (1966); Keefer Gold Medal of the Engineering Institute of Canada "for the best paper on an engineering subject in 1965" (1966).

Died: 11 July, 1966.

Notes

PART 5

CHAPTER 1

[1] Interview with General Pope, October, 1966.
[2] "Chain of Events" file, McNaughton Papers.
[3] Interview with Dr. C. J. Mackenzie, September, 1966.
[4] "Chain of Events" file.
[5] Summary by Lieutenant Gerry Rau (ADC), "Chain of Events."
[6] "Chain of Events" file.
[7] Connolly to Rau, 5 February, 1944, *ibid.*
[8] McNaughton to Mackenzie King, 21 September, 1944, "Retirement from Cdn Army and N.R.C." file, McNaughton Papers.
[9] "Chain of Events" file.
[10] *Ibid.*
[11] Connolly to Rau, 17 February, 1944.
[12] Miscellaneous press clippings, McNaughton Papers (newspaper and date unknown).
[13] J. W. Pickersgill, *The Mackenzie King Record,* Vol. I (Toronto: University of Toronto Press, 1960, pp. 615, 626-8.
[14] Mountbatten to McNaughton, 16 February, 1944, McNaughton Papers.
[15] Dill to McNaughton, 28 December, 1943, *ibid.*
[16] Birdwood to McNaughton, 7 April, 1944, *ibid.*
[17] Miscellaneous correspondence, *ibid.*
[18] Moosomin *World-Spectator,* 15 May, 1944.
[19] Ross Munro, *Gauntlet to Overlord* (Toronto: Macmillan, 1945), p. 34.
[20] Toronto *Evening Telegram,* 15 June, 1944.
[21] *Maple Leaf* (Italy Edition), 1 July, 1944.
[22] Interview, 27 June, 1966.
[23] "Progressive Conservatives, 1944," McNaughton Papers; "Political Letters, July 1944." *Ibid.*
[24] "Retirement from Canadian Army . . ."
[25] *Ibid.*
[26] *Ibid.*
[27] *The Mackenzie King Record,* Vol. 1, p. 74.
[28] "Retirement . . ."
[29] *Ibid.*
[30] *Ibid.*
[31] *Ibid.*
[32] J. W. Pickersgill and D. F. Forster, *The Mackenzie King Record,* Vol. II (Toronto: University of Toronto Press, 1968), p. 119.
[33] *Ibid.* p. 120.
[34] "Retirement . . ."
[35] McNaughton interview, 7 July 1966.
[36] "Retirement . . ."
[37] *The Mackenzie King Record,* Vol. II, p. 121.
[38] "Retirement . . ."
[39] Montreal *Star,* 30 September, 1944.
[40] "Retirement . . ."
[41] *Ibid.*
[42] *Ibid.*
[43] "Progressive Conservatives" file.
[44] *Ibid.*
[45] *Ibid.*
[46] *Ibid.*
[47] *Ibid.*
[48] *The Mackenzie King Record,* Vol. II, p. 123.
[49] *Ibid.,* p. 125.

CHAPTER 2

[1] *Debates, House of Commons,* 14 February, 1944.
[2] File "Reinforcement Crisis, 1944," McNaughton Papers.
[3] Pickersgill and Forster, *The Mackenzie King Record,* Vol. II, p. 111.
[4] "Reinforcement Crisis."
[5] The *Mackenzie King Record,* Vol. II, p. 145.
[6] "Reinforcement Crisis."
[7] *Ibid.*
[8] *Ibid.*
[9] *The Mackenzie King Record,* Vol. II, pp.126-8.
[10] *Ibid,* p.147.
[11] *Ibid,* p.169.

[12] Memo of conversation with Mackenzie King, 1 November, 1944, "Appointment to Cabinet, 2 November, 1944."
[13] Norman Ward, (ed.), *A Party Politician: The Memoirs of Chubby Power* (Toronto: Macmillan, 1966), p. 151.
[14] *Ibid.*
[15] *Ibid.*, p.152
[16] *The Mackenzie King Record* Vol. II, p. 134.
[17] *Ibid.*, p. 151.
[18] *Chubby Power*, p. 152.
[19] *Ibid.*, pp. 152-3.
[20] *Ibid.*, p. 153.
[21] *Ibid.*, p. 154.
[22] R. MacGregor Dawson, *The Conscription Crisis of 1944* (Toronto: Copp Clark, 1963), p. 38.
[23] Information supplied by Dr. J. Mackay Hitsman.
[24] The *Mackenzie King Record*, Vol. II, pp. 157, 161,164.
[25] *Ibid.*, p. 138.
[26] Record of Conversation with P.M., 1 November 1944, "Appointment to Cabinet, November, 1944," McNaughton Papers.
[27] *The Mackenzie King Record*, Vol. II, pp. 174-5.
[28] Conversation with C. D. Howe, 30 October, 1944, "Appointment to Cabinet."
[29] *The Mackenzie King Record* Vol. II, pp. 175-6.
[30] Conversation with Mackenzie King, 31 October, 1944.
[31] Discussion with Prime Minister, "Letters, P.M., 1944-5."
[32] *The Mackenzie King Record* Vol. II, pp. 178-9.
[33] *Ibid.*, p. 177.
[34] *Ibid.*, p. 178.
[35] *Ibid.*, p. 184.
[36] *Ibid.*, pp. 182-3.
[37] Conversation with Mackenzie King, 1 November, 1944, "Appointment to Cabinet."
[38] *Ibid.*
[39] *The Mackenzie King Record*, Vol. II, p. 183.
[40] *Ibid.*
[41] Conversation with Mackenzie King, 1 November, 1942.
[42] *Ibid.*
[43] Letters dated 2 November, 1944, "Progressive Conservatives, 1944."
[44] *The Mackenzie King Record*, Vol. II, p. 187.
[45] *Ibid.*, p. 191.
[46] "Appointment to Cabinet."
[47] *Ibid.*
[48] *Ibid.*
[49] Interview with General Pope, October, 1966.
[50] *The Mackenzie King Record*, Vol. II, p. 195.
[51] *Ibid.*, pp. 195-6.
[52] "Letters, P.M., 1944-5."
[53] *Ibid.*
[54] *Ibid.*
[55] *The Mackenzie King Record*, Vol. II, p. 199-200.
[56] *Canadian Military Gazette*, November, 1944.
[57] *Ibid.*
[58] *The Mackenzie King Record*, Vol. II, p. 202.
[59] *Ibid.*
[60] "Speech to the House, 23 Nov., 44"; file "Appointment to Cabinet."
[61] *Canadian Military Gazette*, November, 1944.
[62] "Speech to the House."
[63] *Ibid.*
[64] Clifford H. Bowering, *Service: The Story of the Canadian Legion 1925-1960* (Ottawa: Dominion Command, Canadian Legion, 1960), pp. 119-20.
[65] Draft Radio address, "Appointment to Cabinet."
[66] "Appointment to Cabinet."
[67] *The Mackenzie King Record*, Vol. II, p. 210.
[68] Interview with General McNaughton, 28 April, 1965.
[69] "Newspaper clippings— unfavourable," McNaughton Papers
[70] *The Mackenzie King Record*, Vol. II, p. 214.
[71] *Ibid.*, p. 217.
[72] Conversations with Mackenzie King, 17 and 20 June, 1945, Letters, P.M.
[73] Interview with R. MacGregor Dawson, 30 January, 1953,

McNaughton Papers.
[74] Conversations with King, 17 and 20 June, 1945.
[75] "Clippings, 1944," McNaughton Papers
[76] *The Mackenzie King Record,* Vol. II, p. 224.
[77] McNaughton to King, 17 June, 1945, "Letters, P.M."
[78] *The Mackenzie King Record,* Vol. II, p. 220.
[79] Dawson, *The Conscription Crisis,* p. 78.
[80] *Chubby Power,* p. 156.
[81] *Ibid.,* pp. 157-9.
[82] *The Mackenzie King Record,* Vol. II, p. 222.
[83] *Ibid.,* p. 223.
[84] *Chubby Power,* p. 159.
[85] *Ibid.,* p. 160.
[86] "Letters, P.M., 1944-5."
[87] *Ibid.*
[88] McNaughton interview, 28 April, 1965.
[89] *The Mackenzie King Record,* Vol. II, p. 229.
[90] McNaughton interview, 28 April, 1965.
[91] Interview with Professor Dawson.
[92] *Chubby Power,* pp. 164-5.
[93] Interview with General Pope.
[94] "Speech to the House, 23 Nov 44."
[95] *The Mackenzie King Record,* Vol. II, p. 241.
[96] File "St. Laurent, Rt. Hon. Louis," McNaughton Papers.
[97] *The Mackenzie King Record,* Vol. II, p. 228.
[98] *Ibid.,* pp. 250-1.
[99] *Debates, Senate,* 30 November, 1944.
[100] McNaughton letter to King, 17 June, 1945.
[101] "Reinforcement Crisis, 1944," McNaughton Papers.
[102] *The Mackenzie King Record,* Vol. II, pp. 252-3.
[103] McNaughton interview, 28 April, 1965.
[104] *Ibid.*

CHAPTER 3

[1] Pickersgill and Forster, *The Mackenzie King Record,* Vol. II, p. 273.
[2] Toronto *Star,* 30 January, 1945.
[3] McNaughton to Ahearn, 8 January 1945, file "T. Franklin Ahearn," McNaughton Papers.
[4] McNaughton to Ahearn, 22 January, 1945.
[5] *Ibid.*
[6] McNaughton to Ahearn, 9 February, 1945.
[7] "Letters, P. M., 1944-5," McNaughton Papers.
[8] *Ibid.*
[9] Ottawa *Journal,* 24 September, 1959.
[10] Information supplied by Dr. J. L. Granatstein.
[11] Correspondence, Roebuck-McNaughton, December, 1944, McNaughton Papers.
[12] *Public Opinion,* January, 1945.
[13] Roebuck-McNaughton correspondence.
[14] *The Mackenzie King Record,* Vol. II, p. 287.
[15] *Ibid.,* pp. 285-6.
[16] *Ibid.,* p. 288.
[17] *Ibid.,* p. 289.
[18] "Press Clippings—Unfriendly" (newspaper unidentified), McNaughton Papers.
[19] Peter C. Newman, *Renegade in Power* (Toronto: McClelland & Stewart, 1963), p. 150.
[20] Toronto *Globe and Mail,* 24 September, 1959.
[21] Hon. John Bracken, "The Truth Overseas," Mc Naughton Papers.
[22] Toronto *Star,* 30 January, 1945.
[23] Ahearn to McNaughton, 29 January, 1945.
[24] *The Canadian Forum,* August, 1966.
[25] File "Letters, P.M., 1944-5."
[26] Ottawa *Journal,* 7 February, 1945.
[27] *The Mackenzie King Record,* Vol. II, p. 292.
[28] Information from Gerry Rau (member of McNaughton's staff).
[29] Draft dated 9 February, 1945, "Letters, P.M."
[30] *The Mackenzie King Record,* Vol. II, p. 311.
[31] "Letters, P.M."

[32] Address to Engineering Institute of Canada, Montreal, 22 February, 1945, McNaughton Papers.
[33] McNaughton to Ahearn, 9 February, 1945.
[34] McNaughton to Ralston, 23 September, 1944, summary contained in "Retirement from Cdn Army . . ." file.
[35] *Ibid.*
[36] *Ibid.*
[37] *The Mackenzie King Record* Vol. II, p. 275.
[38] Colonel C. P. Stacey, *Six Years of War* (Ottawa: The Queen's Printer, 1956), pp. 512-3.
[39] Lt.-Colonel G. W. L. Nicholson, *The Canadians in Italy* (Ottawa: The Queen's Printer, 1956), pp. 657-8.
[40] Brooke, cited by Bryant (ed.), *The Turn of the Tide* (London: Collins, 1957), pp. 57-8.
[41] *Ibid.*, p. 71.
[42] *The Turn of the Tide*, pp. 222, 227.
[43] McNaughton to Nicholson, 13 January, 1953, McNaughton Papers.
[44] "Comments on Canadian Official Historical Sketch," *ibid.*
[45] *The Canadians in Italy*, pp. 660, 663.
[46] Cited by Bryant, *The Turn of the Tide*, pp. 428-9.
[47] "Comments on Canadian Official Historical Sketch."
[48] McNaughton to Alexander, 14 April, 1946, file "F.M. Viscount Alexander," McNaughton Papers.
[49] "Comments on Canadian Official Historical Sketch."
[50] Peter Young, *World War 1939-45: A Short History* (London: Arthur Barker, 1966), p. 151.
[51] Bryant, *The Turn of the Tide*, p. 227.
[52] "Comments on Canadian Official Historical Sketch."
[53] McNaughton address to Economic Club of Detroit, 24 March, 1947, McNaughton Papers.
[54] McNaughton to Alexander, 14 April, 1946.
[55] McNaughton interview, 12 May, 1966.
[56] *The Mackenzie King Record*, Vol. II, p. 351.
[57] Ottawa *Journal*, 9 April, 1945.
[58] Ahearn to McNaughton, 16 May, 1945.
[59] Ottawa *Journal*, 30 May, 1945
[60] *Ibid.*, 1 June, 1945.
[61] *Ibid.*, 2 June, 1945.
[62] Canadian Press despatch, 12 May, 1945.
[63] *Six Years of War*, 514, 517, 519.
[64] McNaughton to Alexander, 14 April, 1946.
[65] Record of meeting with Mackenzie King, 16 June, 1945, "Letters. P.M." file.
[66] Winnipeg *Tribune*, quoted in Ottawa *Journal*, 17 May, 1945.
[67] Letter initialled "E.C.N."; Ottawa *Journal*, 17 May, 1945.
[68] Editorial, Ottawa *Journal*, 18 May, 1945.
[69] Toronto *Globe and Mail*, 9 June, 1945.
[70] Winnipeg *Tribune*, 8 June, 1945.
[71] Ottawa *Journal*, 24 May, 1945.
[72] *Ibid.*
[73] *Ibid.*, 2 June, 1945.
[74] Meeting with Mackenzie King, 16 June, 1945.
[75] Stenographic notes of McNaughton's speech, Ottawa *Journal*, 2 June, 1945.
[76] Ottawa *Journal*, 2 June, 1945.
[77] *Ibid.*
[78] Interview with Mrs. McNaughton, 12 April, 1967.
[79] *Ibid.*
[80] Ottawa *Journal*, 1 June, 1945.
[81] *The Mackenzie King Record*, Vol. II, p. 418.
[82] *Ibid.*, p. 412.
[83] Montreal *Gazette*, 28 June, 1945.
[84] Toronto *Globe and Mail*, 20 June, 1945.
[85] *Ibid.*, 1 August, 1945.
[86] Austin Neville, letter to Ottawa *Citizen*, 7 July, 1945.
[87] W. K. Walker, *ibid.*, 2 July, 1945.
[88] Ottawa *Journal*, 20 June, 1945.
[89] *Globe and Mail*, 18 July, 1945.
[90] Meeting with Mackenzie King.
[91] Memo, dated 24 November, 1945, "Letters, P.M."

[92] Meeting with Mackenzie King.
[93] *The Mackenzie King Record* Vol. II, p. 418.
[94] Norman Hillson, *Alexander of Tunis* (Toronto: W. H. Allen, 1952), pp. 188-91.
[95] Alexander file, McNaughton Papers.
[96] *The Turn of the Tide*, pp. 472, 474.
[97] Cited in Bishop's College School Magazine, June, 1945.
[98] Field-Marshal The Viscount Montgomery, *Memoirs* (London: Collins, 1958), p. 439.
[99] Memo. dated 22 August, 1946, "Maclean Publishing Co., Issues 15 Sept., 46, 1 Oct., 46," McNaughton Papers.
[100] *The Mackenzie King Record*, Vol. II, p. 463.
[101] Meeting with Mackenzie King.
[102] *Ibid.*
[103] McNaughton interview, 12 May, 1966.
[104] *The Mackenzie King Record*, Vol. II, p. 435.
[105] *Ibid.*, p. 439.
[106] *Ibid.*, p. 442.
[107] *Ibid.*, p. 444.
[108] Montreal *Star*, 21 July, 1945.
[109] McNaughton to Mackenzie King, 6 August, 1945.
[110] King to McNaughton, 21 August, 1945.
[111] Toronto *Evening Telegram*, 3 October, 1945.
[112] Ottawa *Journal*, 23 August, 1945.
[113] V. V. McNitt, *The MacNauchtan Saga*, Vol. 1 (Palmer, Mass.: Hampden Hill Press, 1951), p. 166.
[114] Kingston *Whig-Standard* 15 May, 1948.

CHAPTER 4

[1] "Brazil, Mission to," McNaughton Papers.
[2] McNaughton to Kennedy, 14 April, 1960, file "T. Dowsley Kennedy," *ibid.*
[3] Address to Engineering Institute of Canada, Toronto, 8 May, 1947, *ibid.*
[4] *Ibid.*
[5] *Ibid.*
[6] *Ibid.*
[7] McNaughton, in "Atomic Energy," CBC Trans-Canada series broadcast, April-June, 1947.
[8] George Ignatieff, "General A. G. L. McNaughton: A Soldier in Diplomacy," *International Journal*, Summer, 1947.
[9] *Time*, 2 July, 1965.
[10] ". . . A Soldier in Diplomacy."
[11] Frederick Osborn, "The U.S.S.R. and the Atom," *International Organization* (August, 1951).
[12] McNaughton, "National and International Control of Atomic Energy," Winter 1947-8.
[13] "The U.S.S.R. and the Atom."
[14] Arnold Wolfers, "The Atomic Bomb in Soviet-American Relations," *The Absolute Weapon*, Bernard Brodie (ed.) (Yale University, 1946).
[15] Moscow *New Times*, 24 January, 1947.
[16] "National and International Control of Atomic Energy."
[17] *Ibid.*
[18] Interview with General McNaughton, 23 March, 1966.
[19] McNaughton, "Regulation of Armaments and Supply of Forces," *Annual Review of United Nations Affairs 1949* (New York University Press).
[20] Richard G. Hewlett and Oscar E. Anderson, Jr., *A History of the United States Atomic Energy Commission*, Vol. I of *The New World, 1939-1946* (Pennsylvania State University Press, 1962), p. 583.
[21] "Regulation of Armaments and Supply of Forces."
[22] Sir George Thomson, "International Control," CBC "Atomic Energy" series.
[23] "Regulation of Armaments . . ."
[24] "The U.S.S.R. and the Atom."
[25] Group Captain L. G. Cheshire, "Dropping the Bomb," CBC series.
[26] McNaughton, "National and International Control . . ."
[27] "Canada's Part in International Control," CBC series.
[28] *New Republic*, 3 April, 1950.
[29] McNaughton to Solandt,

22 September, 1958, Solandt file, McNaughton Papers.
[30] Information from Mrs. McNaughton.
[31] Interview with General McNaughton, 5 January, 1966.
[32] "The U.S.S.R. and the Atom."
[33] ". . . A Soldier in Diplomacy."
[34] "The U.S.S.R. and the Atom."
[35] P. M. S. Blackett, *Fear, War and the Bomb* (New York: McGraw Hill, 1949), p. 9.
[36] *Ibid.*, p. 95.
[37] *Ibid.*, p. 150.
[38] *Ibid.*, p. 152.
[39] W. J. Hall to Ottawa *Citizen*, 12 June, 1954.
[40] P. M. S. Blackett, *Atomic Weapons and East-West Relations* (Cambridge: Cambridge University Press, 1956), p. 91.
[41] Interview with General McNaughton, 14 July, 1965.
[42] McNaughton to Solandt, 22 September, 1958.
[43] *New Republic*, 3 April, 1950.
[44] Margaret Coit, cited in *Time*, 2 July, 1965.
[45] "The U.S.S.R. and the Bomb."
[46] *Ibid.*
[47] V. V. McNitt, *The MacNauchtan Saga*, Vol. I, p. 159.
[48] Kingston *Whig-Standard*, 15 May, 1948.
[49] Cited McNitt, *The MacNauchtan Saga*, Vol. I, p. 162.
[50] Montreal *Star*, 1 April, 1948.
[51] "Regulation of Armaments. . ."
[52] McNaughton to P.M., 18 July, 1950, Mackenzie King file.
[53] ". . . A Soldier in Diplomacy."
[54] "National and International Control of Atomic Energy."
[55] *Ibid.*
[56] File "Dr. W. B. Lewis," McNaughton Papers.
[57] McNaughton to S. B. Challis, 22 January, 1948, file "Canadian Delegate to U.N."
[58] File note, *ibid.*
[59] "National and International Control . . ."
[60] Canadian Press release, 9 January, 1948.
[61] "McNaughton: Wise, Patient, and what a Worker!"; file of unidentified newspaper clippings, McNaughton Papers.
[62] ". . . A Soldier in Diplomacy."
[63] ". . . Wise, Patient, and what a Worker!"
[64] O'Hearn, cited by Ignatieff, ". . . A Soldier in Diplomacy."
[65] ". . . Wise, Patient, and what a Worker!"
[66] *Ibid.*
[67] ". . . A Soldier in Diplomacy."
[68] *Ibid.*
[69] Address to Canadian Institute of International Affairs, Montreal, 30 January, 1950, McNaughton Papers.
[70] ". . . A Soldier in Diplomacy."
[71] *Ibid.*
[72] F. H. Soward and Edgar McInnis, *Canada and the United Nations* (New York: Manhattan Publishing Co., 1956), p. 109.
[73] *Ibid.*
[74] "Important Issues before the United Nations," 11 August, 1949, McNaughton Papers.
[75] *Canada and the United Nations*, p. 108.
[76] Cited *ibid.*, p. 109.
[77] Text contained in file "Hemming, Harold L.," McNaughton Papers.
[78] McNaughton to Martin, 18 June, 1963, file "Columbia River Treaty—Martin," McNaughton Papers.
[79] Walter O'Hearn, unidentified newspaper clipping of November, 1949, McNaughton Papers.
[80] ". . . A Soldier in Diplomacy."
[81] "Canadian Delegate to U.N." file.
[82] File "Paris trip Sep.-Dec. 48," McNaughton Papers.
[83] John Holmes to Larratt Higgins, comments on draft MS, 26 February, 1969.
[84] Riddell to McNaughton, 22 January, 1949, file "Riddell, G.K.," McNaughton Papers.
[85] Mackenzie King file.
[86] U.N. *Security Council, Verbatim Record*, 29 December, 1949.
[87] Interview given by General McNaughton after consulting his

working file.
[88] *Ibid.*
[89] New York *Times*, 15 March, 1950.
[90] Holmes to Higgins, 26 February, 1969.
[91] London *Times*, 22 February, 1950.
[92] File "IJC—miscellaneous," McNaughton Papers.
[93] *Ibid.*
[94] File "Bevin, Rt. Hon. Ernest," *ibid.*
[95] File "Cadogan, Sir Alexander," *ibid.*
[96] "IJC—miscellaneous."
[97] Holmes to Swettenham, 18 March, 1969.
[98] ". . . A Soldier in Diplomacy."
[99] *[Editor's note, 2nd edition: no reference in source material.]*

CHAPTER 5
[1] McNaughton, "Defence of North America," address to the Council of Foreign Relations, New York, 12 April, 1948, McNaughton Papers.
[2] *Ibid.*
[3] Stanley W. Dziuban, *United States Army in World War II, Special Studies, Military Relations Between the United States and Canada 1939-1945*, p. 34.
[4] *Ibid.*
[5] *Ibid.*, p. 33.
[6] "Defence of North America."
[7] Pickersgill, *The Mackenzie King Record*, Vol. I, p. 234.
[8] *Ibid.*, p. 203.
[9] *United States Army in World War II* . . ., p. 70.
[10] *The Mackenzie King Record*, Vol. I, p. 436.
[11] *Ibid.*, p. 523.
[12] Information from Mrs. McNaughton.
[13] *Debates, House of Commons*, 1944, pp. 41-2.
[14] *The Mackenzie King Record*, Vol. I, p. 671.
[15] *Ibid.*, pp. 636-40.
[16] *Ibid.*, p. 643.
[17] "Defence of North America."
[18] *Ibid.*
[19] Address to joint meeting of the American Society of Mechanical Engineers and the Engineering Institute of Canada in Detroit, 5 October, 1945, McNaughton Papers.
[20] Nicholas Mansergh (ed.), *Documents and Speeches on British Commonwealth Affairs, 1931-1952*, Vol. II (Oxford: Oxford University Press, 1952), pp. 1191-2.
[21] *United States Army in World War II* . . ., p. 336.
[22] Montreal *Gazette*, 22 September, 1947.
[23] "LaGuardia file," McNaughton Papers.
[24] *United States Army in World War II* . . ., p. 337.
[25] Interview with General McNaughton, 9 February, 1966.
[26] Address to A.S.M.E. and E.I.C.
[27] *United States Army in World War II* . . ., p. 337.
[28] McNaughton interview.
[29] *United States Army in World War II* . . ., p. 329.
[30] McNaughton interview.
[31] "Defence of North America."
[32] Louis St. Laurent, "Problems of Canadian Security," Rotary Tri-District Conference, Montreal, 26 April, 1948, McNaughton Papers.
[33] St. Laurent, "Our North American Partnership," St. Lawrence University, Canton, N.Y., 11 June, 1950, *ibid.*
[34] P.M.'s Remembrance Day broadcast, 1948, *ibid.*
[35] *Debates, House of Commons*, 17 March, 1948.
[36] *Ibid.*, 12 February, 1947.
[37] R. G. Robertson to McNaughton, "Resignation from PJBD," McNaughton Papers.
[38] McNaughton to Vice Admiral H. E. Orem, U.S.N., 12 March, 1958, "Orem file," *ibid.*
[39] Dr. O. Solandt to McNaughton, "Solandt file."
[40] *Debates, House of Commons*, 13 November, 1953.
[41] D. Goodspeed (ed.) "*Armed Forces*

of Canada, 1867- 1967' (Ottawa: The Queen's Printer, 1967), p. 225.
[42] *Ibid.*, p 227.
[43] *Ibid.*, p. 226.
[44] McNaughton to Solandt, 22 September, 1958.
[45] Interview with General McNaughton, 23 March, 1966.
[46] Montreal *Gazette*, 16 December, 1948.
[47] "Standardization of Screw Threads," McNaughton Papers.
[48] Morrow to McNaughton, 30 October, 1948, *ibid.*
[49] Morgan to Turner, 11 April, 1951, file "Lieut-Gen Sir Frederick Morgan," McNaughton Papers.
[50] Morrow to McNaughton.
[51] McNaughton interview, 23 February, 1966.
[52] D. J. Goodspeed, *A History of the Defence Research Board of Canada* (Ottawa: The Queen's Printer, 1958), p. 126.
[53] Windsor *Star*, cited Lethbridge *Herald*, 23 August, 1951.
[54] Information from W. H. Barton, May, 1967.
[55] Acheson to McNaughton, 17 October, 1947, "Acheson file," McNaughton Papers.
[56] "Scrapbook 1955," *ibid.*
[57] Fordren to McNaughton, 9 August, 1951, File "Lt Col W. M. Fordren," *ibid.*
[58] "Defence of North America"
[59] Meeting with Robert T. Anderson, U.S. Deputy Secretary of Defense, 6 July, 1954, McNaughton Papers.
[60] Montreal *Gazette*, 21 April, 1954.
[61] "Defence of North America"
[62] *Ibid.*
[63] *Ibid.*
[64] McNaughton interview, 9 February, 1966.
[65] "Defence of North America"
[66] *Ibid.*
[67] McNaughton interview.
[68] *Ibid.*
[69] Meeting with Anderson.
[70] *Maclean's*, 23 June, 1956.
[71] Meeting with Anderson.
[72] Fulton to McNaughton, file "Fulton, Fraser F.," McNaughton Papers.
[73] Letter, C. K. Hurst, *Atlantic Advocate*, October, 1967.
[74] File "R. H. Stokes-Rees," McNaughton Papers.
[75] Meeting with Anderson.
[76] McNaughton, "Mostly Reminiscences," address at University of British Columbia, 19 May, 1960, McNaughton Papers.
[77] McWhorter to McNaughton, file "PJBD (1950)."
[78] Information from W. H. Barton.
[79] Meeting with Anderson.
[80] Leonard W. Brockington, CBC recording.
[81] CBC recording.
[82] Brockington to McNaughton, 12 April, 1954, "Visit to Churchill, 27-28 Mar. 54," McNaughton Papers.
[83] Côté to McNaughton, 28 January, 1963, "National Aviation Museum," *ibid.*
[84] "Resignation from PJBD."
[85] *Ibid.*
[86] *Ibid.*
[87] *Ibid.*
[88] *Ibid.*
[89] *Ibid.*
[90] *Ibid.*
[91] *Ibid.*
[92] Ottawa *Journal*, 21 April, 1954.
[93] "Resignation from PJBD."
[94] McNaughton to H. L. Keenleyside, 6 February, 1961, "Keenleyside file," McNaughton Papers.

CHAPTER 6

[1] L. M. Bloomfield and Gerald F. Fitzgerald, *Boundary Waters Problems of Canada and the United States* (Toronto: Carswell Co. Ltd, 1958), p. 64.
[2] Address, Canadian-U.S. Committee of Chambers of Commerce, Ste Adèle-en-Haut, Quebec, 30 September–1 October, 1954.
[3] Meeting with a group of Sloan Fellows, Massachusetts Institute of Technology, 22 May, 1958.
[4] *Ibid.*
[5] Margaret S. Mattson, "The Adoption of an All-Canadian Seaway Plan 1950-1953: the

Necessity and Impact of the Canadian Initiative" (M.A. major paper, University of Windsor, 1967).

[6] Brooke Claxton, "Canadian-American Arbitration," Department of External Affairs, *Statements and Speeches*, No. 51/15.

[7] Address to Empire Club, Toronto, 6 December, 1951, McNaughton Papers.

[8] McNaughton to External Affairs Committee of the House of Commons, 11 August, 1958.

[9] *Ibid.*

[10] Pickersgill, *The Mackenzie King Record*, Vol. 1, pp. 162-5.

[11] Address to Empire Club.

[12] Meeting with Sloan Fellows.

[13] Address, Royal Roads, Esquimalt, 19 February, 1954, McNaughton Papers.

[14] Address to National Defence College, 15 October, 1951, *ibid.*

[15] Cited in Windsor *Star*, 31 January, 1952.

[16] Editorial, *Saturday Evening Post*, 1 March, 1952.

[17] Cited (U.S.) *St. Lawrence Seaway Manual.*

[18] File "Matthews, Herbert L.," McNaughton Papers.

[19] Toronto *Globe and Mail*, 1 August, 1959.

[20] *Boundary Waters Problems*, p. 199.

[21] Address to Canadian Club of Montreal, 8 March, 1954, McNaughton Papers.

[22] *Ibid.*

[23] Meeting with Sloan Fellows.

[24] U.S. Congress, *Report of the Committee on Foreign Relations . . .* Senate Document No. 441, 83rd Congress, 1st Session, 1953.

[25] *See* ref. 13.

[26] Interview with General McNaughton, 3 April, 1966.

[27] Address to N.D.C., 20 July, 1954.

[28] *Ibid.*

[29] Interview, 3 April, 1966.

[30] Memo of meeting with Robert T. Anderson, U.S. Deputy Secretary of Defense, 6 July, 1954, McNaughton Papers.

[31] *Ibid.*

[32] *Ibid.*

[33] *Ibid.*

[34] *Ibid.*

[35] *Ibid.*

[36] Canadian Note, 17 August, 1954, copy in McNaughton Papers.

[37] McNaughton to Grant, 2 September, 1954, file "Col. L. F. Grant," *ibid.*

[38] McNaughton to External Affairs Committee.

[39] Address to N.D.C., 15 October, 1951.

[40] Ottawa *Journal*, 17 June, 1967

[41] Address to Royal Military College, 27 October, 1952

[42] E. W. Webber at semi-annual meeting of IJC, Washington, 3 April, 1962, copy of proceedings in McNaughton Papers.

[43] IJC terms of reference, cited by Bloomfield and Fitzgerald, *Boundary Waters Problems*, pp. 164-5.

[44] *Ibid.*

[45] *See* ref. 13.

[46] Meeting with Sloan Fellows, M.I.T.

[47] McNaughton to Committee on External Affairs, 4 March, 1953.

[48] *Ibid.*

[49] *Ibid.*

[50] *Ibid.*

[51] "Mostly Reminiscences," Address by McNaughton to University of British Columbia, 19 May, 1960, McNaughton Papers.

[52] Address to N.D.C., 15 October, 1951.

[53] C. D. Bourne, "The Columbia River Controversy," *Canadian Bar Review*, September, 1959, p. 447.

[54] Address to Pacific Northwest Trade Association, Spokane, Washington, 2-3 November, 1953.

[55] Address to R.M.C.

[56] *Financial Post*, 22 September, 1956.

[57] *Canadian Tribune*, 1 August, 1966.

[58] *Financial Post*, 22 September, 1956.

[59] McNaughton to House of Commons Committee on External Affairs, 9 March, 1955.

[60] *Debates, House of Commons*, 4 February, 1955.

[61] To Committee on External Affairs.
[62] *Boundary Waters Problems,* p.177.
[63] *Ibid.,* p. 177-8.
[64] *Ibid.,* p. 179.
[65] *Ibid.,* p. 178.
[66] *Ibid.*
[67] Address to Committee of U.S. and Canadian Chambers of Commerce.
[68] *Boundary Waters Problems,* p.196.
[69] *Ibid.,* p. 192.
[70] Meeting with Anderson, 6 July, 1954.
[71] McNaughton to Standing Committee on External Affairs, Minutes of Proceedings and Evidence, No. 7, pp. 167-8.
[72] Cited by Bourne, "The Columbia River Controversy," p. 465.
[73] "The Columbia River Controversy," p. 465.
[74] *Ibid.*
[75] New York *Times,* 15 July, 1964.
[76] Larratt Higgins: Address to seminar on Canada-U.S. relations, University of Windsor, 30 January, 1967.
[77] Cited by Jacob Austin, ". . . A Study of the History and Influence of the Harmon Doctrine," *Canadian Bar Review,* September, 1959, p. 421.
[78] "The Columbia River Controversy, p. 451.
[79] *Ibid.,* p. 453.
[80] "A Study of the History and Influence of the Harmon Doctrine," p. 440.
[81] Anderson, cited by Bourne, "The Columbia River Controversy", p. 454.
[82] Pugsley, cited by Austin, "A Study of the History and Influence of the Harmon Doctrine," p. 440.
[83] "The Columbia River Controversy," p. 455.
[84] A Study of the History and Influence of the Harmon Doctrine," pp. 400, 408.
[85] M. E. Marts, to Association of Pacific Coast Geographers, 7 April, 1961, "Marts file," McNaughton Papers.
[86] *Debates, House of Commons,* 6 December, 1910.
[87] *Ibid.*
[88] Austin, p. 424.
[89] *Ibid.,* p. 439.
[90] General A. G. L. McNaughton, "Problems of Development of International Rivers," *Fünfte Weltkraftkonferenz, Wien, 1965.*
[91] *Ibid.*
[92] IJC meeting, Washington, 5 April, 1955, McNaughton Papers.
[93] "The Columbia River Controversy," p. 449.
[94] Michael Barkway, recorded interview for CBC radio programme "Andrew McNaughton–Canadian."
[95] *Boundary Waters Problems,* p. 67.
[96] *Ibid.,* pp. 169-70.
[97] "The Columbia River Controversy," p. 461.
[98] Cited *ibid.,* p. 451; International Columbia River Engineering Board Report, p. 33.
[99] "A Study of the History and Influence of the Harmon Doctrine," p. 439.
[100] *Boundary Water Problems,* p. 167.
[101] Bourne, cited by Austin, "A Study of the History and Influence of the Harmon Doctrine," p. 442.
[102] *Ibid.,* p. 442.
[103] "A Study of the History and Influence of the Harmon Doctrine, p. 442.
[104] *Ibid.*
[105] "The Columbia River Treaty and Protocol–a Presentation," *External Affairs,* April 1964, p. 68.
[106] McNaughton to Sloan Fellows, M.I.T., 22 May, 1958.
[107] Committee on External Affairs, 9 March, 1955.
[108] McNaughton to Fulton (Member for Kamloops, B.C.), 13 April, 1956, "Fulton file," McNaughton Papers.
[109] McNaughton to Sloan Fellows.
[110] *Ibid.*
[111] *Ibid.*
[112] Barkway, "Andrew McNaughton–Canadian."
[113] Mansfield, *ibid.*
[114] Jordan and Neuberger, cited in *Financial Post,* 21 January, 1956.
[115] *Wall Street Journal,* 27 December, 1957.
[116] Address to N.D.C., 15 October,

1951.
[117] *Financial Post,* 22 September, 1956.
[118] Barkway, "Andrew McNaughton–Canadian."
[119] *Financial Post,* 15 February, 1958.
[120] *Ibid.,* 22 September, 1956.
[121] Vancouver *Sun,* 25 March, 1957.
[122] "The Columbia River Treaty and Protocol."
[123] "Fulton file," McNaughton Papers.
[124] McNaughton to Ladner, 13 July, 1957, file "Leon Ladner, Q.C.," McNaughton Papers.
[125] McNaughton to Fulton, 13 April, 1956.
[126] McNaughton to Ladner.
[127] *Ibid.*
[128] *Ibid.*
[129] *Financial Post,* 18 October, 1958.
[130] Charlottetown *Evening Patriot,* 10 November 1959.
[131] Toronto *Globe and Mail,* 30 January, 1959.
[132] "Memorandum of meeting 13 Jul 57 concerning attitude of new Government to use of waters of Columbia," McNaughton Papers.
[133] *Financial Post,* 7 February, 1959.
[134] Ottawa *Journal,* 18 December, 1957.
[135] *Electrical News and Engineering,* December, 1957.
[136] Information from Mrs. McNaughton.
[137] Toronto *Globe and Mail,* 6 December, 1958.
[138] *Ibid.,* 11 December, 1958.
[139] *Ibid.*
[140] *Ibid.,* 6 December, 1958.
[141] "The Columbia River Treaty and Protocol."
[142] "The Columbia River Controversy," p. 469.
[143] Presentation by McNaughton to External Affairs Committee, 18 March, 1960.
[144] Vancouver *Sun,* 18 December, 1959.
[145] *Ibid.*
[146] *Financial Post,* 18 December, 1958.
[147] *Electrical News and Engineering,* December, 1957.
[148] *Engineering News-Record,* 12 November, 1959.
[149] "Columbia River Treaty, Protocol and Related Documents," *External Affairs,* February 1904, p. 4.
[150] Ladner to McNaughton, 1 August, 1959.
[151] Address to Canadian Institute of International Affairs, Montreal, 16 February, 1960.
[152] *Financial Post,* 18 October, 1958.
[153] Address to Engineering Institute of Canada, Montreal, 13 June, 1962.
[154] Letter, E. H. Davidson to Victoria *Times,* 14 April, 1962.
[155] Victoria *Times,* 21 October, 1960.
[156] Statement by Columbia River for Canada Committee, 1 August, 1964.
[157] *The Albertan,* 31 December, 1962.
[158] *Ibid.,* 2 January, 1963.
[159] Letter, E. H. Davidson to McNaughton, 23 May, 1963.
[160] Letter, Fulton to Ripley, 17 September, 1962.
[161] Press release, 29 May, 1964, McNaughton Papers.
[162] Cited Vancouver *Province,* 26 January, 1959.
[163] Barkway, for "Andrew McNaughton–Canadian."
[164] Fulton address on Columbia River Treaty, 7 February, 1961, McNaughton Papers.
[165] McNaughton's comments on Columbia River Treaty, *ibid.*
[166] *Ibid.*
[167] *Ibid.*
[168] Columbia River Treaty, Article XIII.
[169] "The Columbia River Treaty and Protocol."
[170] Larratt Higgins, submission to External Affairs Committee, 29 April, 1964, McNaughton Papers.
[171] McNaughton's comments on letter, Fulton to Ripley, 17 September, 1962.
[172] *The Albertan,* 31 December, 1962.
[173] *Electrical News and Engineering,* September, 1962, p. 52.
[174] *Engineering and Contract Record,* September, 1962, p. 39.
[175] Fulton on Columbia River Treaty.
[176] *Engineering and Contract Record,*

September, 1962.
[177] Press release, 29 May, 1964.
[178] "The Columbia River Treaty and Protocol."
[179] Address to E.I.C., 13 June, 1962.
[180] McNaughton's comments on letter, Fulton to Ripley.
[181] Address to E.I.C.
[182] John V. Krutilla, *The Columbia River Treaty: The Economics of an International River Basin* (Baltimore: Johns Hopkins University Press, 1967), p. 99; Higgins to External Affairs Committee.
[183] Fulton on Columbia River Treaty.
[184] *Ibid.*
[185] Comments by M. E. Marts, Washington, McNaughton Papers.
[186] *The Albertan*, 3 January, 1963.
[187] Comments on letter, Fulton to Ripley.
[188] *Ibid.*
[189] *Ibid.*
[190] *Ibid.*
[191] Udall, Hearings on the Columbia River Treaty, *U.S. Senate Foreign Relations Committee*, 8 March, 1961, p. 24.
[192] Comments on letter, Fulton to Ripley.
[193] *Debates, House of Commons*, 3 January, 1963.
[194] Cited in *The Albertan*, 3 January, 1963.
[195] *Debates, House of Commons*, 12 December, 1962.
[196] Comments on letter, Fulton to Ripley.
[197] *Ibid.*
[198] Keenleyside at annual dinner of First World War veterans, Saanichton, 12 May, 1962, McNaughton Papers.
[199] McNaughton's comments on Keenleyside address.
[200] Barkway, "Andrew McNaughton—Canadian.
[201] *Debates, House of Commons*, 12 December, 1962.
[202] Address to E.I.C.
[203] Fulton on Columbia River Treaty.
[204] *Ibid.*
[205] *Varsity Graduate* (University of Toronto), Christmas, 1961.
[206] Writing to Vancouver *Sun*, 23 September, 1963.
[207] Release by Fulton, 25 July, 1961, McNaughton Papers.
[208] Interview for "Andrew McNaughton—Canadian."
[209] Larratt Higgins at University of Windsor.
[210] Barkway, "Andrew McNaughton—Canadian.
[211] Ripley, *ibid.*
[212] Toronto *Globe and Mail*, 7 April, 1962.
[213] Toronto *Star*, 1 April, 1962.
[214] *Debates, House of Commons*, 12 December, 1962.
[215] *Ibid.*
[216] Pearson to McNaughton, 11 December, 1962, "Davies file," McNaughton Papers.
[217] File "York," *ibid.*
[218] Peter C. Newman, *Renegade in Power* (McClelland & Stewart), pp. 264, 266-7.
[219] Toronto *Globe and Mail*, 30 April, 1962.
[220] Ripley, "Andrew McNaughton—Canadian."
[221] Larratt Higgins at University of Windsor.
[222] "Meeting, Hon. Paul Martin, 18 June 63," "Martin file," McNaughton Papers.
[223] "The Columbia River Treaty and Protocol."
[224] "Martin file," McNaughton Papers.
[225] *Ibid.*
[226] "Davidson file," McNaughton Papers.
[227] Draft dated 4 August, 1964, "Martin file."
[228] McNaughton to Martin, 12 December, 1963.
[229] Pearson to Mrs. Elizabeth Wood, Secretary, Columbia River for Canada Committee, 25 July, 1963, McNaughton Papers.
[230] Pearson to Cady, 25 July, 1963, in Trail *Daily Times*, 2 August, 1963.
[231] McNaughton to Davidson, 1 September, 1963.
[232] "Martin file," McNaughton Papers.
[233] *The Albertan*, 2 June, 1964.

[234] *Ibid.*
[235] Vancouver *Sun,* 3 October, 1963.
[236] Toronto *Globe and Mail,* 14 March, 1964.
[237] *The Albertan,* 15 May, 1964.
[238] Calgary *Herald,* 15 May, 1964.
[239] L. Austin Wright to External Affairs Committee, 7 August, 1963.
[240] Bartholomew to McNaughton, 15 March, 1963, McNaughton Papers.
[241] Statement by Columbia River for Canada Committee, 1 August, 1963.
[242] Press release, 29 May, 1964, McNaughton Papers.
[243] *Ibid.*
[244] Draft letter, McNaughton to Martin, 4 August, 1964.
[245] Ralph W. Johnson, "The Columbia Basin," Garretson, Hayton, Olmstead, *The Law of International Drainage Basins* (New York: New York University School of Law, 1967), p. 232.
[246] *Financial Post,* 25 May, 1964.
[247] Larratt Higgins at University of Windsor.
[248] Institute of Electric and Electronic Engineers *Spectrum,* September, 1965.
[249] Economic analysis by Larratt Higgins, tabled with External Affairs Committee by McNaughton on 15 May, 1964.
[250] *Debates, House of Commons,* 12 May, 1967.
[251] Vancouver *Sun,* cited *ibid.,* 18 May, 1967.
[252] Toronto *Globe and Mail,* 5 September, 1967.
[253] "Notes on Power," McNaughton Papers.
[254] McNaughton, cited by Walter Stewart, "The Soldier Who Fights For Canada's Precious Water," *Star Weekly,* 26 February, 1966.

CHAPTER 7

[1] "Canada's Water," address to Engineering Institute of Canada, 28 May, 1965.
[2] *Saturday Evening Post,* 23 April, 1966.
[3] *Ibid.*
[4] "Canada's Water."
[5] Larratt Higgins, address to Woodsworth Foundation, 12 November, 1966.
[6] Cited in *Congressional Record,* 4 February, 1966.
[7] *Ibid.*
[8] *Ibid.*
[9] *Ibid.*
[10] *Ibid.*
[11] Higgins to Woodsworth Foundation.
[12] Reston, cited Ottawa *Journal,* 16 March, 1965.
[13] *Ibid.*
[14] T. C. Douglas to Saskatchewan Provincial Convention, 19 November, 1965
[15] *Ibid.*
[16] Higgins, address to Seminar on Canada-U.S. Relations, University of Windsor, 30 January, 1967.
[17] Higgins to Woodsworth Foundation.
[18] Higgins, recorded interview for CBC radio programme, "Andrew McNaughton—Canadian."
[19] "Canada's Water."
[20] James Ripley, for "Andrew McNaughton—Canadian."
[21] *Ibid.*
[22] "Canada's Water."
[23] *Ibid.*
[24] Address to Canadian Club, Montreal, 4 October, 1965, McNaughton Papers.
[25] *Ibid.*
[26] *Ibid.*
[27] *Ibid.*
[28] *Ibid.*
[29] *Debates, House of Commons,* 31 January, 1966, H. W. Herridge.
[30] *Ibid.*
[31] *Ibid.*
[32] *Ibid.*
[33] *Ibid.*
[34] *Ibid.*
[35] *Ibid.*
[36] *Ibid.*
[37] *Ibid.*
[38] *Ibid.*
[39] *Ibid.*
[40] Ottawa *Journal,* 3 November, 1966.
[41] Cited *ibid.,* 10 June, 1966.

[42] Dillon O'Leary in *The Reporter*, cited *Debates, House of Commons*, 31 January, 1966.
[43] *Debates, House of Commons*, 31 January, 1966.
[44] *Saturday Night*, December, 1965.
[45] Herridge, 31 January, 1966.
[46] Claude E. Dolman (ed.), *Water Resources of Canada* (Toronto: University of Toronto Press, 1967), p. 25.
[47] Cited by McNaughton in address to Royal Society of Canada, Sherbrooke, 8 June, 1966, McNaughton Papers.
[48] Address to Royal Society.
[49] *Ibid.*
[50] *Ibid.*
[51] *Ibid.*
[52] *Ibid.*
[53] Cited, *ibid.*
[54] Cited, Higgins to Woodsworth Foundation.
[55] *Water Resources of Canada.*
[56] *Debates, House of Commons*, 31 January, 1966.
[57] *Ibid.*
[58] Ottawa *Journal*, 22 August, 1966.
[59] *Ibid.*
[60] *Ibid.*, 27 February, 1967.
[61] *Ibid.*, 8 December, 1966.
[62] Various newspapers, including Moose Jaw *Times-Herald*, 26 July, 1966.
[63] Ottawa *Journal*, 6 September, 1966.
[64] *Ibid.*, 7 September, 1966.
[65] Debates, House of Commons, 31 January, 1966.

CHAPTER 8

[1] 1.Côté to McNaughton, 28 January, 1963, McNaughton Papers.
[2] Ballard to McNaughton, 18 January, 1965, *ibid.*
[3] Dr. C. J. Mackenzie, interview for CBC programme, "Andrew McNaughton—Canadian."
[4] Dr. L. E. Howlett, *ibid.*
[5] James Ripley, *ibid.*
[6] Walter Stewart, "The Soldier Who Fights For Canada's Precious Water."
[7] McNaughton to Ross Munro, 27 November, 1959, McNaughton Papers.
[8] Fullerton to McNaughton, 16 March, 1962, "Fullerton, James T.," *ibid.*
[9] File, "Forde, Capt. Eric," *ibid.*
[10] *Congressional Record*, U.S. Senate, 12 July, 1966.
[11] Ottawa *Journal*, 12 July, 1966.
[12] Letters of condolence in the possession of Mrs. McNaughton.
[13] Alex O. McMurtry, *ibid.*
[14] Hon. J. M. Macdonnell, *ibid.*
[15] C.O., 11th Field Artillery Regiment, Royal Canadian Artillery Association, *ibid.*
[16] J. M. Harrison, Assistant Deputy Minister (Research), Department of Mines and Technical Surveys, *ibid.*
[17] A. D. McLean, *ibid.*
[18] President, Canadian Aeronautical and Space Institute, *ibid.*
[19] Dr. Guy Ballard, *ibid.*
[20] Secretary of State for Commonwealth Relations, *ibid.*
[21] G. E. Beament, *ibid.*
[22] R. H. Stokes-Rees to Mrs. McNaughton, 28 January, 1969.
[23] Cardinal Roy, letters of condolence.
[24] Archbishop O'Neill.
[25] Douglas LePan.
[26] Joseph Scannell.
[27] Philip Jessup.
[28] Dr. David A. Keys.
[29] William Wight.
[30] Air Marshal Reyno.
[31] Robert Moses.
[32] Eric Reford.
[33] R. Douglas Stuart.
[34] Jean Lesage.
[35] Pierre Dansereau.
[36] Ian M. Bell-Irving.
[37] Alfred Archibald.
[38] British Columbia Research Council.
[39] Lt.-General E. L. M. Burns.
[40] Henry Davis.
[41] Maj.-Gen. W. J. Megill.
[42] Isobel Wilson.
[43] Solandt, interview for "Andrew McNaughton—Canadian."
[44] Sir Saville Garner.
[45] Address to Loyola College, Montreal, 1 June, 1957.

INDEX

PUBLISHER'S ACKNOWLEDGMENTS

Grateful acknowledgment is made to the following for permission to quote from copyrighted material:

THE AUTHORS, *Boundary Waters Problems of Canada and the United States* by L. M. Bloomfield, Q.C., and Dr. Gerald F. Fitzgerald.

COLLINS PUBLISHERS, *The Turn of the Tide* by Sir Arthur Bryant, and *The Memoirs of Field-Marshal Viscount Montgomery.*

THE LITERARY EXECUTORS OF THE LATE MACKENZIE KING and THE UNIVERSITY OF TORONTO PRESS, *The Mackenzie King Record,* Volume I, by J. W. Pickersgill; *The Mackenzie King Record,* Volume II, by J. W. Pickersgill and D. F. Forster.

THE MACMILLAN COMPANY OF CANADA LIMITED, *A Party Politician: The Memoirs of Chubby Power,* edited by Norman Ward.

MANHATTAN PUBLISHING COMPANY, 225 Lafayette Street, New York, N.Y. 10012, *Canada and the United Nations* by F. H. Soward and Edgar McInnis.

THE ROYAL CANADIAN LEGION, *Service: The Story of the Canadian Legion, 1925-1960* by Clifford H. Bowering.

TURNSTILE PRESS LTD., *Fear, War and the Bomb* by Professor P. M. S. Blackett.

UNIVERSITY OF TORONTO PRESS, *Water Resources of Canada,* edited by Claude E. Dolman.

WATER POWER, London, England, *Water Power* by Professor M. E. Marts, which appeared in the journal's issue of February, 1961.

THE TEXT OF THIS BOOK WAS SET IN 11 POINT BASKERVILLE OLD FACE TYPE WITH THE DISPLAY MATERIAL SET IN OREGON LDO.

About IEEE Canada

General A. G. L. McNaughton (1887–1966) was an exceptional Canadian electrical engineer (B.A. in 1910 and M.Sc. in 1912, both from McGill University in Montreal), scientist, army officer, President of the National Research Council of Canada, Minister of National Defense, diplomat and President of the UN Security Council. The late British-Canadian historian John Alexander Swettenham (1920–1980) authored a three-volume biography of General McNaughton, first published by Ryerson Press in 1968. The publication was a successful attempt to keep the legacy of an outstanding Canadian engineer, inventor and war hero alive until it went out of print during the first decade of the 21st century.

In 2015, the author's son John (Junior) collaborated with IEEE Canada (the Canadian arm of the Institute of Electrical and Electronics Engineers, the largest technical professional organization in the world) and IEEE Canadian Foundation (the philanthropic partner of IEEE Canada) on the McNaughton Biography Reprint Project. This project was an initiative undertaken by the IEEE Canada History Committee in order to preserve an important chapter of Canadian history. Many volunteers contributed to the reprint project and I wish to acknowledge their contributions with heartfelt gratitude. I thus extend a special thank-you to Dave Kemp (2016 IEEE Canada History Committee Chair), David Whyte (2013-2021 IEEE Canadian Foundation President), Branislav Djokic (2014-2016 IEEE Canada Publications and Communications Chair) and Sarah Van Sickle (volunteer typographer and book advisor) for their generous and invaluable assistance and support.

In 2016 a McNaughton plaque dedication ceremony was held in honour of the General in Ottawa's Beechwood Cemetery on October 16, following the IEEE Canada Fall Board Meeting. Many of the IEEE Canada volunteers, the IEEE USA President, the IEEE Canada President, and several of the General's relatives, including the now retired Lieutenant-General, the Honourable Andrew Leslie, attended. IEEE Canada has recently placed a second plaque in the National Research Council McNaughton Building in Ottawa, recognizing A. G. L. McNaughton's service as National Research Council President.

The significant contributions of Canadian electrical engineers who took major strides for well over a century towards advancing technology for the benefit of humanity, which IEEE is today dedicated to, were honoured by IEEE's Milestones Program. This impressive list of achievements dates back to the 1866 Landing of the Transatlantic Cable—recognized by IEEE in 1985. The innovative inventions of Canadians attest to the significant contributions this country made to the industry overall, and underscores the Canadian engineers' recognition by IEEE in particular. Unfortunately, one of the most innovative engineers of the era, General A. G. L. McNaughton, has been, until now, somewhat overlooked.

In memory of General McNaughton, IEEE Canada has created, since 1979, over 40 IEEE McNaughton Learning Resource Centres intended to enhance the learning experiences of IEEE student members at Canadian Universities and Colleges. Each centre keeps a copy of the McNaughton biography which helps inform, educate and inspire many generations of students.

The past has shown how it can take just one or a few, often narrow-minded, people to start a war, but requires thousands of smart, often broad-minded, people to end it. Thankfully General McNaughton, one of the wisest men of his era, contributed significantly to the efforts which ended the First World War through his compelling leadership skills and technological innovations. Today's generation of scientists and engineers is once again being called upon to address major world problems, such as global warming for example. It will come as no surprise that a number of future Canadian leaders, echoing General McNaughton's exemplary accomplishments, will emerge from new generations of IEEE members.

Dr. Amir Aghdam, Ph.D., P.Eng.

2014–2015 President, IEEE Canada
Professor of Electrical and Computer Engineering
Concordia University

About the Author

John Alexander Swettenham was born in 1920 in England. When he was a young teenager, his father died from health complications resulting from service in World War I. Convinced by Hitler's rhetoric that another war was imminent, and resolved to serve, John undertook the education and training required to enter the Army as an Officer Engineer by 1939, when war was declared. He served as an Officer throughout the war. The least enjoyed part of his wartime career was defusing mines on Mediterranean beaches. The most enjoyed, also in the Mid-East, was partaking in the hospitality of Sheiks and oil company executives while laying explosives to destroy oil infrastructure in the event of potential Nazi invasion.

After the war, John served with the British in the liberation of occupied Norway, where he met his wife Ellinor Mørch. They were married in 1948 and moved shortly thereafter to Germany where John served with the provisional Government and was in charge of a displaced persons camp. Many in the camp were refugees from the Baltic States and their stories inspired John to write his first book, *The Tragedy of the Baltic States* published in England (later republished in the US). John had found homes in Canada for many displaced persons and in turn decided to emigrate to Canada himself, arriving in Nova Scotia in 1952.

After an unsuccessful attempt at farming in the Annapolis Valley, John joined the Canadian Army—again as an Engineer. This took John and Ellinor across the country which they fell in love with and fully adopted. John also learned first-hand the competence and expertise of the Canadian Army and of its significant and often overlooked contribution to the two World Wars.

Having written one book and now passionate about the Canadian military, John yearned to write more and was afforded the opportunity when he joined the Canadian Army's Department of History and later the Canadian War Museum. His many books were largely about the Canadian military including: *To Seize the Victory: The Canadian Corps in World War I*; *Allied Intervention in Russia and the Part Played by Canada*; *D-Day*; *Canada and the First World War*; *The Evening of Chivalry* (on medieval armour and warfare); *Serving the State* (with David Kealy, a history of the Professional Institute of Canada); *Silent Witnesses* (Canadian War Graves); *Valiant Men: Canada's Victoria Cross and George Cross Winners*; *Canada's Atlantic War*; and of course *McNaughton*, the three volume biography of the great Canadian General.

John Alexander Swettenham died suddenly from a heart attack in 1980. At the time he was completing another book, *Canada's Clandestine War*, which remains unpublished. Ellinor Margarete Swettenham died in 2015. They are survived by their three children: Victoria, Julie and John.

AUTHOR'S PHOTOGRAPH
BY THOMAS STUDIO, OTTAWA

Made in United States
Troutdale, OR
11/03/2026

56847573R00239